Introduction to the physics of nanoelectronics

## Related titles:

*Electromigration in thin films and electronic devices: Materials and reliability*
(ISBN 978-1-84569-937-6)
Electromigration is a significant problem affecting the reliability of microelectronic devices such as integrated circuits. Recent research has focused on how electromigration affects the increasing use by the microelectronics industry of lead-free solders and copper interconnects. Part one reviews ways of modelling and testing electromigration, part two discusses electromigration in copper interconnects and part three covers solder and wirebonds.

*Silicon–germanium (SiGe) nanostructures: Production, properties and applications in electronics*
(ISBN 978-1-84569-689-4)
Nanostructured silicon–germanium (SiGe) provides the prospect of novel and enhanced electronic device performance. This book reviews the materials science and technology of SiGe nanostructures including crystal growth, fabrication of nanostructures, material properties and applications in electronics.

*Advanced piezoelectric materials: Science and technology*
(ISBN 978-1-84569-534-7)
Piezoelectric materials produce electric charges on their surfaces as a consequence of applying mechanical stress. They are used in the fabrication of a growing range of devices such as transducers, actuators and pressure sensor devices and increasingly as a way of producing energy. This book provides a comprehensive review of advanced piezoelectric materials, their properties, methods of manufacture and applications. It covers lead zirconate titanate (PZT) piezo-ceramics, relaxor ferroelectric ceramics, lead-free piezo-ceramics, quartz-based piezoelectric materials, the use of lithium niobate and lithium in piezoelectrics, single crystal piezoelectric materials, electroactive polymers (EAP) and piezoelectric composite materials.

Details of these books and a complete list of titles from Woodhead Publishing can be obtained by:

- visiting our web site at www.woodheadpublishing.com
- contacting Customer Services (e-mail: sales@woodheadpublishing.com; fax: +44 (0) 1223 832819; tel.: +44 (0) 1223 499140 ext. 130; address: Woodhead Publishing Limited, 80 High Street, Sawston, Cambridge CB22 3HJ, UK)
- in North America, contacting our US office (e-mail: usmarketing@woodheadpublishing.com; tel.: (215) 928 9112; address: Woodhead Publishing, 1518 Walnut Street, Suite 1100, Philadelphia, PA 19102-3406, USA)

If you would like e-versions of our content, please visit our online platform: www.woodheadpublishingonline.com). Please recommend it to your librarian so that everyone in your institution can benefit from the wealth of content on the site.

# Introduction to the physics of nanoelectronics

Seng Ghee Tan and Mansoor B. A. Jalil

WOODHEAD PUBLISHING

Oxford Cambridge Philadelphia New Delhi

Published by Woodhead Publishing Limited,
80 High Street, Sawston, Cambridge CB22 3HJ, UK
www.woodheadpublishing.com
www.woodheadpublishingonline.com

Woodhead Publishing, 1518 Walnut Street, Suite 1100, Philadelphia,
PA 19102-3406, USA

Woodhead Publishing India Private Limited, G-2, Vardaan House,
7/28 Ansari Road, Daryaganj, New Delhi – 110002, India
www.woodheadpublishingindia.com

First published 2012, Woodhead Publishing Limited


British Library Cataloguing in Publication Data
A catalogue record for this book is available from the British Library.

Library of Congress Control Number: 2012901492

ISBN 978-0-85709-511-4 (print)
ISBN 978-0-85709-588-6 (online)

The publisher's policy is to use permanent paper from mills that operate a sustainable forestry policy, and which has been manufactured from pulp which is processed using acid-free and elemental chlorine-free practices. Furthermore, the publisher ensures that the text paper and cover board used have met acceptable environmental accreditation standards.

Typeset by RefineCatch Limited, Bungay, Suffolk

# Contents

# Author contact details

(* = main contact)

*Professor Seng Ghee Tan
Data Storage Institute
A*STAR
Singapore

and

National University of Singapore
Singapore

Email: TAN_Seng_Ghee@dsi.a-star.edu.sg

Professor Mansoor B. A. Jalil
National University of Singapore
Singapore

# Foreword by S. Murakami

Our everyday life is supported by countless electronic products, built with silicon complementary metal–oxide semiconductor (CMOS) technologies. Thanks to efforts to achieve high integration, the transistor density at the minimum cost has been doubling every two years – a trend known as Moore's law, which has continued for over half a century. Moore's law has provided us with a roadmap for semiconductor technology. Nevertheless, as the chip size becomes smaller, we encounter many obstacles, such as the production of heat in the circuit in operation, leakage current and difficulties in the fabrication process. A breakthrough for this limit for silicon CMOS is being strongly sought. One possibility for a breakthrough is to utilize internal degrees of freedom of electrons.

The degree of freedom we utilize for electronics is the charge. Meanwhile, quantum mechanics tells us that the electrons are waves; they also have spin degrees of freedom. This wave nature and spins offer us new possibilities to provide each electron with internal degrees of freedom. Among them are spintronics and graphene, both currently attracting considerable interest. Spintronics is the field in electronics using the spin degrees of freedom of electrons. Because the spins obey the profound rules of quantum mechanics, there are various and non-trivial ways to manipulate electron spins, in comparison with conventional electronics. These various possibilities of spin manipulation show great promise.

Graphene is a single sheet of graphite, a common material. Nevertheless, the electronic band structure of graphene is unique and is made from a cone-shaped dispersion. This band structure endows electrons with internal degrees of freedom, such as pseudospin and valley-spin. These internal degrees of freedom also serve like electron spins, offering various possibilities for electron manipulation. Spintronics and graphene physics have developed rapidly in recent years, thanks to the progress in theory, nanofabrication and measurement techniques.

This book reviews basic knowledge for spintronics and graphene nanoelectronics. An introduction to the relevant mathematics, quantum mechanics and fundamental aspects of electron transport in solids is given. The reader is then introduced to the central issues of the book – spintronics and graphene. In these topics we notice with great surprise that the electrons behave so differently and elegantly in these materials. We can also see in this book that, in the fields of spintronics and graphene, science and technology are close to each other; fancy ideas from pure theory can be measured directly or even used in applications. The role of gauge theory and topological structure is emphasized in this book. Actually, due to the electronic band structure in solids, the space where an electron travels becomes curved. The gauge theory describes how this space is 'curved'. This curvature sometimes gives 'topological structure' to the space. By formulating the physics of electron transport in terms of gauge theory and topology, one can get new ideas for novel physical phenomena like the spin Hall effect, topological insulators and so forth. These abstract concepts hold great interest for condensed matter physicists because they can indeed be found in nature and have great promise for potential applications.

*Professor Shuichi Murakami*
*Tokyo*

# Foreword by B. Luk'yanchuk

Nanoscale electronics and photonics are becoming increasingly important areas of research which have excited politicians and attracted a large number of scientists, engineers and graduate students. Recent developments in nanoscale photonics (e.g. plasmonics, near-field photonics and optical metamaterials) might lead to new applications such as high-resolution lithography and heat-localized nanoparticles for cancer treatment. Likewise, on the electronics front, spintronics, graphene electronics and topological insulators might lead to high-density memory and high-speed transistors. By extrapolation, the future of nanoscale electrophotonics could be on the horizon.

Nanoscale electronics and photonics are modern contexts in which renewed interest is found in physical effects discovered much earlier (e.g. the Rashba and spin Hall effect in spintronics, and surface wave propagation in plasmonics). This book is dedicated to the field of nanoscale electronics, one of the twin engines that power modern research in applied physics. It is a good introduction to spintronics, single electronics, molecular electronics, Hall effects, carbon nanostructures, graphene and topological-based electronics. This book is of a theoretical nature, and modern approaches such as non-equilibrium Green's function (NEGF) and topological methods are introduced to discuss physical concepts such as the spin torque effect due to Rashba spin orbit coupling.

*Professor Boris Luk'yanchuk*
*Data Storage Institute, Singapore*
*Honorary Professor at Johannes Kepler University, Linz, Austria*

# Endorsements

*'This book will enable physics and engineering students to understand the most important underlying physics of modern nanoelectronics.'*

Professor Jian-Ping Wang, Director of the Center for Micromagnetic and Information Technologies (MINT), University of Minnesota, USA

*'We can see in this book that in the fields of spintronics and graphene, science and technology are close to each other; fancy ideas from pure theory can be directly measured or even be used for applications. The role of gauge theory and topological structure is emphasized in this book.'*

Professor Shuichi Murakami, Tokyo Institute of Technology, Japan

*'An excellent contribution to the fast-expanding field of nanoelectronics. It is my pleasure to recommend this book to all researchers and students who are working in the field.'*

Professor Yong Jiang, University of Science and Technology Beijing, China

*'This book is a good introduction to spintronics, single electronics, molecular electronics, Hall effects, carbon-nanostructures, graphene and topological-based electronics.'*

Professor Boris Luk'yanchuk, Data Storage Institute, Singapore (Honorary Professor, Johannes Kepler University, Linz, Austria)

*'This book provides a thorough overview of the theoretical aspects of nanoscale electronic devices.'*

Professor Kouichiro Inomata, Emeritus Fellow of National Institute of Material Science, Tsukuba, Japan

*'An indispensable reference for anyone interested in the latest advances in nanoelectronics – very highly recommended.'*

Alexander A. G. Driskill-Smith, VP Business Development, Grandis, Inc., Milpitas, CA 95035, USA

# Preface

Due to its long history, the term 'electronics' is well known to laymen and experts alike. But to prefix it with 'nano' is something that would baffle the laymen and set the experts off on an endless debate. The main motivation for writing this book arises from our distress at the confusion and the desire to seek an appropriate interpretation. And we decided that this should be done in the context of modern applied physics, an area which is exciting and relevant to many scientists worldwide. It is a broad area which involves condensed matter, optical and atomic, devices and nanotechnologies. We therefore think that nanoelectronics should be about the discovery and implementation of new physics in electronic devices that are approaching the nanometer scale. It should thus encompass, among others, the modern topics of spintronics, topological insulators, carbon-based electronics (principally carbon nanotubes and graphene), single electronics, quantum Hall systems and devices.

There have been many good-quality condensed matter physics textbooks, ranging from solid state, many-body physics to, more recently, mesoscopic physics. On the other hand, there is no shortage of books on devices, particularly semiconductor devices, which introduce the workings of bipolar, diodes and complementary metal–oxide semiconductors (CMOS) transistors. Nonetheless, it is hard not to notice that condensed matter physics has been rather decoupled from the physics of electronic devices. Modern books on mesoscopic physics and non-equilibrium Green's functions (NEGFs) might have taken physics a few steps closer to the world of devices, but clear links remain elusive. The main intention of this book is to extend previous efforts to an extent that future efforts to complete the link between physics and electronic devices will be greatly simplified. Nanoelectronics is, however, an area so diverse and inclusive that a complete study of it would require inter-disciplinary fields that would certainly overfill a book this size. We thus remind readers that reading this book might need to be complemented by other specialized textbooks.

This book contains seven chapters and focuses on the theoretical aspects of nanoscale devices. Chapter 1 provides a revision of mathematics and quantum mechanics, which will be of use in subsequent chapters. Chapter 2 introduces the general idea of nanoelectronics as well as the many new contemporary topics that define this concept. The concept of energy and many-body physics in condensed matter system is introduced as a prelude to the more important topics of electron and spin dynamics.

Chapters 3 and 4 provide the foundational study of electron charge and spin dynamics in condensed matter and device systems. Hence, the concepts of interaction, disturbances and non-equilibrium become important here. Discussions begin with an equilibrium condensed matter system. The non-equilibrium formalism is introduced for device systems with clear boundaries and external fields. Chapter 5 is exclusively focused on spintronics and introduces the different theoretical topics related to quantum spin transport and individual spin manipulation. Thus, there is a much greater emphasis on spin orbit coupling, spin gauge theory and spin Hall effects than is found in existing books on magnetism or metal-based spintronics which, owing to the history of the subject, tend to emphasize the semi-classical approach. Nonetheless, the spin drift-diffusive theory of giant magnetoresistance (GMR) and tunneling magnetoresistance (TMR) effects are also covered in this chapter.

Chapter 6 introduces electronics based on carbon materials such as carbon nanotubes and graphene. Monolayer and bilayer graphene are discussed and the forms of their Hamiltonian are studied. The concept of gauge potential induced by strain or graphene deformation is also introduced. Particular attention is paid to the extra degrees of freedom in graphene, namely the pseudospin and the valley-spin, which share the spinor physics of electron spin. Novel graphene behaviors like localization, Klein tunneling and minimal conductivity are also discussed. The integer quantum Hall effect (IQHE) in graphene is discussed in relation to the semiconductor IQHE.

Chapter 7 provides a systematic introduction to the field and gauge theoretical methods, with a focus on their application in nanoscale systems (e.g. spin orbit coupling systems and graphene). This is a departure from the usual treatment of gauge theory and topology in existing books, which mainly emphasize the development of these theories in high-energy physics. In this book, field and gauge theoretical methods are used alongside a phenomenological and statistical approach, with an emphasis on applications in nanoelectronics.

The writing of this book began with the rearrangement of the lecture notes for a graduate class we taught at the National University of Singapore. Nevertheless, this apparently straightforward process has taken us three years to complete. This task would not be possible without the help and intensive checking by our doctoral students and colleagues. We would like to thank Zhuobin Siu, Takashi Fujita, Yuan Li, Congson Ho, Mingjun Xing, Jie Guo and Nyuk Leong Chung for reading the manuscripts intensively and correcting our mistakes.

What we hope to achieve in this book is that readers will generally recognize the importance of integrating the physics of condensed matter, atom, electronic devices and particle dynamics in one book. In this way, a book like this one is able to discuss effectively the emerging technologies of spintronics, graphene, single electronics, topological insulators, plasmonics and many more that we believe will continue to emerge and dominate the development of physics in the twenty-first century.

*Seng Ghee Tan*
*Mansoor B. A. Jalil*

# 1 Physics and mathematics for nanoscale systems

**Abstract:** This chapter contains a number of useful mathematical subjects that might be relevant for the study of the physics of electron transport on the nanoscale level. This chapter is intended to be a fast-forward through basic mathematics and quantum mechanics topics. Useful topics covered in the chapter include vector calculus, Fourier transform and Dirac delta functions, basic quantum mechanics, Green's function and second quantization as a method for keeping account of the electron.

**Key words:** quantum mechanics, vector calculus, Fourier transform, Dirac delta, Green's function, second quantization.

## 1.1 Introduction

The rapid development in the 1990s of nanoscale technologies and devices which span material and electronic engineering, semiconductor physics, device fabrication and characterization technologies, theoretical and computational physics, optics and chemistry turns this field into a sort of assembly hall that congregates cross-disciplinary expertise. We want to provide in this chapter a revision of the mathematics and physics that many (apart from perhaps physicists) may have forgotten but are necessary for a better appreciation of the many qualitative discussions in this book. This book is intended to be easy to read; its derivations may thus appear long-winded for those who are already familiar with them. Field theoretic, many-body and topological concepts are introduced 'gently' to readers to minimize psychological aversion.

This chapter contains a number of exercises, with the solutions given in most of them. It presents a number of useful mathematical subjects that we think are most relevant to the study of electron transport and nanoscale physics. As these are well-established topics that can be found in many textbooks, our presentation here will be brief, exercise based and summarized in tables.

The main purpose of this chapter is to provide a quick revision, a reminder of what most readers may have learnt in their younger (undergraduate) years and a quick look-up for equations or standard techniques (e.g. integration) that might come in handy for understanding the rest of the book.

## 1.2 Vector calculus

Vector calculus is usually covered in undergraduate mathematics and electrodynamics courses. As mentioned earlier, most are not repeated here.

*Table 1.1* Summary of useful vector identities

| | Useful identities |
|---|---|
| 1. | $\boldsymbol{A}.(\boldsymbol{B}\times\boldsymbol{C}) = \boldsymbol{B}.(\boldsymbol{C}\times\boldsymbol{A}) = \boldsymbol{C}.(\boldsymbol{A}\times\boldsymbol{B})$ |
| 2. | $\nabla.(\boldsymbol{A}\times\boldsymbol{B}) = \boldsymbol{B}.(\nabla\times\boldsymbol{A}) - \boldsymbol{A}.(\nabla\times\boldsymbol{B})$ |
| 3. | $\nabla\times(\boldsymbol{A}\times\boldsymbol{B}) = (\boldsymbol{B}.\nabla)\boldsymbol{A} - \boldsymbol{B}(\nabla.\boldsymbol{A}) - (\boldsymbol{A}.\nabla)\boldsymbol{B} + \boldsymbol{A}(\nabla.\boldsymbol{B})$ |
| 4. | $\boldsymbol{A}\times(\boldsymbol{B}\times\boldsymbol{C}) = (\boldsymbol{A}.\boldsymbol{C})\boldsymbol{B} - (\boldsymbol{A}.\boldsymbol{B})\boldsymbol{C}$ |
| 5. | $(\boldsymbol{A}\times\boldsymbol{B})\times\boldsymbol{C} = (\boldsymbol{A}.\boldsymbol{C})\boldsymbol{B} - (\boldsymbol{B}.\boldsymbol{C})\boldsymbol{A}$ |
| 6. | $\nabla\times(\nabla\phi) = 0$ |
| 7. | $\nabla.(\phi\boldsymbol{A}) = (\nabla\phi).\boldsymbol{A} + \phi(\nabla.\boldsymbol{A})$ |
| 8. | $\nabla\times(\phi\boldsymbol{A}) = (\nabla\phi)\times\boldsymbol{A} + \phi(\nabla\times\boldsymbol{A})$ |
| 9. | $\nabla.\nabla\phi = \nabla^2\phi$ |
| 10. | $\nabla.(\nabla\times\boldsymbol{A}) = 0$ |

However, some useful identities will be handy for solving advanced quantum mechanics or engineering problems, and these are summarized in Table 1.1. Here $\boldsymbol{A}$, $\boldsymbol{B}$ and $\boldsymbol{C}$ are differentiable vector functions, while $\Phi$ is a scalar function. The following is the vector calculus of some common expressions presented in the form of exercises and solutions.

### Exercise 1.1

***A, B*** are differentiable vector functions and show that:

(a) $\nabla\times(\nabla\times\boldsymbol{A}) = \nabla(\nabla.\boldsymbol{A}) - \nabla^2\boldsymbol{A}$

(b) $\nabla(\boldsymbol{A}.\boldsymbol{B}) = (\boldsymbol{B}.\nabla)\boldsymbol{A} + (\boldsymbol{A}.\nabla)\boldsymbol{B} + \boldsymbol{B}\times(\nabla\times\boldsymbol{A}) + \boldsymbol{A}\times(\nabla\times\boldsymbol{B})$

### Exercise 1.2

(1) Prove that:

$$\nabla\left(\frac{1}{r}\right) = \frac{-\boldsymbol{r}}{r^3}$$

(2) With the above where $\delta^3(r) = \delta(x)\delta(y)\delta(z)$, prove that:

$$\nabla^2\left(\frac{1}{r}\right) = -4\pi\,\delta^3(r)$$

***Solution***

(1) $$\nabla\left(\frac{1}{r}\right) = \left(\frac{\partial r}{\partial x}\frac{\partial}{\partial r}\boldsymbol{i} + \frac{\partial r}{\partial y}\frac{\partial}{\partial r}\boldsymbol{j} + \frac{\partial r}{\partial z}\frac{\partial}{\partial r}\boldsymbol{k}\right)\frac{1}{r} = \left(\frac{x}{r}\boldsymbol{i} + \frac{y}{r}\boldsymbol{j} + \frac{z}{r}\boldsymbol{k}\right)\frac{-1}{r^2} = \frac{-\boldsymbol{r}}{r^3}$$

(2) Let $R^2 = r^2 + v^2$

$$\nabla^2\left(\frac{1}{r}\right) = \lim_{v\to 0}\nabla^2\left(\frac{1}{R}\right) = \lim_{v\to 0}\left(\frac{-R^3+3x^2R}{R^6}+\frac{-R^3+3y^2R}{R^6}+\frac{-R^3+3z^2R}{R^6}\right)$$

$$= \lim_{v\to 0}\frac{-3v^2}{(r^2+v^2)^{\frac{5}{2}}} = A\delta^3(r)$$

Note that $\partial_x\left(\frac{1}{r}\right) = \frac{-x}{r^3}$ and applying volume integral to both sides:

$$\iiint \nabla^2\left(\frac{1}{r}\right) d^3r = \iiint A\delta^3(r)\, d^3r$$

One derives by divergence theorem that:

$$\rightarrow A = \iiint \nabla\cdot\nabla\left(\frac{1}{r}\right) dV = \iint \nabla\left(\frac{1}{r}\right)\cdot \boldsymbol{dS}$$

Recalling that $\nabla\left(\frac{1}{r}\right) = \frac{-\boldsymbol{r}}{r^{3,}}$, one now has:

$$A = \iint \frac{-\boldsymbol{r}}{r^3}\cdot r^2 \sin\theta\, d\theta\, d\phi\, \boldsymbol{a}_r = \iint -\sin\theta\, d\theta\, d\phi = -4\pi$$

---

## 1.3 Fourier transform and Dirac delta functions

The mathematics of Fourier transform has extensive use in engineering and physics and can be found in most second year textbooks for engineering and physics undergraduate courses. Here we dole out the quick reminder that, if $f(x)$ is periodic with period $L$, it can be written in Fourier series as follows:

$$f(x) = \sum_{-\infty}^{+\infty} a_n\, e^{ik_n x} \qquad [1.1]$$

where $k_n = \frac{2\pi n}{L}$, then:

$$a_n = \frac{1}{L}\int_0^L f(x)e^{-ik_n x}\, dx. \qquad [1.2]$$

In the case where $f(x)$ is not periodic but a function, one can find its spectrum $g(k)$ using the Fourier transform with which one has:

$$g(k) = \frac{1}{\sqrt{2\pi}}\int f(x)e^{-ikx}\, dx$$

$$f(x) = \frac{1}{\sqrt{2\pi}}\int g(x)e^{ikx}\, dk \qquad [1.3]$$

where $g(k) = \mathfrak{F}\{f(x)\}$. Note that we have followed the convention of unitary, angular frequency. Table 1.2 summarizes some useful Fourier transform (FT) identities.

The Dirac delta function is an important mathematical object that simplifies calculations required for the studies of electron motion and propagation. It is not really a function but a symbol for physicists and engineers to represent some calculations. It can be regarded as a shorthand notation for some complicated limiting processes. Once the role it plays, for example, as a distribution function is implicitly understood, many derivations can be greatly simplified. Table 1.3 contains a few useful identities.

*Table 1.2* Summary of useful Fourier transform identities

| | Function $f(x)$ | Unitary, angular FT $g(k)$ |
|---|---|---|
| 1. | $\partial_x^n f(x)$ | $(ik)^n g(k)$ |
| 2. | $f(x-a)$ | $e^{-iak} g(k)$ |
| 3. | $f(ax)$ | $\frac{1}{\lvert a \rvert} g\left(\frac{k}{a}\right)$ |
| 4. | $(f * h)x$ | $\sqrt{2\pi} g_f(k) g_h(k)$ |
| 5. | $f(x)\, h(x)$ | $\frac{(g_f * g_h)k}{\sqrt{2\pi}}$ |

*Table 1.3* Common relations and identities involving the Dirac delta functions

| | Useful identities |
|---|---|
| 1. | $\int f(x)\delta(x-x_0)dx = f(x_0)$ |
| 2. | $\int \delta(x-x_1)\delta(x-x_2)dx = \delta(x_1-x_2)$ |
| 3. | $\delta(x-x_0) = \frac{1}{2\pi}\int e^{ik(x-x_0)}\,dk$ |
| 4. | $\delta(x) = \frac{d}{dx}\theta(x)$ where $\theta(x) = 1$ for $x>0$; 0 for $x<0$; $\frac{1}{2}$ for $x=0$ |
| 5. | $\delta(ax) = \frac{\delta(x)}{\lvert a \rvert}$ |

**Exercise 1.3**

Show that $\delta(x-x_0)=\frac{1}{2\pi}\int e^{ik(x-x_0)}\,dk$

***Solution***

Let $f(x)$ be $\delta(x)$. One then has:

$$g(k)=\frac{1}{\sqrt{2\pi}}\int \delta(x)e^{-ikx}dx=\frac{1}{\sqrt{2\pi}}$$

$$\delta(x)=f(x)=\frac{1}{\sqrt{2\pi}}\int g(k)e^{ikx}dk=\left(\frac{1}{\sqrt{2\pi}}\right)\left(\frac{1}{\sqrt{2\pi}}\right)\int e^{ikx}dk$$

**Exercise 1.4**

Prove the identity $\delta(ax)=\frac{\delta(x)}{|a|}$

***Solution***

Let $u=|a|x$. It is straightforward that:

$$\int \delta(ax)\,dx=\int \delta(u)\frac{du}{|a|}=\frac{1}{|a|}$$

leading to:

$$|a|\int \delta(ax)\,dx=1=\int \delta(x)dx.$$

By comparing integrand, one has:

$$\delta(ax)=\frac{\delta(x)}{|a|}$$

Strictly speaking, the above is an integrand identity. By quick inspection, it is clear that the Dirac delta is an even function where $\delta(x)=\delta(-x)$. With some ingenuity, the Dirac function can be represented explicitly by mathematical functions that are truly functions. One very useful and common example is:

$$\delta_c(x)=\begin{cases}\frac{1}{c} & |x|\le\frac{c}{2}\\ 0 & |x|>\frac{c}{2}.\end{cases} \quad [1.4]$$

Now $\delta_c(x)$ is a true function. The conventional limiting process with Dirac delta should therefore be written in a mathematical manner, that is:

$$\int \delta(x)\,dx = 1 \rightarrow \lim_{c\to 0} \int \delta_c(x)\,dx = 1 \qquad [1.5]$$

$$\int f(x)\,\delta(x)\,dx = f(0) \rightarrow \lim_{c\to 0} \int f(x)\delta_c(x) = f(0) \qquad [1.6]$$

If interchanging the order of the limiting process is allowed, then one can safely say $\delta(x) = \lim_{c\to 0} \delta_c(x)$. Let us prove the second equation:

$$\lim_{c\to 0} \int f(x)\delta_c(x)\,dx = \lim_{c\to 0} \int_{-\infty}^{-\frac{c}{2}} + \int_{-\frac{c}{2}}^{+\frac{c}{2}} + \int_{+\frac{c}{2}}^{+\infty} f(x)\delta_c(x)dx = \lim_{c\to 0} \int_{-\frac{c}{2}}^{+\frac{c}{2}} f(x)\delta_c(x)\,dx$$

$$= \lim_{c\to 0} \frac{1}{c} \int_{-\frac{c}{2}}^{+\frac{c}{2}} f(x)\,dx = \lim_{c\to 0} \frac{1}{c} f(\underline{0})c = f(0) \qquad [1.7]$$

where $\underline{0}$ takes on any value of $x$ that lies within $-\frac{c}{2} \le x \le +\frac{c}{2}$.

Table 1.4 provides a summary of three Dirac delta representations commonly used in Green's functions and the physics of nanoelectronics.

*Table 1.4* Representation of the Dirac delta by smooth functions. More representations can be found in standard textbooks

| | Dirac delta representations $\delta(x) \rightarrow \lim_{c\to 0} \delta_c(x)$ |
|---|---|
| 1. | $\delta_c(x) = \begin{cases} \frac{1}{c} & \lvert x\rvert \le \frac{c}{2} \\ 0 & \lvert x\rvert > \frac{c}{2} \end{cases}$ |
| 2. | $\delta_c(x) = \frac{1}{c\sqrt{\pi}} e^{-x^2/c^2}$ |
| 3. | $\delta_c(x) = \frac{1}{\pi} \frac{c}{x^2 + c^2}$ |

**Exercise 1.5**

The function $\lim_{c\to 0} \frac{1}{A} \frac{c}{x^2 + c^2}$ clearly plays the role of $\lim_{c\to 0} \delta_c(x) = 0$ everywhere except at $x = 0$, but there seems to be a freedom in choosing $A$. Find $A$.

***Solution***

$$\int_{-\infty}^{\infty} \lim_{c\to 0} \frac{c}{c^2 + x^2}\, dx = \int_{-\infty}^{\infty} A\delta(x)\, dx$$

which leads to:

$$A = \left[ \tan^{-1} \frac{x}{c} \right]_{-\infty}^{\infty} = \pi.$$

In fact $A$ has a fixed value.

## 1.4 Basic quantum mechanics

Standard quantum mechanics textbooks normally consist of the following topics:

1. Wave mechanic description – well-known in Schrödinger's equation
2. State vector formalism based upon Dirac's bra-ket notations
3. Green's function formalism
4. Principle of gauge and symmetry
5. Second quantization
6. Relativistic quantum mechanics.

In this book, emphasis is given to the applications of these well-known theories and formalism in the new context of electronics that are now fast approaching the nanoscale dimension. The foundational and mathematical aspects of quantum mechanics are not discussed in detail in this book. Instead we refer readers to textbooks[1,2] for foundational knowledge.

### 1.4.1 Vector spaces

In this section we show that, under the generalization of vector spaces, the physical states of a system can be described by state vectors in so-called bra-ket formalism, as well as by wave mechanic representation in Schrödinger's wavefunctions. In an $n$-dimensional space, one can choose a set of $n$ linearly independent vectors $|u_i\rangle$ ..... $|u_n\rangle$ as the basis vectors for a particular state vector. These basis vectors span the vector space and form a complete set of vectors as shown in Table 1.5 where the standard bra-ket notations are now used.

An electron in a momentum state with momentum $p$ is often denoted by $|p\rangle$. On the other hand, a spin up state with spin quantization axis chosen along $z$ by convention is denoted by $|\uparrow\rangle$. For simplicity in nanoelectronics one often just refers to an electron as spin up or down, where it is understood that the spin quantization axis has already been defined. One explicit representation of the spin states of up and down are the column vectors of $(1 \quad 0)^T$, $(0 \quad 1)^T$, respectively, as these vectors are linearly independent and they are complete in two dimensions, thus satisfying the requirements to represent a single particle's quantum spin states.

*Table 1.5* Bra-ket representation of quantum states in discrete and continuous pictures

| | System | Quantum state vectors $\|\psi\rangle$ | Resolution of identity |
|---|---|---|---|
| 1. | Discrete | $\|\psi\rangle = \sum a_i \|u_i\rangle$ | $\sum_i \|u_i\rangle\langle u_i\| = 1$ |
| 2. | Continuous | $\|\psi\rangle = \int a_u \|u\rangle du$ | $\int \|u\rangle\langle u\| du = 1$ |

One wonders who might have imposed these requirements of linear independence and completeness. The answer lies of course in the law of nature. Since this is not a quantum mechanics or mathematical physics book, it suffices to accept that these are mere rules imposed by nature. In applied physics, all one needs to do is find the mathematical methods or objects that satisfy these rules. In the above, $|u_i\rangle$ has been introduced as state vectors that are linearly independent. But linear independence does not imply orthogonality or orthonormality (e.g. $\langle u_i|u_i\rangle = \delta_{ij}$), a feature we will use later. For example, $\frac{1}{\sqrt{2}}\left(|\uparrow\rangle + \langle\downarrow|\right)$ and $|\uparrow\rangle$ are linearly independent but they are not orthogonal.

**Exercise 1.6**

Prove that $\Sigma_i|u_i\rangle\langle u_i| = 1$.

***Solution***

$$|\psi\rangle = \sum a_i |u_i\rangle \quad \rightarrow \quad \langle u_j|\psi\rangle = \langle u_j|\sum a_i |u_i\rangle = \sum_i \delta_{ij} a_i = a_j.$$

Thus one can write:

$$|\psi\rangle = \sum a_i |u_i\rangle \;=\; \sum \langle u_i|\psi\rangle |u_i\rangle = \sum |u_i\rangle\langle u_i|\psi\rangle.$$

In the following, we discuss the coordinate and momentum representations in slightly more detail as these two pictures will be suitable for different types of nanostructures. In nanoelectronics and mesoscopic physics, knowledge of the flow or distribution of electrons in nano-sized structures or devices is required to understand some important measurables (e.g. conductivity, spin or charge accumulation). Let us now write the position operator $X^v$ in vector space (bra-ket) notation, and write the state vector as $|\psi\rangle$. The position operator could provide at least 'rudimentary' insights into the whereabouts of an electron. In coordinate representation, the position eigenvectors { $|x_i\rangle$ } are chosen as the basis set. Since:

$$X^v|x\rangle = x|x\rangle \qquad [1.8]$$

one can find $X^v$ in the function space, i.e. $X^f$. Since:

$$\langle x|X^v|\psi\rangle = x\langle x|\psi\rangle = x\,\psi(x). \qquad [1.9]$$

Defining:

$$X^f\psi(x) = x\,\psi(x) \qquad [1.10]$$

one arrives at $X^f\langle x \mid \psi\rangle = \langle x|X^v|\psi\rangle$. On the other hand, if the above is to be carried out in the momentum representation, the basis set will be the momentum eigenvectors $|p\rangle$ satisfying:

$$P^v|p\rangle = p|p\rangle. \qquad [1.11]$$

---

**Exercise 1.7**

(1) Prove that

$$\varphi^3(x) = \langle \boldsymbol{x}|\hbar \boldsymbol{k}\rangle = \frac{1}{(2\pi\hbar)^{\frac{3}{2}}}e^{i\boldsymbol{k}\cdot\boldsymbol{x}}$$

where the following convention is followed:

$\varphi^3(x) = \varphi(x)\varphi(y)\varphi(z)$; $\delta(x) = \delta^3(\boldsymbol{x}) = \delta(x)\delta(y)\delta(z)$.

(2) An electron can be regarded as a wave propagating in the crystal structure of materials that we use to make nanoscale devices. The position operator helps to locate the electron's whereabouts. Prove that the position operator in function space is represented in momentum space by $X_\mu^f = i\dfrac{\partial}{\partial k_\mu}$, where $\mu = x, y$, or $z$ and $k$ is the wavevector of the electron.

***Solution***

(1) Since $\langle p|p'\rangle = \langle \hbar k|\hbar k'\rangle = \delta_{pp'}$ or $\delta(p-p')$ and $\delta(\boldsymbol{w}) = \left(\dfrac{1}{2\pi}\right)^3 \int e^{i\,\boldsymbol{w}\cdot\boldsymbol{x}}\,d^3x$, one now has:

$$\delta(\boldsymbol{p}-\boldsymbol{p}') = \left(\frac{1}{2\pi}\right)^3 \int e^{i(\boldsymbol{p}-\boldsymbol{p}')\cdot\boldsymbol{x}}\,d^3x = \left(\frac{1}{2\pi\hbar}\right)^3 \int e^{i(\boldsymbol{k}-\boldsymbol{k}')\cdot\hbar\boldsymbol{x}}\,d^3(\hbar x)$$

$$= \left(\frac{1}{\hbar^3}\right)\left(\frac{1}{2\pi}\right)^3 \int e^{i(\boldsymbol{k}-\boldsymbol{k}')\cdot\boldsymbol{s}}\,d^3s \rightarrow \delta(\boldsymbol{p}-\boldsymbol{p}') = \frac{1}{\hbar^3}\delta(\boldsymbol{k}-\boldsymbol{k}')$$

Recall the above is consistent with the identity $\delta(ax) = \frac{\delta(x)}{|a|}$ proven earlier. With the general form of $\langle \boldsymbol{x}|\hbar k\rangle = ce^{i\boldsymbol{k}.\boldsymbol{x}}$, now one can also write, using the closure relation that:

$$\delta(\boldsymbol{p}-\boldsymbol{p}') = \int \langle \hbar\boldsymbol{k}'|\boldsymbol{x}\rangle\langle \boldsymbol{x}|\hbar\boldsymbol{k}\rangle\, d^3x$$
$$= \int cc^* e^{i(\boldsymbol{k}-\boldsymbol{k}')\cdot x}\, d^3x = cc^*(2\pi)^3\delta(\boldsymbol{k}-\boldsymbol{k}').$$

One could thus deduce that $c = \frac{1}{(2\pi\hbar)^{\frac{3}{2}}}$. Note that $\langle \hbar k|x\rangle = \langle x|\hbar k\rangle^*$ and that in the convention we follow, $\langle \boldsymbol{x}|\hbar\boldsymbol{k}\rangle = \langle x|\hbar k_x\rangle\,\langle y|\hbar k_y\rangle\,\langle z|\hbar k_z\rangle$. Therefore:

$$\varphi^3(x) = \langle \boldsymbol{x}|\hbar\boldsymbol{k}\rangle = \frac{1}{(2\pi\hbar)^{\frac{3}{2}}} e^{ik\cdot x}$$

(2) **Hint:** Use the fact that $X_\mu^f\,\psi(k) = \langle \hbar k|X_\mu^v|\psi\rangle$.

---

### Exercise 1.8

(1) With the above, deduce that $\psi(\boldsymbol{k}) = \frac{1}{(2\pi\hbar)^{\frac{3}{2}}}\int \psi(x)e^{-ik\cdot x}d^3x.$

(2) The locality of an electron can be described by a vector $|\psi\rangle$ in an infinite-dimensional, abstract vector space. The component of this vector $\psi(x)$ (itself a continuous function) can be interpreted as the probability amplitude of finding the electron in the continuous position of $x$.

(a) Show in coordinate representation that the inner product of two such vectors is:

$$\langle \psi_2|\psi_1\rangle = \int_{-\infty}^{+\infty} \psi_2^*(x)\psi_1(x)dx$$

(b) Show that:

$$\langle x|[X, P^v]|x'\rangle = (x - x')\langle x|P^v|x'\rangle$$

where $X$ is the position operator and $P^v$ is the momentum operator.

**Hint:** Use the continuous closure relation.

(c) Show that $\int x\frac{d}{dx}\delta(x)\,dx = -\int \delta(x)\,dx.$

**Hint:** Use integration by parts.

(d) Hence show that $\langle x|P^v|x'\rangle = P^f\,\delta(x - x')$

where $P^v$ is the momentum operator in the vector space and $P^f$ is its coordinate representation in function space.

**Hint:** $[X, P^v] = i\hbar$.

Thus to find an operator in different spaces, one can define:

$$\left\langle x\middle|A^{v}\middle|\psi\right\rangle \equiv A^{f}\psi(x) \tag{1.12}$$

where superscripts $v$ and $f$ indicate vector and function spaces, respectively.

We will now study in more detail how to explicitly write $A^f$ such that one translates abstract mathematics to clear, quantifiable engineering physics. This will better illustrate and enumerate the dynamics of the motion of an electron. Let us begin with the Hamiltonian:

$$\left\langle x\middle|H^{v}\middle|\psi\right\rangle = H^{f}\psi(x). \tag{1.13}$$

In order for the above to be true:

$$\left\langle x\middle|H^{v}\middle|\psi\right\rangle = \int\left\langle x\middle|H^{v}\middle|x'\right\rangle\left\langle x'\middle|\psi\right\rangle dx' = \int H^{f}\delta(x-x')\left\langle x'\middle|\psi\right\rangle dx'. \tag{1.14}$$

By comparing integrand, one can deduce that:

$$\left\langle x\middle|H^{v}\middle|x'\right\rangle = H^{f}\delta(x-x'). \tag{1.15}$$

Alternatively, one can begin by finding the expression for the momentum by writing $\langle x|P^v|x'\rangle = \int \langle x|P^v|p\rangle\langle p|x'\rangle dp = \int p\,\langle x|p\rangle\langle p|x'\rangle dp$. Since, by definition, $P^f\langle x|p\rangle = p\,\langle x|p\rangle$, one now has:

$$\left\langle x\middle|P^{v}\middle|x'\right\rangle = \frac{1}{2\pi}\int P^{f}e^{ip(x-x')}dp = P^{f}\delta(x-x') \tag{1.16}$$

and subsequently:

$$\left\langle x\middle|P^{2v}\middle|x'\right\rangle = \int \left\langle x\middle|P^{v}\middle|x''\right\rangle\left\langle x''\middle|P^{v}\middle|x'\right\rangle dx'' = (P^{f})^{2}\delta(x-x'). \tag{1.17}$$

Hence, for a Hamiltonian operator $H$, which depends on positive powers of the momentum and position operators, we have $\langle x|H^v|x'\rangle = H^f\,\delta(x - x')$. But what is $P^f$ explicitly in the coordinate representation? In fact it can be found that $P^f = -i\hbar\dfrac{\partial}{\partial x}$.

## 1.4.2 Introduction to Green's function

The Green's function method is particularly useful for studies of electron transport. The propagator in quantum electrodynamics and chromodynamics are based on perturbative expansion of the Green's functions. In condensed matter physics, Green's functions are used in a similar fashion to study the many-body effects[3,4] (e.g. the Matsubara Green's function considers the temperature effect on electron dynamics). The Kubo–Greenwood formula was developed for conductivity. In modern times, the Keldysh modification of Green's function becomes particularly useful for nanoscale devices where electrical bias drives the electron channel into a state of non-equilibrium. The descriptions above are meant only to give an idea of the importance of Green's function. This section introduces the mathematical techniques.

*Complex integral*

We start by deriving an important identity used frequently in the Green's function description of electron propagation:

$$\lim_{\eta\to 0}\frac{1}{x\pm i\eta}=P\left(\frac{1}{x}\right)\mp i\pi\delta(x) \qquad [1.18]$$

where $P$ means the principal value. In order to avoid the pole at $x = 0$, integration should be carried out as shown by the arrow. This leads to:

$$\int_{-\infty}^{\infty}\lim_{\eta\to 0}\frac{f(x)}{x-i\eta}dx=\int_{-\infty}^{\infty}\frac{f(x)}{x}dx=\left[\int_{-\infty}^{-\eta}\frac{f(x)}{x}dx+\int_{+\eta}^{\infty}\frac{f(x)}{x}dx\right]+\int\frac{f(z)}{z}dz$$
$$=p\int_{-\infty}^{\infty}\frac{f(x)}{x}dx+\int\frac{f(z)}{z}dz \qquad [1.19]$$

where $\int\frac{f(z)}{z}dz$ is the integration along the contour of the semicircle taking a detour to the south instead of passing through the origin as shown in Fig. 1.1:

$$\int\frac{f(z)}{z}dz\approx f(0)\int\frac{1}{z}dz=f(0)\int_{-\pi}^{0}i\,d\theta=i\pi\,f(0). \qquad [1.20]$$

Alternatively, one can also prove that $\lim_{\eta\to 0}\frac{1}{x\pm i\eta}=p\left(\frac{1}{x}\right)\mp i\pi\delta(x)$ by decomposing $\frac{1}{x\pm i\eta}$ into $\frac{x}{x^2+\eta^2}\mp\frac{i\eta}{x^2+\eta^2}$.

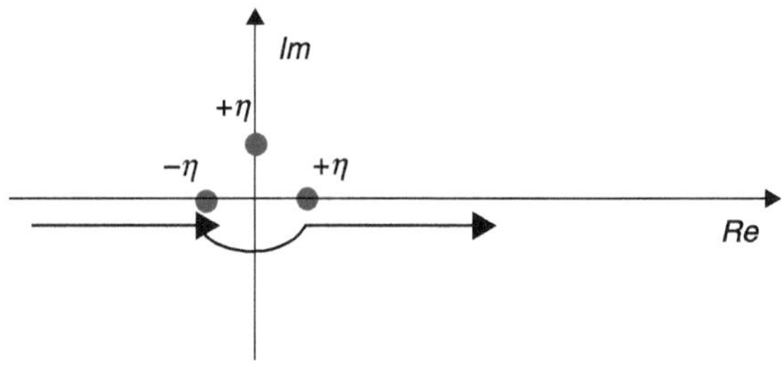


*1.1* A complex diagram to illustrate the integration of the function in Eq. 1.20.

*Electrical potential due to a charge density*

Green's function can be a useful method for solving a differential equation. For example, in electronics, one is often required to find the electrical potential for a fixed charge distribution:

$$\nabla_r^2 \phi(r) = -\frac{1}{\varepsilon}\rho(r). \qquad [1.21]$$

Making use of the Dirac delta function, one can write:

$$\nabla_r^2 \phi(r) = -\frac{1}{\varepsilon}\int \delta(r-r')\,\rho(r')\,dr' = -\frac{1}{\varepsilon}\int \nabla_r^2 G(r-r')\rho(r')\,dr'. \qquad [1.22]$$

In the above, a Green's function has been defined as $\nabla_r^2 G(r-r') = \delta(r-r')$. Thus, one can now deduce that:

$$\phi(r) = -\frac{1}{\varepsilon}\int G(r-r')\rho(r')\,dr'. \qquad [1.23]$$

Thus the electrical potential distribution can be found by solving $\nabla_r^2 G(r-r) = \delta(r-r')$. But solving for $G(r)$ may not be trivial. One needs the relationship $\mathfrak{F}\{\partial_x^n f(x)\} = (ik)^n g(k)$ mentioned above, where $g(k) = \mathfrak{F}\{f(x)\}$. Thus:

$$\mathfrak{F}\{\partial_r^2 G(r)\} = (ik)^2 G(k) \qquad [1.24]$$

and one is led to:

$$A\int \delta(r)\,e^{-ikr}\,dr = (ik)^2 G(k)$$

$$G(k) = A\frac{-1}{k^2} \qquad [1.25]$$

where $A$ is the constant in the Fourier transform. Thus $G(r)$ can be found by inverse Fourier transform of $G(k)$, i.e.

$$G(r) = B\int d^3k\, e^{ik.r} G(k) = \int d^3k\, \frac{-e^{ik.r}}{k^2} AB = \frac{1}{4\pi r} \qquad [1.26]$$

where $B$ is the constant in the inverse Fourier transform. As is shown in Section 1.3, $AB = \left(\frac{1}{2\pi}\right)^3$. Thus one sees that if $A = \left(\frac{1}{\sqrt{2\pi}}\right)^3$, $B = \left(\frac{1}{\sqrt{2\pi}}\right)^3$, or if for convenience $A = 1$, then $B = \left(\frac{1}{2\pi}\right)^3$.

---

### Exercise 1.9

The physics of the electric field in nanoscale electronics is often described in differential equations that can be solved with integration techniques.

(1) Show, using Laplace transform or other methods that $\int_0^\infty \frac{\sin y}{y}dy = \frac{\pi}{2}$.

(2) For $\nabla^2 G(r) = \delta(r)$, show that its inverse Fourier transform is $G(k) = \frac{-1}{k^2}$, where $k$ is the wavevector. Eventually, making use of results in (1), prove that $G(r) = 1/4\pi r$.

(3) In free space, the electric potential $V(r)$ due to the presence of charge density $\rho(r)$ is given by $\nabla^2 V = \rho(r)$. Show that:

$$V(r) = \int \frac{\rho(r')}{4\pi(r - r')} dr'.$$

(4) Thus for a point charge located at $r = 0$, show that the solution for the electric potential is $V(r) = \dfrac{1}{4\pi r}$.

(5) In a many-electron system (e.g. a device channel), the effects of screening is important because it determines the effect of the external electric field on electron motion. The potential–charge relation is modified to $\nabla^2 V = q^2 e^{-qr} + 4\pi\delta(r)$. Show that the solution will then be:

$$V(r) = \frac{1}{r} e^{-qr}.$$

(6) Using the fact that $G(r) = 1/4\pi r$, show that $\nabla^2 \left(\frac{1}{r}\right) = -4\pi\delta^3(r)$.

***Solution***

(1) Differentiate with respect to an auxiliary parameter. Define $I(a) = \int_0^\infty e^{-ay} \frac{siny}{y} dy$,, so that $\frac{dI}{da} = -\int_0^\infty e^{-ay}\ sinydy$. The above can continue straightforwardly using by parts integration, but we speed up with the following $\frac{dI}{da} = -\mathrm{Im}\int_0^\infty e^{(i-a)y} dy = -Im\frac{1}{i-a} = -\frac{1}{1+a^2}$.

Then $I(b) = \int_{-\infty}^{b} \frac{dI}{da} da = -\int_{-\infty}^{b} \frac{1}{1+a^2} da = \frac{\pi}{2} - tan^{-1}\ b.$

Hence $\int_0^\infty \frac{siny}{y} dy = lim_{a\to 0} I(a) = \frac{\pi}{2}$.

**Hint:** $e^{ik.r} = e^{ikrcos\theta}$; $r$ is taken to be the $z$ axis, symmetry allows $r$ to be taken in any direction.

**Hint:** Rewrite the differential equation $\nabla^2 V = \rho(r)$ into an integral equation with the help of $\nabla^2 G(r)$.

---

The retarded Green's function is particularly useful for describing electron propagation and many other properties related to electron propagation in condensed matter or nanoscale devices. The retarded Green's function by definition is:

$$G^R(xx', tt') = -i\left\langle x \middle| e^{-iH'\tau} \middle| x' \right\rangle \theta(\tau) \qquad [1.27]$$

where $\tau = t - t'$. Choosing the eigenstates $|\alpha\rangle$ as basis vectors, one can also write:

$$iG^R(xx', tt') = \theta(\tau)\sum_\alpha \phi_\alpha(x)\phi_\alpha^\dagger(x')e^{-iE_\alpha\tau}. \qquad [1.28]$$

Fourier transform with respect to spatial difference can be performed on $G^R(xx', tt')$, taking $\phi_\alpha(x) = e^{ik_\alpha x}$. The eigenstates of a non-interacting system are plane waves due to translational invariance.

$$
\begin{aligned}
iG^R(k,\tau) &= \int \sum_\alpha \phi_\alpha(x)\phi_\alpha^\dagger(x')e^{-iE_\alpha\tau}\theta(\tau)e^{-ik(x-x')}d(x-x') \\
&= \int \sum_\alpha e^{-i(k-k_\alpha)(x-x')}e^{-iE_\alpha\tau}\,\theta(\tau)d(x-x') \\
&= \sum_\alpha \delta(k-k_\alpha)e^{-iE_\alpha\tau}\theta(\tau) = e^{-iE_k\tau}\theta(\tau).
\end{aligned} \quad [1.29]
$$

In the context of nanoelectronics, $iG^R(k,\tau)$ can be viewed as the coherent, continued propagation of electron. Its Fourier transform with respect to time difference is:

$$
iG^R(k,E) = \int e^{iE\tau}\; e^{-iE_k\tau}\theta(\tau)\,d\tau = \int_0^\infty e^{i(E-E_k)\tau}\,d\tau = \frac{1}{(E-E_k+i\eta)}. \quad [1.30]
$$

Thus:

$$
iG^R(xx',E) = \sum_\alpha \frac{\phi_\alpha(x)\phi_\alpha^\dagger(x')}{(E-E_\alpha+i\eta)}. \quad [1.31]
$$

We have dealt with the explicit Green's function, but what is its origin? Retarded Green's function in quantum mechanics is a solution to a differential equation related to the Schrödinger equation, i.e.

$$
(i\partial_t - H^f)G^R(xx',tt') = \delta(x-x')\delta(t-t'). \quad [1.32]
$$

Compare this to the earlier expression where $\nabla^2 G(r) = \delta(r)$. The Green's function above is a solution to a differential equation of the above. We have seen that $G(r)$ is a useful correlator for functions that solve $\nabla^2\phi(r) = \rho(r)$. Similarly $G^R(xx', tt')$ will be a useful correlator for functions that solve $(i\partial_t + \nabla^2)\,\psi(r) = 0$, where $\psi(r)$ is the wavefunction of the system. In perspectives:

$$
\begin{aligned}
&(i\partial_t - H^f)G^R(xx',tt') = \delta(x-x')\delta(t-t') \\
&(i\partial_t + \nabla^2)\psi(r) = 0.
\end{aligned} \quad [1.33]
$$

$\psi(r)$ is the wavefunction that contains all the electron's information, but it is difficult to solve for. $G^R(rr', tt')$ is the correlation function that contains information related to the electron's dynamics. The Green's function might contain less information on the system but it is easier to solve for and is therefore a more powerful tool with respect to electron dynamics.

Green's function is a correlator in general. But we will be interested in a more specific interpretation of the physical significance of various Green's functions in nanoelectronics and nanotechnologies. The retarded Green's function can be regarded as a propagator which describes the coherent dynamic of an electron in the electronic device. The Fourier transformed $G^R$ with respect to time represents the strength of propagation at different energy levels. The mathematical expression $G^R(rr', tt') = -i\langle r|e^{-iH^V\tau}|r'\rangle\,\theta(\tau)$ can be regarded as the probability amplitude (chance) that an electron found at earlier time $t'$ in location $r'$ will be found at a later time $t$ in location $r$. Similarly one can construct a function that describes the 'chance' that an electron in state $|n'\rangle$ at earlier time $t'$ will be found in state $|n\rangle$ at a later time $t$.

$$G^R(nn',tt') = -i\left\langle n\left|e^{-iH^v\tau}\right|n'\right\rangle\theta(\tau) \quad [1.34]$$

where $|n_i\rangle$ belongs to a complete set of basis states. Recalling that $\tau = t - t'$, one can write:

$$G^R(xx',tt') = -i\left\langle x\left|e^{-iH^v\tau}\right|x'\right\rangle\theta(\tau) = \sum_{nn} -i\langle x\,|\,n\rangle\left\langle n\left|e^{-iH^v\tau}\right|n'\right\rangle\langle n'\,|\,x'\rangle\,\theta(\tau)$$
$$= -i\sum_{nn}\phi_n(x)G^R(nn',tt')\phi^\dagger_{n'}(x')\theta(\tau). \quad [1.35]$$

In the absence of scattering, it is certain that an electron will continue to propagate in its initial eigenstate, accumulating only phase changes to its wavefunction. This, as is proved later, is evident in the expression $G^R(kk', tt') = G^R(k, tt')\delta_{kk'}$. Thus, Eq. 1.35 can be written as:

$$G^R(xx',tt') = -i\sum_{nn'}\phi_n(x)\phi^\dagger_{n'}(x')\,G^R(n,tt')\delta_{nn'}\theta(\tau). \quad [1.36]$$

If $|n\rangle$ and $|n'\rangle$ are eigenstates of the Hamiltonian, one then has:

$$G^R\left(xx',tt'\right) = -i\sum_{nn}\phi_n(x)\phi^\dagger_{n'}(x')\,e^{-iE_{n'}\tau}\delta_{nn'}\,\theta(\tau)$$
$$= -i\sum_{n}\phi_n(x)\phi^\dagger_n(x')e^{-iE_n\tau}\,\theta(\tau). \quad [1.37]$$

An important function that can be derived from the studies of the Green's function is the spectral function, which by definition is:

$$A(k,E) = -2\,Im\,G^R(k,E). \quad [1.38]$$

We have learnt that $G^R(k,E) = \dfrac{1}{E - E_k + i\eta}$ and that $\dfrac{1}{\omega + i\eta} = P\dfrac{1}{\omega} - i\pi\delta(\omega)$.

Thus the explicit expression of the spectral function is:

$$A(k,E) = 2\pi\delta(E - E_k). \quad [1.39]$$

The spectral function is a delta function which says that injecting an electron into the system can generate excitation only when the energy of the electron is $E_k$. It is thus related to the density of states at a given energy.

$$\int A\frac{dE}{2\pi} = 1. \quad [1.40]$$

In the absence of scattering and in the $k$ eigenstate of the system:

$$G^R(k,tt') = -i\theta(\tau)e^{-iE_k\tau}$$
$$G^A(k,tt') = i\theta(-\tau)e^{-iE_k\tau}$$
$$G^>(k,tt') = -i(1 - n_k)e^{-iE_k\tau}$$
$$G^<(k,tt') = in_k e^{-iE_k\tau}. \quad [1.41]$$

Note that the above Green's functions spawn 12 other versions with Fourier and inverse Fourier transforms of the space and time parameters. Physically, $G^>(k,tt')$ is related to the propagation of electrons because it is proportional to empty states; $G^<(k,tt')$ is related to the propagation of holes because it is proportional to the electron filling those empty states. In fact it is more intuitive to view $iG^>(k,tt')$ as a measure of the chance/strength of electron scattering into a point in k-space, while $-iG^<(k,tt')$ as a measure of the chance/strength of electron scattering out, or holes scattering in as shown in Fig. 1.2.

Since $G^<(k,tt') = in_k e^{-iE_k\tau}$, it makes sense to deduce that $-iG^<(k,0) = n_k$ represents the number of electrons at a particular $k$. In fact, in many-body physics, the above is simply one example of an equal time field operation, i.e.

$$-iG^<(k,0) = \langle c_k^\dagger(t)c_k(t)\rangle = n_k. \quad [1.42]$$

In the equilibrium condition, $n_k = f_k$ where $f_k$ is the Fermi Dirac distribution function. By inverse Fourier transform, $-iG^<(k,\tau) = -i\int \frac{dE}{2\pi} G^<(k,E)e^{-iE\tau}$. Note that $B = 1/2\pi$ and $A = 1$ is normally followed in inverse Fourier transform and Fourier transform, respectively. One can deduce the same for the equal time correlation function as follows:

$$-iG^<(k,0) = -i\int \frac{dE}{2\pi} G^<(k,E) = \int \frac{dE}{2\pi} n_k 2\pi\, \delta(E - E_k) \quad [1.43]$$

which merely states that the equal time lesser correlation function is the electron density. In nanoscale electronic devices with leads playing the function of electron reservoirs, one can make use of the kinetic equation of:

$$-iG^<(k,0) = \sum_\alpha G^R f_\alpha \Gamma_\alpha G^A. \quad [1.44]$$

From the above, one can thus define a $A_C^\alpha = G^R\Gamma_\alpha G^A$, which will now carry the physical meaning of the density of states of the central region due to lead $\alpha$ where $\alpha$ is the lead index. Electron density in the central region is $n_c = f_\alpha A_c^\alpha$ with $A_c^\alpha$ playing the role of something like the transfer density of states. The

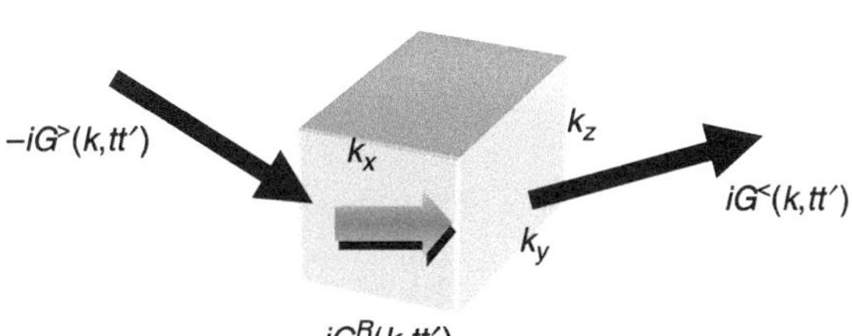


*1.2* Schematic illustration of the physical meanings of the various Green's functions in the context of electron transport in nanoscale electronic devices.

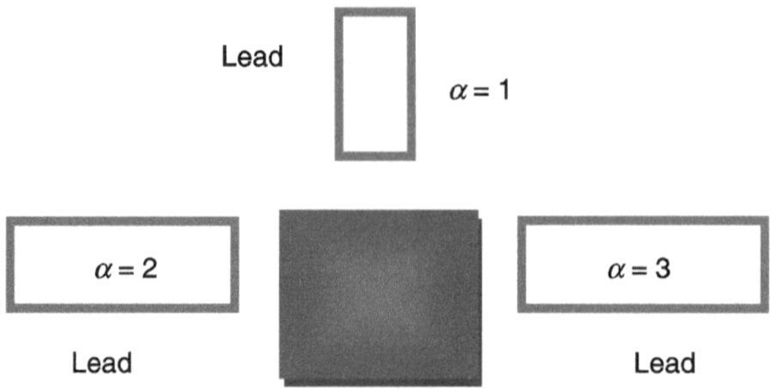


*1.3* Schematic illustration of a multi-part nanoscale device with the central region being where non-equilibrium electron transport is the main interest in this book.

non-equilibrium effect enters via $f_\alpha$ which measures the distribution ('chance') of electron presence in the individual leads as shown in Fig. 1.3.

---

**Exercise 1.10**

(1) Broadening of the spectral function occurs in the presence of interaction. For $G^R(k,\tau) = -i\theta(\tau)e^{-iE_k\tau}\,e^{-\tau/t_d}$, prove that the spectral function is broadened, i.e. $A(k,E) = \dfrac{2/t_d}{(E-E_k)^2 + (1/t_d)^2}$.

(2) The retarded Green's function is explicitly $G^R(xx',tt') = -i\langle x|e^{-iH\tau}|x'\rangle\,\theta(\tau)$. Show that the following is true:

$$(i\partial_t - H_0)\,G^R(xx',tt') = \delta(x-x')\delta(t-t').$$

**Hints:** $\dfrac{d}{dx}\theta(x) = \delta(x)$. Use chain rules in differentiation. Alternatively, one can use the equation of motion method.

(3) (a) Explain the physical significance of the various Green's functions and the spectral functions with respect to electron dynamics.
(b) Why is Green's function a more powerful tool than the wavefunction with respect to electron dynamics?
(c) Derive the energy version of the following Green's functions:
$G^R(k,tt') = -i\theta(\tau)e^{-iE_k\tau}$
$G^A(k,tt') = i\theta(-\tau)e^{-iE_k\tau}$
$G^>(k,tt') = -i(1-n_k)e^{-iE_k\tau}$
$G^<(k,tt') = in_k e^{-iE_k\tau}$.

---

## 1.5 Second quantization for electron accounting

Nanoscale electronic devices consist of a reasonably large number of electrons that interact to give rise to correlation effects on electron transport which cannot be neglected, nor can these effects be hidden under simple macroscopic manifestation (e.g. conductivity and mobility). Any reliable formalism for describing them must embrace many-body considerations, which is usually done via 'second quantization'. In essence, it involves the electron creation operators ($c_i^\dagger$) and the annihilation operators ($c_i$) that keep track of the number of electrons in the state of the system.

In this section, we choose the lattice sites as the basis states. For instance, let us define $|a, b, c\rangle$ as the state with $a$ particles on spatial site 1, $b$ particles on site 2 and $c$ particles on site 3. Here $a$, $b$, and $c$ are either 0 or 1 because of the Pauli exclusion principle.

$$c_2^\dagger \,|\, a,b,c\rangle = |\, a,b+1,c\rangle; \qquad c_2^\dagger \,|\, a,0,c\rangle = |\, a,1,c\rangle \qquad [1.45]$$

meaning that $c_2^\dagger$ is a creation operator that adds one electron to site 2, and:

$$c_1 \,|\, a,b,c\rangle = |\, a,-1,b,c\rangle; \qquad c_1|\, 1,b,c\rangle = |\, 0,b,c\rangle \qquad [1.46]$$

since $c_1$ is an annihilation operator that subtracts one electron from site 1. The famous Pauli exclusion principle states that no two electrons (in general fermions) can occupy the same state. This is because fermionic states are anti-symmetric under an exchange of particles. In our context, it implies that we cannot have two or more electrons in the same site. Hence, at any site $i$:

$$c_i^\dagger c_i^\dagger = 0; \qquad c_i c_i = 0. \qquad [1.47]$$

Actually, this fact can be generalized to $\{c_i^\dagger, c_j^\dagger\} = c_i^\dagger c_j^\dagger + c_j^\dagger c_i^\dagger = 0$, which means that adding two electrons in an opposite order (which amounts to swapping them) gives a state with the opposite sign. Likewise, $\{c_i, c_j\} = 0$. The operator $c_i^\dagger c_i$, however, has a special significance: it means the number of electrons on site $i$. Let us consider the effect of $c_3^\dagger c_3$ on spatial site 3. If there are no electrons on site 3, $c_3^\dagger c_3|\, a, b, 0\rangle = c_3^\dagger (c_3|\, a, b, 0\rangle) = 0$ because there is nothing to annihilate at site 3. But $c_3^\dagger (c_3|\, a, b, 1\rangle) = c_3^\dagger|\, a, b, 0\rangle = +\,|a, b, 1\rangle$. We thus see that $c_3^\dagger c_3$ is an operator on the state of the system with eigenvalues +1 or 0, depending on whether an electron is present. In general, the operator:

$$N = \sum_i c_i^\dagger c_i \qquad [1.48]$$

is known as the number operator that counts the total number of electron in the system. The above method, which keeps track of electrons in discrete spatial sites, is particularly suited to nanoscale electronic systems that contain discrete parts (e.g. semi-infinite contact reservoirs and a finite-sized central region).

We will show its relation to the continuous spatial formalism commonly used in condensed matter many-body physics. The quantum field associated with an electron is:

$$\psi(x)=\sum_k c_k\phi_k(x);\ \psi^\dagger(x)=\sum_k c_k^\dagger\phi_k^*(x) \qquad [1.49]$$

where $k$ is the momentum index. In momentum eigenstates instead of states labeled by sites, index $i$ will be replaced by the momentum variables $k$, $q$ or $p$. The total number of electrons in the system is:

$$\int\psi^\dagger(x)\psi(x)\,dx=\int\sum_{k\,k'}c_{k'}^\dagger\phi_{k'}^*(x)c_k\phi_k(x)\,dx=\sum_{k\,k'}c_{k'}^\dagger c_k\delta_{kk'}=\sum_k n_k. \qquad [1.50]$$

The above is the sum of all electrons over their momentum states, which will yield the total number of particles $N$ in the system. This leads one to deduce that:

$$\psi^\dagger(x)\psi(x)=\rho(x) \qquad [1.51]$$

which is the particle density operator. Table 1.6 shows the identities involving the use of these electron operators.

*Table 1.6* Below is a list of useful identities for second quantized fermionic operators

| | Useful identities |
|---|---|
| 1. | $[c_j, n_j] = c_j$ |
| 2. | $[c^\dagger_j, n_j] = -c^\dagger_j$ |
| 3. | $\{c_i, c_j\} = 0$ |
| 4. | $\{c^\dagger_i, c^\dagger_j\} = 0$ |
| 5. | $\{c_i, c^\dagger_j\} = \delta_{ij}$ |
| 6. | All commutations between electron and number operators denoted by $i$, $j$ are zero, e.g. $[a_i, b_j] = 0$ |

**Exercise 1.11**

We will frequently need to evaluate the commutations of operators involving more than one creation/annihilation operator.

(1) Prove the identity $[AB, C] = A\,[B, C] + [A, C]\,B$.
(2) Derive a similar identity involving $[A, BC]$.
(3) Prove the Jacobi identity $[A, [B, C]] + [B, [C, A]] + [C, [A, B]] = 0$.

Note that $[A, B] = AB - BA$.

**Exercise 1.12**

Show that $[n_j n_k, c_j] = -n_k c_j$ where $n_j = c_j^\dagger c_j$.

***Solution***

$$[n_j\, n_k, c_j] = n_j\, n_k\, c_j - c_j\, n_j\, n_k = -c_j\, n_j\, n_k = -n_k\, c_j\, n_j.$$

The above can be written as:

$$-n_k\, c_j\, n_j = -n_k\, c_j\, c_j^\dagger c_j = -n_k\,(1 - c_j^\dagger c_j)c_j = -n_k\, c_j.$$

---

In retrospect, many-body physics began with the many-body wavefunction which gives the probability amplitude of finding $N$ number of particles in a system. But to find the explicit solution for the many-body wavefunction is an immense task. This is not surprising as the wavefunction is the probability amplitude of the whereabouts of all the particles! Thus the Green's function, which means the probability of finding a particle at place $r'$ at time $t'$ given it was found at $r$ at time $t$, is probably easier to find. In fact the Green's function for a particle, even in the presence of interaction due to many other particles, is handier and more useful than the wavefunction – especially with respect to electron dynamics. This is the major motivation behind the elaborate use of Green's function to understand electron propagation and scattering in nanoscale devices. In field theoretic condensed matter physics, the Green's function has been used to study the effect of interactions on electron energy, transport, and so forth. Thus, the Green's function is also a natural tool to include many-body physics in nanoelectronics. The incorporation of the effects of electron–electron and electron–phonon scattering are particularly useful. In the more recent developments where additional degree of freedoms are studied (e.g. spintronics, graphene electronics), the Green's function provides a platform for including spin orbit coupling and spin–spin interaction.

In nanoelectronics, one may describe electron dynamics in the conduction band only. This point will be apparent in the many-body description, where:

$$\langle \psi \,|\, c_k^\dagger c_k \,|\, \psi \rangle \qquad [1.52]$$

really depends on what the many-body state vector is. In nanoelectronics, it makes sense to keep analysis to one band; for example, $|\psi = C\rangle$ refers to the conduction band only. However, in a more general context, $|\psi = G\rangle$ is the ground state that represents all filled states below the Fermi level and all unoccupied states above it at zero temperature. Furthermore, the operator $c_k$ morphs an electron at above the Fermi level to a hole at below the Fermi level. Thus, $\langle G|c_k^\dagger c_k|G\rangle = \langle G|n_k|G\rangle$ is always zero because in the state of $|G\rangle$ there will be no electron above the Fermi level and no hole below the Fermi level:

$$\langle G \,|\, c_k^\dagger c_k \,|\, G\rangle = \langle G \,|\, n_k^e \,|\, G\rangle\, \theta(k - k_F) + \langle G \,|\, n_k^h \,|\, G\rangle\, \theta(k_F - k). \qquad [1.53]$$

Since $|G\rangle$ is the filled Fermi sea at zero temperature:

$$\langle G \,|\, c_k^\dagger c_k \,|\, G\rangle = f_{T=0}\, \theta(k - k_F) + (1 - f_{T=0})\, \theta(k_F - k) = 0. \qquad [1.54]$$

By contrast in nanoelectronics, $c_k$ is the operator for electron. Thus:

$$\langle C \mid c_k^\dagger c_k \mid C \rangle = \langle C \mid n_k^e \mid C \rangle. \qquad [1.55]$$

Since $|C\rangle$ refers to states above the Fermi level, at finite temperature:

$$\langle C \mid c_k^\dagger c_k \mid C \rangle = f_{T\neq 0}\, \theta(k-k_F) = \frac{\theta(k-k_F)}{1+\exp\left(\dfrac{E-E_F}{kT}\right)} < 1. \qquad [1.56]$$

Second-quantized representation of a measurable quantity in a system (e.g. momentum, spin, energy) is important when one needs to derive the Green's function to describe the motion of electron in that system. For example, the spin angular momentum in second-quantized form is:

$$S_z^{(2)} = \sum_{k\,k'\sigma\,\sigma'} \langle k\sigma \mid S_z \mid k'\sigma' \rangle\, c_{k\sigma}^\dagger c_{k'\sigma'} \qquad [1.57]$$

where $k$ is the momentum and $\sigma$ the spin quantum number. One can easily show that the above leads to:

$$S_z^{(2)} = \sum_k (c_{k\uparrow}^\dagger c_{k\uparrow} - c_{k\downarrow}^\dagger c_{k\downarrow}). \qquad [1.58]$$

It is also important to note the following relation for a general measurable quantity in second-quantization:

$$O^{(2)} = \sum_{k\,k'\,\sigma\,\sigma'} \langle k\sigma \mid O^v \mid k'\sigma' \rangle c_{k\sigma}^\dagger c_{k'\sigma'} = \int \psi^\dagger(x) O^f \psi(x)\, dx. \qquad [1.59]$$

## 1.6 References


[1] Ballentine L E (2003), *Quantum Mechanics: A Modern Development*, World Scientific, New York.

[2] Merzbacher E (1998), *Quantum Mechanics*, John Wiley & Sons, New York.

[3] Bruus H and Flensberg K (2004), *Many-Body Quantum Theory in Condensed Matter Physics*, Oxford University Press, Oxford.

[4] Fetter A L and Walecka J D (2003), *Quantum Theory of Many-Particle Systems*, Dover Publications, Mineola, NY.

# 2
# Nanoscale physics and electronics

**Abstract:** This chapter defines an important area of physics we call the physics of nanoscale electronics. The core concepts of non-equilibrium, size effects and neighboring perturbation are introduced and a quick run through the emerging topics including metal spintronics, semiconductor spintronics, single electronics and quantum dot, molecular electronics, carbon nanotube transistors and graphene electronics is provided. Lastly, we discuss the electronic environment in which all carrier transport takes place. The effect of ionic potential results in the bandstructure of metal, semiconductor and insulator. The effect of carrier interaction gives rise to electron gas, liquid and solid.

**Key words:** non-equilibrium, size effects, neighboring perturbation, spintronics, quantum dot, single electronics, carbon nanotube, graphene, interaction, electron gas.

## 2.1 Introduction to nanoscale electronics

What is 'nanoscale electronics' and how does it differ from the conventional electronics with which we are familiar? The term 'electronics' is related to electric current, that is, the flow of electrons driven by, for example, an externally applied voltage across a device channel. Why then does one need to prefix it with 'nano'? To begin with, an electron is a point particle which might not even have a well-defined volume. The 'nano' prefix would only be a superfluous term, wouldn't it? These are common questions that arise when one is introduced to this novel, exciting and rapidly developing branch of electronics – 'nanoelectronics'. Nanoelectronics is an area which encompasses, among others, subtopics such as spintronics, topological insulators and carbon-based electronics (principally carbon nanotubes and graphene).

Indeed, electron interaction occurs over the average distance between individual electrons, which of course is on the nanometer scale. We are also familiar with the fact that electrons, like other fundamental particles, are governed by the axioms or rules of quantum mechanics. However, more often than not in electronic physics, the rules of quantum mechanics are implicit. The degree of freedom (DOF) relevant to observation/measurement is normally macroscopic. Underlying these macroscopic observables or manifestations are microscopic DOFs which can be treated as constituent effects of the macro manifestation.

In conventional electronics, one usually does not need to consider quantum effects, other than the fact that electrons obey the Fermi–Dirac distribution. The transport and dynamics of these electrons can be adequately described by the semiclassical drift and diffusion equations, which depend on macroscopic properties such as mobility and conductivity. In fact, in micron-sized devices such as the conventional metal–oxide–semiconductor field-effect transistor (MOSFET), the Fermi–Dirac distribution and the semiclassical drift–diffusion equations are all that is required to describe the physics of electron transport under, for example, the effect of a biased electrochemical potential. In other words, condensed matter physics has been rather decoupled from the device and electron transport physics. This also explains why the in-depth study of condensed matter physics has traditionally been confined mainly to superconductivity, matter phase transition and other strongly correlated phenomena, but not to the ubiquitous semiconductor devices. In fact, with hindsight, it is rather odd to think that propagators (Feynman diagrams) and gauge dynamics, so well established in high energy physics and superconductivity, have had so little application in the physics of electron current and dynamics in conventional electronics.

This scenario is about to undergo a drastic change: nanoelectronics will mark the beginning of the application of advanced physics in devices and electronic systems. The mesoscopic aspects of condensed matter physics become more important due to the following reasons:

1. In nanoscale systems, the boundary effects are becoming prominent, giving rise to sub-band and edge effects, and tunneling phenomena. For example, in single electronics, the quantization of electronic energy and the absence of all continuous states in the zero-dimensional quantum dot result in a novel correlated transport process where the flow of electrons no longer obeys the usual ohmic current–voltage (I–V) characteristics. Other examples include a two-dimensional electron gas (2DEG) in high-electron mobility transistors (HEMTs) and nanowires in carbon nanotube devices, where the presence of sub-bands and circumferential boundary conditions, respectively, alters the electron transport behavior.
2. Many of the new bandstructures or sub-bands that arise due to geometrical modification or reduction in the device generate momentum-dependent internal DOFs (e.g. spin–orbit coupling, graphene pseudospin DOF). Some of these may lead to topological properties in the transport behavior (e.g. quantum Hall effect, quantum spin Hall effect and others).
3. Due to the shrinking size of devices, tunneling current (e.g. across some oxide barrier), which is a purely quantum mechanical phenomenon, can no longer be ignored.
4. As devices become small, it is also possible for electrons to travel through an entire device without being scattered. Ballistic transport becomes important and electron motion can be investigated in terms of wavefunction or non-interacting Green's function. In fact, in the absence of interaction, electron dynamics can be directly described with basic quantum mechanics.
5. As the interaction effect becomes more arbitrary and less intense, internal DOFs which normally survive over a short length scale would manifest

themselves. For example, the spin coherence of an ensemble of electron spin can be observed in a nanoscale device shorter than the spin diffusion length.

6. For a device channel length approaching the screening length, the potential drop across the channel becomes significant. Thus specific corrections might be needed to control the on/off states of transistor devices.
7. Interaction will not totally disappear even in small devices. In fact, the effect of interaction becomes more complicated as the number of electrons is too small for sufficient averaging that would lead to statistical macroscopic DOFs. One simply cannot rely on the statistical effects to protect the macroscopic observables from microscopic interactions. In other words, small device size means fewer electrons with less intense interactions but, paradoxically, more detectable results (due to the lack of statistical averaging).

In the following, we discuss one of the most important consequences of device miniaturization in the nanoscale range, that is, the effects of boundary which gives rise to sub-bands. The sub-band configuration superimposes itself on the intrinsic energy band arising due to the periodic ionic core potential. Rapid advances in nanofabrication technologies have enabled us to fabricate lower (two, one or even zero) dimensional structures and devices via either the bottom-up or, more commonly, the top-down approach. If we regard a three-dimensional bulk structure as an infinitely large system for the electron, fabricating nanostructures means imposing artificial boundaries to the erstwhile infinite structure.

### 2.1.1 Three-dimensional nanostructures

A three-dimensional (3D) system consists of a periodic crystalline structure and macroscale geometrical boundaries which define its shape. Such a system lacks any boundaries at the nanoscale, and thus, one would not expect additional energy bands or sub-bands apart from its own intrinsic bandstructure. Typical metal-based devices treated as 3D systems include spin valves and magnetic tunnel junctions (MTJ). The former have been widely used in recording technologies such as hard disk drives. The latter are the crucial storage element of the magnetic random access memory (MRAM). These devices are also known as current-perpendicular-to-plane devices. Although they appear to be two-dimensional (2D) devices, with one dimension (their thickness) being much smaller than the other two lateral dimensions, electron transport should still be treated as in a 3D system, owing to the small Fermi wavelength in metals. Because of the translational symmetry in the in-plane directions, the electron transport in these spin valve multilayers or MTJs can be modeled by assuming a simple one-dimensional (1D) treatment. Figure 2.1 provides a pictorial depiction of a 3D electronic system.

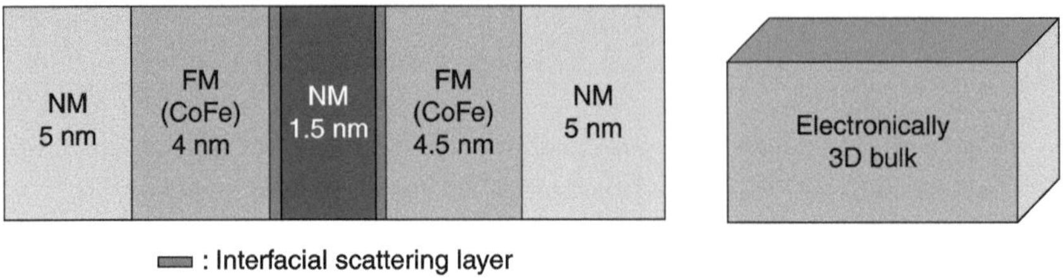


*2.1* Schematic diagram of a multilayer spin valve. Although it appears visually as a two-dimensional system, electronically, it can be treated as a three-dimensional bulk system. NM, non-magnetic; FM, ferromagnetic.

## 2.1.2 Two-dimensional nanostructures

A two-dimensional nanostructure is confined in one dimension, normally referred to as the vertical *z*-axis. An electron is free to move in two dimensions contained in a plane perpendicular to the confinement dimension. Semiconductor 2D electron gas is a typical example of a 2D system. Such a 2D electron gas system was first realized in the inversion layer of silicon-based MOSFET. Subsequently, we showed in Fig. 2.2 a 2DEG which was realized in $GaAs/Ga_xAl_{1-x}As$ heterostructures, where long mean free paths exceeding 10 μm can be achieved. This makes 2DEG heterostructures important for high-speed electronics.

The next most commercially important 2D system is graphene. Graphene is a monolayer of carbon atoms packed into a 2D honeycomb lattice. Given its thickness of just one atomic layer, it is the first truly 2D system. Thus, monolayer graphene is the basic building block of all carbon-based nanostructures in higher dimensions. The effect of boundary in 2D systems, however, becomes more complicated when magnetic field is applied along the vertical dimension (e.g. due to the formation of Landau levels).

Since in a 2D system an electron is restricted in one direction but free to move around in the other two, quantization along the confined direction gives rise to an energy sub-band in addition to the usual energy bands arising due to the 2D periodic core potential. The effect on the conduction electron of the periodic core potential is captured by the effective electron mass. The single-electron Hamiltonian of a 2DEG in the presence of perpendicular magnetic fields and linear spin orbit coupling (SOC) effects is given by:

$$\frac{-\hbar^2}{2m}\phi''(z)\psi(x,y)+ezF_z\phi(z)\psi(x,y)$$
$$+\left[\frac{(p_x+eA_x)^2}{2m}+\frac{p_y^2}{2m}+U+H_Z+H_{SO}\right]\phi(z)\psi(x,y)$$
$$=E\phi(z)\psi(x,y) \qquad [2.1]$$

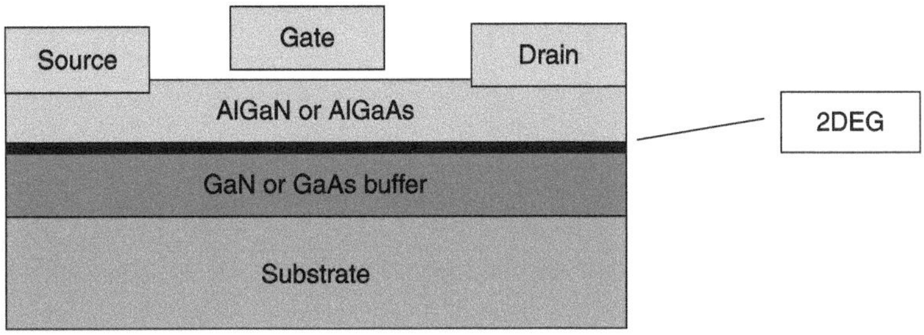


*2.2* Schematic of a typical 2DEG device, which has become an essential component in high-speed electronics as well as in future spintronic devices.

where $F_z$, $H_{SO}$, $H_z$ and $U$ are terms corresponding to the 2DEG triangular confining potential, the linear spin orbit coupling, the Zeeman split and the applied electric potential, respectively. The solution to the bound wavefunction $\phi(z)$ is given by a linear combination of Airy functions, i.e.

$$\phi(z) = e^{-\frac{bz}{2a}} Ai\left[\frac{b^2 + 4ac - 4aE_n + 4adz}{4a^{4/3}d^{2/3}}\right]C_1$$
$$+ e^{-\frac{bz}{2a}} Bi\left[\frac{b^2 + 4ac - 4aE_n + 4adz}{4a^{4/3}d^{2/3}}\right]C_2 \qquad [2.2]$$

where $C_1$ and $C_2$ are constants, and $a = \frac{\hbar^2}{2m}, b = 0, c = \frac{p_x^2}{2m} + \frac{(p_y - eA_y)^2}{2m}$ $+ U + E_Z + E_{SO}$, and $d = eF_z$ ($E_z$ and $E_{SO}$ are the Zeeman and spin orbit energies). Solving for the Hamiltonian at $z = 0$, the lowest energy eigenvalue of the system is given by:

$$E = \frac{(p_x + eA_x)^2}{2m} + \frac{p_y^2}{2m} + U + E_Z + E_{SO} + 2.338\left(\frac{\left[eF_z\hbar\right]^2}{2m}\right). \qquad [2.3]$$

Thus, the energy of the lowest sub-band due to vertical confinement is $E_S = 2.338\left(\frac{\left[eF_Z\hbar\right]^2}{2m}\right)$, assuming the triangular well approximation for the confining potential.

## 2.1.3 One-dimensional nanostructures

To date, the most important one-dimensional system in nanoelectronics is the carbon nanotube (CNT). The CNT is a normal graphite sheet rolled up into a

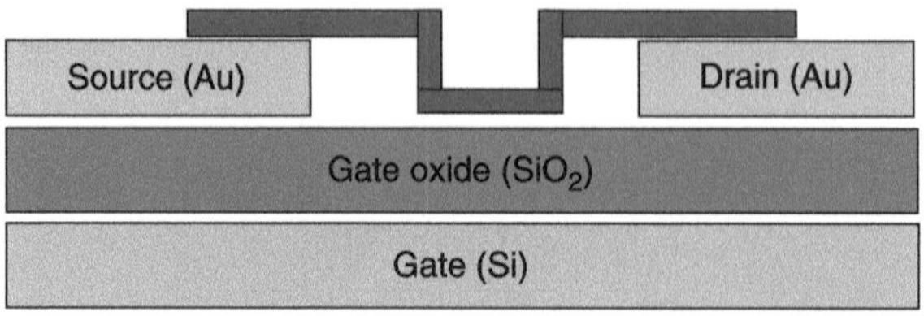


*2.3* Transistor device modified to incorporate the carbon nanotube as the channel.

cylinder with typical dimensions of radius 2 nm and length at least a thousand times larger. Electron motion in CNTs is confined in all except one dimension, which is along the cylindrical axis. Figure 2.3 shows the incorporation of the CNT to a transistor device.

The wavefunction for an electron in the CNT is normally expressed in the cylindrical coordinate:

$$\psi_k^\sigma(r) = \frac{1}{\sqrt{V}} e^{ik.r} \chi_\sigma \rightarrow \psi_{k_x}^{n,l,\sigma}(r) = \frac{1}{\sqrt{L}} e^{ik_x x} R_{n,l}(r) Y_l(\phi) \chi_\sigma, \qquad [2.4]$$

from which the eigenenergy can be found as follows:

$$E_{k_x,n,l} = E_0^R + E_l^\phi + \frac{\hbar^2 k_x^2}{2m}. \qquad [2.5]$$

Electrons are confined in two dimensions, i.e. azimuthal angle $\phi$ and $R$, but free in $x$. In the following, we estimate the sub-band energy due to the confinement. For a typical CNT diameter, the sub-band due to confinement along $R$ is $\Delta R = 0.1\ nm \rightarrow E_0^R \sim \frac{\hbar^2}{2m(\Delta R)^2} \sim 10\text{eV}$. Likewise, the sub-band energy due to confinement along the circumferential direction $\phi$ can be estimated as $\lambda_l = \frac{2\pi R_0}{l} < 2\,nm \rightarrow E_l^\phi \sim \frac{\hbar^2}{2m(\lambda_n)^2} \sim 1\text{eV}$.

## 2.1.4 Zero-dimensional nanostructures

It is not hard to deduce a further reduction to the lowest possible dimensionality, that is, a zero-dimensional (0D) nanostructure in which the electron is confined in all three spatial dimensions as shown in Fig. 2.4. In other words, the electron is not free to travel anywhere within such a geometry. In nanoelectronics, such zero-dimensional nanostructures usually take the form of quantum dots, in which electron states exist as discrete momentum states. Due to the small capacitance of the quantum dots, Coulomb blockade or single electron charging effects will be significant in these devices.

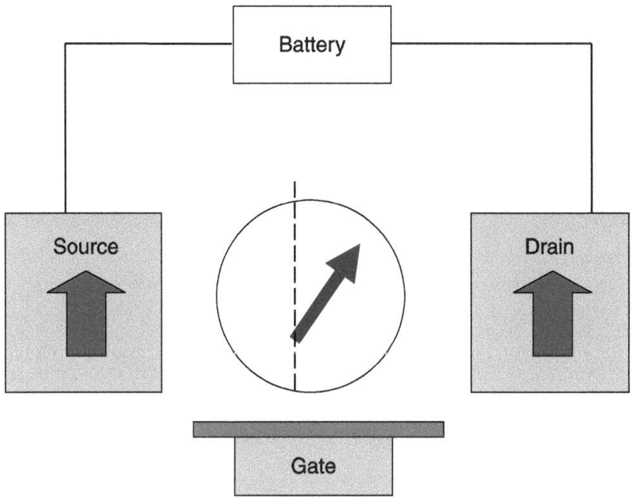


*2.4* Metal island or quantum dot structure with source, drain and gate to control electron or spin flux conduction.

### 2.1.5 Relevant areas of knowledge

Having provided an introduction to the low-dimensional structures that underpin nanoelectronic devices, we now consider the different knowledge sets required for understanding the many current and emerging aspects of nanoelectronics. Without doubt, new advances in nanoelectronics have provided a platform and test bed for the application of advanced theoretical physics beyond its traditional grounds of cosmology and high-energy physics. The trend became evident since the early 1990s when the Green's function propagator and gauge physics started to be applied to the new 'pasture' of condensed matter physics and nanoelectronics. These powerful theoretical tools could now play very significant roles in advancing the understanding of electron transport in the realms of nanoscale electronics. Nevertheless, nanoelectronics is a very diverse and inclusive field, and a complete study of it would require a diverse set of knowledge and skills ranging from conventional electronics to advanced cutting-edge physics. This fast-expanding field cannot be understood in a holistic manner without interdisciplinary knowledge that cuts across device physics, nanofabrication and technologies, electron transport, statistical physics, many-body physics and quantum mechanics. Methods and approaches aplenty, we attempt to summarize the required skill and knowledge sets in Table 2.1 based on the methods applied in the various fields over the last half-century.

The knowledge required to understand the full breadth of recent development in nanoelectronics is too wide to be covered in this book alone. We will assume pre-existing knowledge on the part of the reader of Topics 1 and 6 in Table 2.1; these

*Table 2.1* Studies of nanoelectronics require a diverse set of knowledge ranging from electronic to condensed matter physics: summary of the different fields relevant to nanoelectronics

| Topic | Required knowledge in different fields |
|---|---|
| 1. | Boltzmann-based semiconductor physics[1,2]<br>Phenomenological condensed matter physics |
| 2. | Quantum mechanics[6,7] |
| 3. | Equilibrium many-body physics[8,9] |
| 4. | Non-equilibrium Green's function device physics[10,11] |
| 5. | Topology and gauge theoretic electron dynamics[12,13] |
| 6. | Electronic devices and MOSFET technologies[3]<br>Nanofabrication and characterization technologies[4,5] |

topics should be familiar to electronic engineering graduates. For Topic 1, we recommend standard solid states physics textbooks (see, for example, references 1 and 2), while Topic 6 is usually covered by most device physics and nanotechnology reference books. For example, topics related to semiconductor physics such as *p-n* junctions, bipolar junction transistors, the junction gate field-effect transistor (JFET) which utilizes the *p-n* junction, the metal–semiconductor field-effect transistor (MESFET) which utilizes the Schottky barrier, MOSFET which utilizes the metal–insulator junction, tunnel devices and semiconductor-based photonic devices are routinely covered in many semiconductor textbooks.[3] One can rely on modern references or books on nanotechnologies[4,5] to give detailed description of the working principles of various nanofabrication and processing systems, which form the mainstay of modern clean rooms, such as multi-target sputtering machine, molecular beam epitaxy (MBE), electron beam lithography, optical mask aligner, focused ion beam writer, ion miller, as well as characterization equipment such as atomic/magnetic force microscopes, transmission electron microscopes, vibrating sample magnetometer, X-ray diffraction (XPD), X-ray photoelectron spectroscopy (XPS), alternating gradient field magnetometer, ellipsometer, torque magnetometer, Fourier transform infra-red (FTIR) spectroscopy, micro-Raman system, hard disk media tester, magneto-optical Kerr effect, etc. Topics 1 and 6 are used or implied throughout this book, without further elaboration on references or their origins. Lastly, we shall assume knowledge of Topic 2 which comprises basic quantum mechanics.[6,7] This is widely covered in undergraduate textbooks for quantum mechanics. Deeper topics on single-particle quantum mechanics are also covered in reference books on quantum information.

This book focuses on theoretical topics that are not normally covered in textbooks on microelectronics. These principally consist of Topics 3, 4 and 5. Topics 3 and 5 contain subject matter that is covered in field theoretic or many-body condensed matter physics.[8,9] Chapters 3 and 4 of this book provides a pedagogical description of the non-equilibrium Green's function method[10,11] and how it is developed for the microscopic transport studies of nanoscale devices under arbitrary voltage bias. Topic 5 on gauge symmetry and topology is a recent addition to the theoretical 'armory' of nanoscale condensed matter physics and electronics. These theories were first developed in cosmology and high-energy particle physics. Their application in mesoscopic or nanoscale condensed matter physics only began in earnest in the 1980s with the discovery of the quantum Hall effect, a topic which cannot be adequately described by traditional condensed matter physics. The last chapter of this book (Chapter 7) provides a systematic introduction to the field and gauge theoretic methods,[12–14] with the focus being their application in nanoscale electronics. This is a departure from the usual treatment of gauge theory and topology in existing books and the literature, which mainly emphasized the development and application of these theories in the more traditional areas of high-energy physics. In this book, field and gauge theoretic methods are used alongside phenomenological and statistical approach in a complementary manner.

### 2.1.6 Overview of nanoelectronics

We now begin our quest into the various fields of nanoelectronics, as well as the underlying physics which unite these seemingly disparate fields. Figure 2.5 is a pictorial description of the various emerging nanoelectronic fields of research and

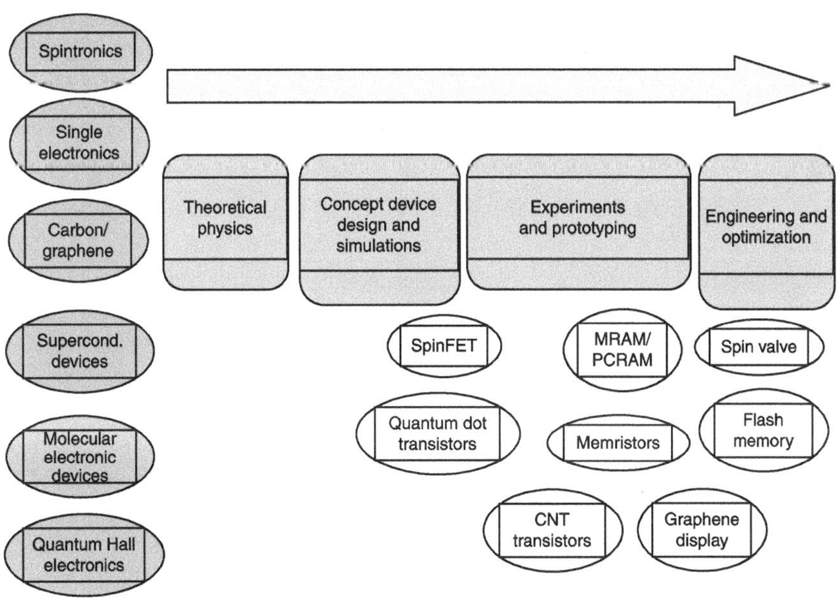


*2.5* Schematic description of the various fields of nanoelectronics and the evolution from theoretical concepts to numerical computation, experimental tests and prototyping, and finally to engineering optimization.

application, and the general path which occurs in all these domains starting from the fundamental physics of the underlying phenomena onto the device concepts and design, and finally to the world of applications, devices and industrial commercialization.

The main underlying physics is nanoscale condensed matter physics, and the most powerful theoretical methods to describe this are Green's function and the topology and gauge theoretic approaches. These theories have led to the prediction and description of new condensed matter phenomena in meso- and nanoscale systems (e.g. quantum Hall effects, novel spin dynamics, electron correlation effects and topological transport behavior), properties which are only now being explored for possible applications in nanoscale devices.

Once the theory of the underlying physics has been formulated, the next step is numerical study and computational simulation to confirm the theoretical predictions and to evaluate the robustness or reliability of these effects for practical device applications. If the results are promising, experiments will be carried out to confirm the physical phenomena and then to resolve any practical difficulties to achieve a basic or crude prototype as a proof of concept. Finally, the last stage is device engineering and optimization. At this point, experimentalists will have shown the theoretical benefits to be feasible (e.g. in terms of robustness and room temperature operation). Furthermore, various environmental and human factors (e.g. cost and scalability in manufacturing) will have been analyzed. At this stage, engineers are focused on refining and improving the practical features of the devices and applications so as to turn them into successful commercial products.

## 2.2 Nanoelectronics and nanoscale condensed matter physics

Our earlier introduction to the various fields of nanoelectronics suggested its strong linkage to condensed matter physics. One can start by visualizing a condensed matter system as something akin to a mini-universe to the electrons with a periodic potential that pervades the entire bulk of the material. Advances in nanotechnologies have led to spatial shrinkage of this universe so that its boundary effects become apparent, giving rise to many properties unique to nanoscale condensed matter systems.

In Section 2.1, we discussed the many mesoscopic or nanoscale condensed matter physics concepts relevant to nanoelectronics as a consequence of its shrinking size and concomitant boundary effects. But we have yet to make explicit the distinction between nanoelectronics and pure nanoscale condensed matter physics. One of the most important effects that distinguishes the physics of nanoelectronic devices from mesoscopic or nanoscale condensed matter physics is the influence of neighboring perturbations. Nanoelectronic devices normally contain a few discrete parts, comprising a central channel and adjacent leads. From the perspective of the central channel, the adjacent leads can be viewed as perturbations. The most common types of leads are the source and drain electrodes, and as in conventional electronics (e.g. in a MOSFET device), the electrical voltage is applied from source to drain across the channel.

Thus, if one were to construct an equation that describes a carrier's propagation in the channel, this equation tends to require local parameter input (e.g. local potential and site energies that extend from the channel onto the leads indefinitely). It may seem that we have an open system with divergent effects, which extend far from the central channel region under consideration. However, in a homogeneous condensed matter system, such an extension to infinity can be made convergent by some renormalization process. In a nanoelectronic device with a channel which is treated as the central region, it makes sense to treat the channel as a closed system with all the effects of perturbation being absorbed into some physical observables of the carrier (electron) in the channel. For example, the lead perturbation that extends semi-infinitely can be re-normalized into a finite self-energy which affects electron propagation in the channel. The device can now be viewed not as an open system that is attached to adjacent regions like leads, but as a closed one with electron propagation in a modified energy landscape.

A consequence of the lead perturbations (under the application of source–drain bias) is the induction of non-equilibrium electron density in the channel. By taking into account the non-equilibrium electron density as a function of some global lead parameters, the channel can be considered as a closed system with modified electron density. One can see this if we assume the ohmic relation for current in the channel:

$$I = \frac{V_S - V_D}{R} \qquad [2.6]$$

where ($V_S$ − $V_D$) are the global lead parameters (i.e. their respective applied voltage) which can also be considered as a property of the closed channel system which reflects its non-equilibrium electron density. While Eq. 2.6 is a rather crude way of incorporating the effect of leads on the non-equilibrium distribution of an electron in momentum space, it is accurate in bulk systems, in which the gradient of the non-equilibrium electron density in the central channel is determined by the difference in the lead voltage. In this way, a device with three discrete units (central channel and two leads) can be viewed as a single closed system without leads, but with a modified electron density (Fig. 2.6).

A more improved approach suitable for microscopic system is to derive the Kubo conductivity, which plays the role of the inverse resistance ($1/R$) in Eq. 2.6. In the linear response, one assumes the change of electron density to have minimal effect on the conductivity, so that current is obtained via direct multiplication with the small voltage difference. Thus, a major challenge in nanoelectronic physics is to find clever means to incorporate the seemingly extensive and divergent perturbation effects due to neighbors into some physical properties of the channel system, such that the channel can be treated as a closed system with modified properties.

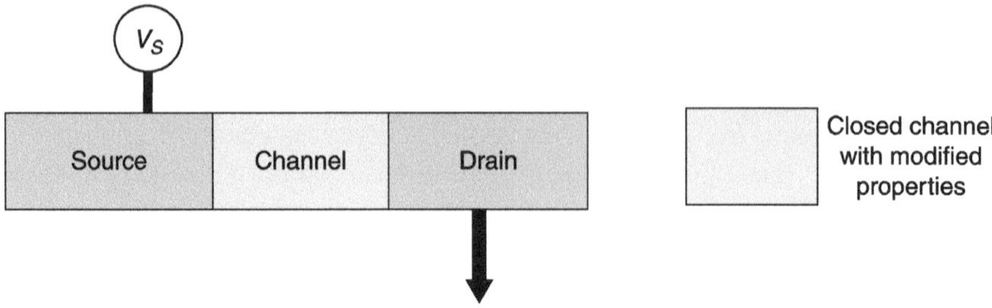


*2.6* A standard device with three discrete regions can be viewed as a closed system with lead perturbations absorbed in terms of a modified electron energy and electron distribution within the channel.

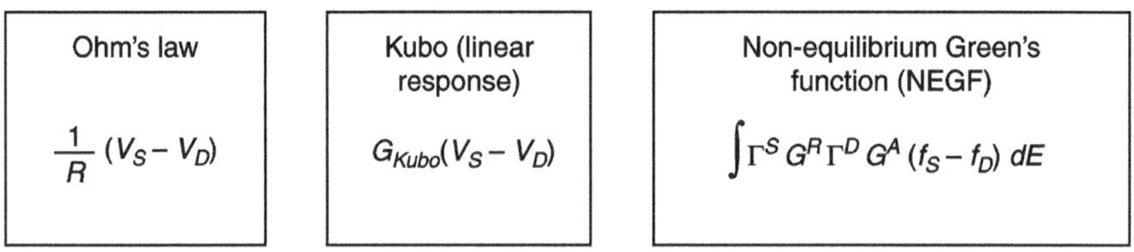


*2.7* Various ways of taking into account the lead effects on the non-equilibrium electron density in the central region.

Different methods of modeling open systems with leads as modified closed systems are shown in Fig. 2.7. These are three effectively closed systems which relate the lead voltage to the non-equilibrium electron density distribution in the central channel. Methods 1 and 2 are straightforward. Device 3 makes use of the so-called non-equilibrium Green's function (NEGF) technique to model the effect of the lead voltage on the non-equilibrium distribution in the channel in a much more involved manner. It involves linewidth functions and Green's functions of the central channel region, which incorporate the coupling strengths to the leads and self-energy due to the leads, respectively. This is discussed further in Chapter 3.

It is now clear that nanoelectronics acquires additional complexities because of the following effects:

1. Nanoscale size effects – modified equilibrium system properties
2. Source–drain applied voltages – non-equilibrium effects
3. Contact proximity – interaction with channel
4. External applied fields – topological effects.

## 2.3 Emerging nanoelectronic devices and systems

Considering the diverse nature of nanoelectronic fields, it is as yet not possible to develop a single formalism suitable for all devices. While the non-equilibrium

Green's function allows lead perturbation effects to be absorbed into the coherent propagation and non-equilibrium electron density, it still needs a separate mechanism to calculate the electron density distribution within the channel. If modification to dimension is not drastic (i.e. if the channel can be considered as a quasi-3D structure), then the electronic background as a function of spatial separation ($E - r$) and in $k$-space ($E - k$) can be determined based on pure condensed matter physics (i.e. disregarding the effects of boundaries).

In the event of single-electronic or quantum dot structures, CNT and 2DEG, mesoscopic or nanoscale condensed matter physics is required to determine the sub-bands. However, in nanoelectronic systems with multiple parts, the electronic background will be modified further. As we know, transport equations are based on the specific background under which they are derived. In quantum dots, electron gas as described in $E - r$ relation, where $r$ is related to electron separation, cannot be assumed. Electron transport is correlated and the transport equation has to be modified to reflect this (see Section 2.3.3 on single electronics below). In semiconductor devices with interfaces like metal–semiconductor or *p-n* junctions, local band-bending occurs at the interfaces and this alters the dispersion relation or the $E - k$ property of the device in the affected parts. These changes and modifications cannot be reflected automatically in most transport physics, although the bandstructure effect could modify the NEGF equations via the modification to the Hamiltonian. It is thus important to understand these physics separately. For more details of the device physics and interfaces, readers can refer to books on devices.[3] For our purpose, it suffices to classify them in terms of their relevance to the new nanoelectronic devices and systems, many of their designs being based on the well-established semiconductor devices.

In the following, we introduce and specifically discuss several important types of nanoelectronic systems. These are spintronics which encompasses metal spintronics, hybrid and semiconductor spintronics, single electronic systems (e.g. the quantum dot and metal Coulomb blockade devices), molecular electronics, carbon nanostructure transistors and graphene electronics.

## 2.3.1 Metal spintronics

Many theoretical and simulation studies have been carried out to investigate the giant magnetoresistance (GMR) and spin transfer effects in both the current perpendicular-to-plane (CPP) and current-in-plane (CIP) configurations. Phenomenological (semiclassical) physics plays an important role, as it describes experimental observations to a rather accurate degree. The main semiclassical transport theory to describe the GMR effect is one based on the Boltzmann equation, which is a model understood by both theorists and experimentalists alike. There is a large body of published works and books about the Boltzmann transport model and its specific applications in spin electronics. Magnetoresistance (MR) based devices are commonly known as readers or recording heads by the hard disk drive storage community.

*Giant magnetoresistance (GMR) and spin valves*

The most prominent and useful effect in metal-based spintronics is giant magnetoresistance (GMR). The GMR effect refers to a large fractional change in resistance induced by an applied external magnetic field, which changes the relative magnetization orientations of the two ferromagnetic (FM) layers separated by a non-magnetic (NM) spacer from the parallel to the anti-parallel configuration.

In fact the entire development of the emerging spin-based electronics was based initially on the success of GMR devices, which are composed primarily of ferromagnetic metal multilayers. The MR response is largely a macroscopic effect arising due to the collective spin-dependent scattering experienced by electrons in the bulk and at interfaces.

GMR effects were first reported in multilayer devices[15,16] in the early 1990s (e.g. Fe/Cr, Co/Cu and Ag/Co). The early GMR devices existed in CIP form where electron flow is within the plane of the device. Present GMR devices exist in CPP form. Both theoretical simulations and experiments have confirmed that the CPP GMR device can achieve a higher MR ratio than a CIP one. In addition, CPP sensors also possess the engineering advantages of requiring a smaller shield-to-shield gap, thus enabling a higher areal storage density of the hard disk drive (HDD). In a CPP GMR element, the resistance difference $\Delta R$ (between the parallel and anti-parallel configurations) scales with the shrinking of the sensor area. This is compatible with future needs as the track density in the recoding medium increases. Figure 2.8 shows a CPP-type GMR device.

*Tunneling magnetoresistance (TMR) and spin valves*

At the turn of the twenty-first century, interest in MR research expanded to encompass an effect with a more pronounced quantum feature, namely tunneling magnetoresistance (TMR).[17,18] This is due in part to the rising demand for HDD

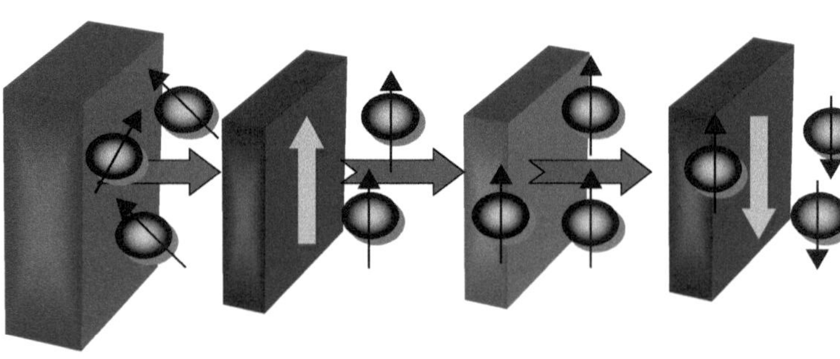


*2.8* Schematic of a CPP-type GMR device, which is the most common field-detecting element in present spin-valve based recording heads.

storage. Increasing bit density in the media platter of the HDD has placed unprecedented demand on the magnetic field sensor to detect weak magnetic field change in order to distinguish one binary state from another. For instance, the use of MgO tunnel barriers has resulted in MR (in the 'pessimistic' definition) well in excess of 75% while most CPP GMR based spin valves deliver around 5%. The advantage was, however, tempered by the high resistance of TMR devices. While a typical CPP GMR device has an overall areal resistance of 100 $m\Omega\mu m^2$, that of TMR hovers at 1–2 $\Omega\mu m^2$, introducing a large impedance mismatch between the magnetic sensor and the pre-amplifier, which results in electrical loading of the detection circuitry.

*Half-metal spin valve*

The CPP GMR device with an MR of around 3–5%, even in the advanced form of an exchange-biased CPP spin valve, is far from adequate to meet the requirements imposed by ultra-high density storage of several terabits per square inch. The solution to the problem of high resistance area in TMR devices remains elusive. In around 2003–2006, some interest also shifted back to all-metal GMR devices but with half-metals (HM) taking over the normally ferromagnetic (FM) contacts (see Fig. 2.9). Materials with a high bulk polarization ratio (90–100%), such as half-metallic (HM) Heusler alloy,[19,20] were studied for their potential to increase magnetoresistance. However, the use of these new materials in CPP sensors faced difficulties in terms of achieving high-quality thin film structure, as well as unknown interfacial effects.

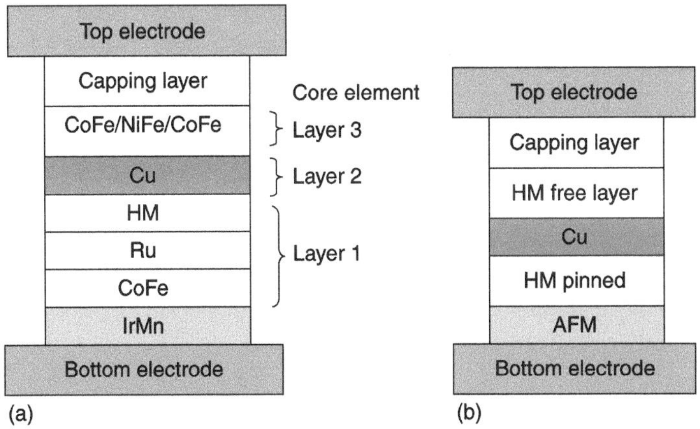


*2.9* (a) Schematic of a modern CPP spin-valve based on half-metal. (b) Schematic with functional descriptions of the different layers of device (a).

*More complicated spin-valve structures*

Many of the spin-valve GMR devices – be they FM GMR, HM GMR or TMR spin valves – have since undergone further technological development and improvement. The basic pseudospin valve structure has spawned more complex structures, such as exchange-biased spin valve, synthetic antiferromagnetic spin valves, half-metal dual spin valve, etc. Readers interested in the specialized field of spin-valve devices can refer to the many papers published in engineering and applied physics journals, as well as proceedings of the main hard-disk related conferences, such as the Magnetism and Magnetic Materials and the International Magnetic conferences.

*Magnetic tunnel junction and spin torque devices*

Two other important aspects of metal spintronics are the tunneling magnetoresistance (TMR) and spin transfer torque. The magnetic tunnel junction has similar physics and working principles to a TMR spin valve but it is used as a storage element instead of as a field sensor. The use of the magnetic tunnel junction introduces the possibility of non-volatile solid state memory. Using such non-volatile memory, the possibility of an instant-on computer is no longer far-fetched. Magnetic random access memory (MRAM),[21] which consists of a lattice of MTJ cells, has since become a leading candidate for spintronic-based non-volatile memory. The reality of solid state memory was advanced further with the realization that a spin-polarized electron flux could interact with a local magnetic moment to effect magnetization switching. The underlying physics of this phenomenon is known as the spin transfer torque, first studied independently by L. Berger and J. C. Slonczewski.[22] Their theoretical studies were later supported by experiments.[23]

This was great news to the MRAM community which hitherto had to rely on tiny current-carrying wires to generate magnetic (Oersted) field-to-effect magnetization switching in the magnetic tunnel junction. With the advent of spin transfer torque switching, the 'bulky' wires (relative to the tunnel junction) could be dispensed with. However, early experimental studies revealed that a relatively large critical current density of the order of $10^8$ A/cm$^2$ is required to trigger current-induced magnetization switching (CIMS) via the spin transfer torque effect. Such excessive current density may cause electromigration and damage electronic devices. Thus much attention is now focused on reducing the required current density for CIMS. Current state-of-the-art results[24] reported for CoFeB/MgO/CoFeB MTJ show high TMR over 120%, high thermal stability at dimension 40 nm and switching current as low as 49 μA.

The physics of spin torque is an interesting topic by itself. Since the GMR effect is the resistance modulation in multilayer ferromagnetic structures arising due to the scattering of itinerant electron spins by local magnetization, it would be natural to conceive that by reciprocity, a high concentration of electron spins may exert some form of spin-current related torque on the local magnetization, leading to possible switching or CIMS. Additionally, spin transfer torque can effect large-angle magnetization precession in the free layers of spin valves and magnetic tunnel junctions, giving rise to microwave oscillations.[25]

Many theoretical methods have been developed to study CIMS and current-induced spin oscillations in a variety of structures, ranging from ferromagnetic spin valves to Coulomb blockade transistors.[26] The theoretical methods range from non-collinear spin drift diffusion (SDD), which imposes phenomenological continuity of spin flux at the multilayer interfaces, to quantum mechanical method with boundary matching of wavefunctions, NEGF and gauge theoretic methods. Except for the latter method of gauge theoretic, which will be introduced in Chapter 7, we will refer readers to the works in the literature for the other methods.

Figure 2.10 provides a lateral view of a standard array design for today's MRAM. Each cross point addresses one MRAM cell, and access to these cells is performed by decoders selecting the lines and the transistors.

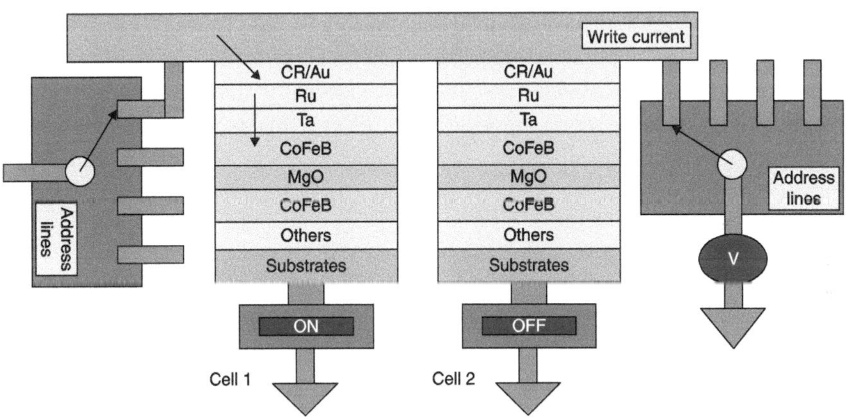


*2.10* Typical MRAM array in which the writing process is based on the CPP spin transfer torque on the free layer, while the reading process of the stored information is effected via CPP magnetoresistance.

### 2.3.2 Hybrid and semiconductor spintronics

*Bipolar spin-valve transistors*

In the mid to late 1990s there were exhilarating breakthroughs in the field of metal spintronics in terms of spin-valve and magnetic memory applications. These sparked interest in the design of hybrid spintronic devices which combined semiconductor with metal spintronics. The spin valve transistors[27,28] were first

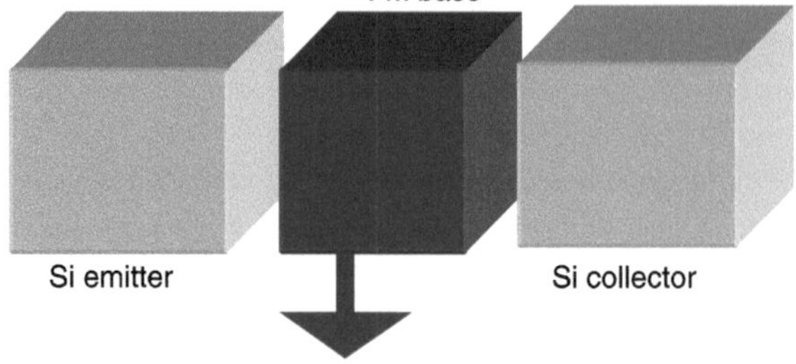


*2.11* Schematic of a basic spin-valve transistor.

conceived in the form of an all-metal bipolar device, before evolving to the semiconductor–metal bipolar system fabricated to the structure of Si–MultilayerFM–Si in 1995 (Fig. 2.11). Schottky barriers, which form at the Si–FM interfaces, act as buffers which set the FM region free from electric fields applied to the terminal pair. Electrons crossing the FM multilayers lose more energy under the anti-parallel configuration than the parallel configuration of magnetizations. The FM configurations can be controlled by applying external magnetic fields. Thus, collector current can be externally modulated via an applied magnetic field. It is, however, important to note that this does not yet constitute a semiconductor spintronics device. Spin conductance was modulated entirely by the all-metal spin valve via the GMR effects. The semiconductor regions merely provide the Schottky barriers and facilitate the injection of spin-neutral hot electrons into the spin-valve base region.

Modifications were made to the above by introducing the injection of spin-polarized instead of neutral current into the base. The base region can also be modified to play the role of a spin filter and detecting spin current. If the filter rejects the spin current, most of the current will be diverted to the collector.

### *Metal–semiconductor spin injection transistor*

Key prerequisites for the functioning of a truly semiconductor spintronic device are the generation of spin-polarized current and long spatial and temporal spin coherence in the semiconductor, which is normally paramagnetic. Since normal semiconductors are non-magnetic, spin current needs to be generated via spin injection from a ferromagnetic material. Current passing through ferromagnetic materials becomes spin polarized due to exchange coupling (due to asymmetry in the density of states) as well as bulk scattering (due to spin asymmetry in mobility). The idea here is to inject the spin polarized current into the non-magnetic semiconductor, the latter being a well-established medium for the control of electronic conductance.

The long spin diffusion length in semiconductors such as GaAs is tremendously helpful for such a purpose. But there is one major obstacle – in diffusive transport, spin polarized current cannot be easily injected into semiconductors. The low dosage of spin injection has been attributed to a factor[29] related to the conductance difference between the metal ferromagnetic and the semiconductor media. However, it seems that the incorporation of a spin-asymmetric tunnel or Schottky barrier can improve spin injection greatly via quantum mechanical tunneling. Besides the intrinsic problem for low spin injection, interfacial spin flip is also crucial for destroying the proper delivery of spin-polarized flux to the semiconductor. In general, the spin injection device can normally generate spin current, which cannot be transported to where it matters (namely the semiconductor region), thus precluding its further exploitation.

*Magnetic–electric field transistor*

Figure 2.12 shows that the external field device is particularly suitable for the design of a semiconductor MRAM or a programmable logic device.[30] In the memory application, the ferromagnetic gates are used to store information. Detection of the stored memory states of the FM gates can be achieved by passing current through the conduction channel directly below them. The edge fields emanating from the gates can be correlated to the orientation of the magnetic moment of the FM gate. This device can rely on either field-induced switching via the write line method (similar to that used in MRAM) or the current-induced-magnetization-switching (CIMS) effect to switch the gate magnetization.

Tunneling through a delta barrier is given by the following:

$$\left[\frac{\hbar^2}{2m^*}\nabla^2 + U\delta(x)\right]\phi(x) = E\phi(x) \qquad [2.7]$$

where $U$ has the units of eV m. Over a length scale of 10–50 nm between the barriers, the boundary effects on electron transport will be significant. Below, we

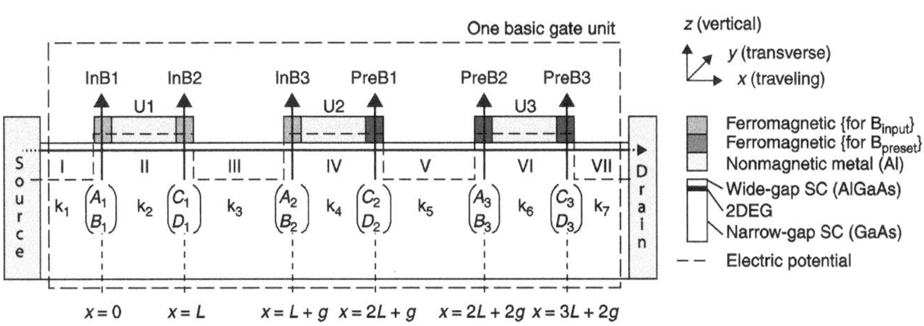


*2.12* Magnetic–electric field MOSFET with multiple ferromagnetic gates to realize functions of non-volatile storage and programmable logic.

describe the ballistic tunneling process which allows the wave amplitude of electron at each region to be determined. Taking $x = 0$ as the interface, and assuming infinitesimal thickness $\pm\varepsilon$ of the $\delta$-function which represents the edge field from the FM gate, we have:

$$\int_{-\varepsilon}^{+\varepsilon}\left[\frac{-\hbar^2}{2m^*}\nabla^2 + U\delta(x)\right]\phi(x)\,dx = \int_{-\varepsilon}^{+\varepsilon} E\phi(x)\,dx \approx 2\varepsilon\, E\,\phi(0) = 0$$

$$\frac{-\hbar^2}{2m^*}(\phi'(+\varepsilon) - \phi'(-\varepsilon)) + U\phi(0) = 0. \qquad [2.8]$$

In the case where the barrier is magnetic, $U = S\dfrac{eg^*}{2m_0}\dfrac{\hbar}{2}B$ is used. To compute spin transport, the standard material parameters for GaAs are used: $m^* = 0.067\ m_0$, $g^* = 0.44$, Fermi energy $E_F = 3.55\ meV$, corresponding to charge density of $n_e = 10^{11} cm^{-2}$, and $B_0 = 0.2$ T.

*Rashba spin orbit coupling spin field-effect transistor (FET)*

Spin orbit coupling (SOC) can be regarded as a form of effective magnetic field 'seen' by the spin of the electron in the rest frame. Based on the notion of effective magnetic field, it will be straightforward to conceive that spin orbit coupling can be a natural, non-magnetic means of generating spin-polarized electron current. It is thus natural to conceive that SOC within a tunnel barrier can be utilized to achieve efficient spin filtering.[31] Despite practical difficulties, it has been accepted that, in nanoscale devices, spin current can at least be generated at one end of the device and detected on the other. The combined effect of the tunnel barriers and the SOC can be used to achieve a spin transistor function. Electrical voltage ($V_g$) is applied to the Schottky gate to alter the Rashba coupling strength within the InAlAs–InGaAs 2DEG and thus modulate the spin state of the conduction electrons in the semiconductor channel. This, in turn, is translated into a conductance modulation by incorporating a ferromagnetic drain electrode to the contact to act as the detector for the electron spin.

Although experiments[32] have confirmed the working principles of such devices, most have fallen short of demonstrating large conductance modulation. The prospect of semiconductor spintronics thus remains unclear for the foreseeable future. Spin orbit coupling based spintronics is still fraught with difficulty due to the poor 'survival' of measurable spin current amidst such effects as temperature, impurity scattering, mode averaging, non-uniformity of spin orbit strength, as well as interfacial spin flips. Temperature smears the electron distribution over the double-conic Rashba bandstructure at the Fermi energy, greatly reducing ideal spin polarization. Spin relaxation length is rather long in a ballistic semiconductor, but impurity scattering can weaken spin current considerably. Mode-averaging can be partially overcome by designing devices that are spatially constrained in the transverse direction, or via an electron wave-guiding action effected by regions with different doping densities. Both of these will result in wavevector (mode) selection, where only a portion of the Fermi surface is probed for spin detection. Note that SOC systems are time-reversal symmetric, and so there will be no net spin polarization if the measurement is averaged over the entire Fermi surface.

### 2.3.3 Single electronics and quantum dot

Electron transport through a device in which the central electrode is of nanometer scale would encounter the so-called Coulomb blockade effect. The size of the CB effect depends on the size of the central region and the material system it is made of. The discovery of single electron tunneling (SET)[33] marked the arrival of a new class of nanoelectronic devices known as single electronic devices. In their simplest form, these devices consist of the central island contacted to left and right leads – acting in the language of MOSFET as the source and drain electrodes, respectively. The device can be in an all-metallic[34] or semiconductor[35] configuration, that is, the central island can be in the form of a metal or semiconductor quantum dot[36] with metal contact electrodes. In single electronic devices, the coupling between the central device and the leads is assumed to be weak so that the tunneling term in the Hamiltonian can be regarded as a perturbation. However, in the case of a molecular island, the coupling effect becomes strong and a different transport mechanism occurs. This is reviewed in Section 2.3.4 on molecular electronics.

Here we introduce the concept of Coulomb blockade (CB), which is crucial in single electronics. If there is already an extra unbalanced electron in the nanostructure, the addition of an electron leads to a charging energy of:

$$E_a = E_C + E_{k'}. \qquad [2.9]$$

In Eq. 2.9, $E_C$ refers to the electrostatic charging energy arising from the Coulomb repulsion of an electron unbalanced by the background positive ionic charge, while $E_k$ is the energy change due to the electron occupying a different quantum level. If the first electron in the nanostructure is of state $k_{1\uparrow}$, then the next electron can be either $k_{1\downarrow}$ *or* $k_{2\uparrow}$. The charging energies involved are $E_a = E_C$ *or* $E_a = E_C + (E_{k2} - E_{k1})$, respectively. In the case of a metallic central island, the small Fermi wavelength implies continuous kinetic energy states. Thus, $(E_{k2} - E_{k1}) \approx 0$ for metal; the charging energy is the same for both cases. However, the scenario would be different in the case of a semiconductor quantum dot as the distinct state quantization implies $(E_{k2} - E_{k1})$ can be large. Therefore, if the first electron is of state $k_{1\uparrow}$, then by the Fermi exclusion principle, the next incoming one can only be $k_{1\downarrow}$. This forms the basis of the Anderson model, in which the charging energy term is expressed by:

$$\widehat{H}_C = E_C\, \hat{n}_{k\uparrow}\, \hat{n}_{k\downarrow}. \qquad [2.10]$$

*Single electron spintronics*

On the other hand, the contact electrodes can be made magnetic to implement spintronics in single electronic system. With magnetic leads, spin injection can be implemented leading to the design of a hybrid SET–spinFET device,[37] an example of which is a SET device based on a self-assembled InAs quantum dot. The interplay between the CB effect and spin-dependent tunneling gives rise to several novel magnetotransport effects such as magneto-Coulomb oscillations, enhanced magnetoresistance due to cotunneling and the spin blockade effect. Figure 2.13 provides a schematic representation of a hybrid SET–spinFET showing magneto-Coulomb effect and a basic single electronic system based on quantum dot with quantized energy levels.

One of the advantages of single electronic device over MOSFET is its high packing density as single electronic devices are scalable down to 1–10 nm. In MOSFET, high power consumption is becoming an increasing problem as device dimension continues to shrink and leakage current occurs across the thin oxide layer via quantum tunneling, draining energy and wasting power. In a single electronic device system, power consumption is generally low because of the low number of electrons involved. Single electronic devices also deliver a higher switching speed as opposed to MOSFETs, which require time to charge up the capacitor. The disadvantages include fabrication on the nanometer scale where most single electronic devices require sequential e-beam patterning, which is a time-consuming process, as opposed to optical lithography for CMOS devices. Controlled SET fabrication with repeatable properties at room temperature is still difficult to achieve. Furthermore, SET devices are particularly sensitive to random background charges. Presently, due to these practical difficulties, single electronics is confined to a few niche applications such as:

- Metrology – very accurate current standard
- Thermometry – absolute temperature measurement based on fundamental constants
- Very high resolution surface charge probes.

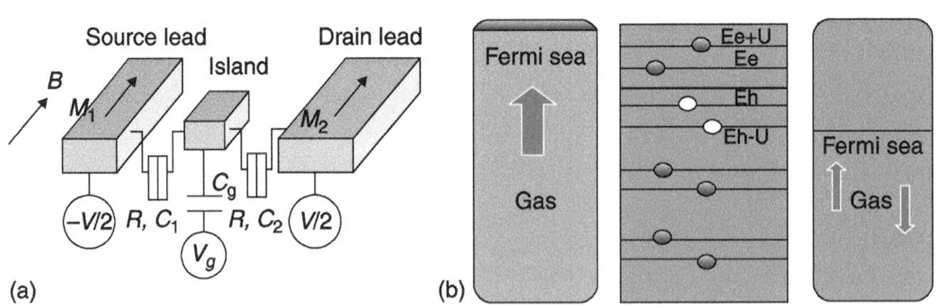


*2.13* (a) Hybrid SET–spinFET device exhibiting magneto-Coulomb effect which combines the physics of single electronics and spintronics to yield new transport properties. (b) Single electronic system based on semiconductor quantum dot.

*Modeling of single electron tunneling transport*

Since electron transport in a single electronic system occurs in a sequential manner and individual tunneling events are independent uncorrelated events, the quantum wavefunction tunneling method cannot be applied here. Furthermore, we assume that equilibrium is re-established after each tunnel event. Thus, single charge tunneling can be treated as Markovian events, so that the overall transport can be modeled via the Master equation approach. Because of charge accumulation in these structures, electron tunneling through these structures or junctions depends on the type of accumulation present on the island or quantum dot (e.g. electron, hole, spin and so forth).

The Master equation method consists of a series of linear equations which characterize the rate of change of the state probability of the SET system:

$$\begin{aligned}\frac{d\langle n_\sigma\rangle}{dt} &= \Gamma^{in}_{n\sigma}\left[1-\langle e_\sigma\rangle-\langle e_{\bar\sigma}\rangle-\langle h_\sigma\rangle-\langle h_{\bar\sigma}\rangle+\langle e_\uparrow e_\downarrow\rangle+\langle h_\uparrow h_\downarrow\rangle\right]\\ &+\bar{\bar\Gamma}^{in}_{n\sigma}\left[\langle n_{\bar\sigma}\rangle-\langle n_\uparrow n_\downarrow\rangle\right]-\Gamma^{out}_{n\sigma}\left[\langle n_\sigma\rangle-\langle n_\uparrow n_\downarrow\rangle\right]\\ &-\bar{\bar\Gamma}^{out}_{n\sigma}\langle n_\uparrow n_\downarrow\rangle-\Gamma^{flip}_{n\sigma}\left[\langle n_\sigma\rangle-\langle n_{\bar\sigma}\rangle\right]\end{aligned} \qquad [2.11a]$$

$$\begin{aligned}\frac{d\langle n_\uparrow n_\downarrow\rangle}{dt} &= \bar{\bar\Gamma}^{in}_{n\uparrow}\left[\langle n_\downarrow\rangle-\langle n_\uparrow n_\downarrow\rangle\right]+\bar{\bar\Gamma}^{in}_{n\downarrow}\left[\langle n_\uparrow\rangle-\langle n_\uparrow n_\downarrow\rangle\right]\\ &-\left(\bar{\bar\Gamma}^{out}_{n\uparrow}+\bar{\bar\Gamma}^{out}_{n\downarrow}\right)\langle n_\uparrow n_\downarrow\rangle\end{aligned} \qquad [2.11b]$$

where $\langle n_\sigma\rangle$ represents the probability of single occupancy of a carrier (electron or hole) with spin $\sigma$ in the island of the SET. Carrier species would have been clearly specified if $\langle e_\sigma\rangle$ and $\langle h_\sigma\rangle$ are used instead. The variable $\langle n_\uparrow n_\downarrow\rangle$ is the probability of double occupancy by two carriers of opposite spin in the island. $\Gamma^{in}_{n\sigma}$ is the tunneling rate of a carrier into an empty island. $\Gamma^{out}_{n\sigma}$ is the tunneling rate out of a singly occupied island. $\bar{\bar\Gamma}^{in}_{n\sigma}$ is the tunneling rate into a singly occupied island. $\bar{\bar\Gamma}^{out}_{n\sigma}$ is the tunneling rate out of a doubly occupied island. The tunneling rates $\Gamma^{in}_{n\sigma}$, $\bar{\bar\Gamma}^{in}_{n\sigma}$, $\Gamma^{out}_{n\sigma}$ and $\bar{\bar\Gamma}^{out}_{n\sigma}$ under different conditions are illustrated in Fig. 2.14, while $\Gamma^{flip}_{n\sigma}$ is the spin flip rate within the central island.

The occupation probabilities of electron within the central island are variables to be determined by solving Eq. 2.11. The tunneling rate equation can be derived by treating the tunnel event as a perturbation and applying the Fermi golden rule. For SET systems with a metallic island, the rate equation (commonly known as the 'orthodox' rate equation) is given by:

$$\begin{aligned}\Gamma^{\alpha}_{NM} &= \frac{1}{e^2 R^{\alpha}_t} g(E_c(N)-E_c(M)+\mu_i-\mu_\alpha),\\ g(E) &= \frac{E}{e^{E/kT}-1}\end{aligned} \qquad [2.12]$$

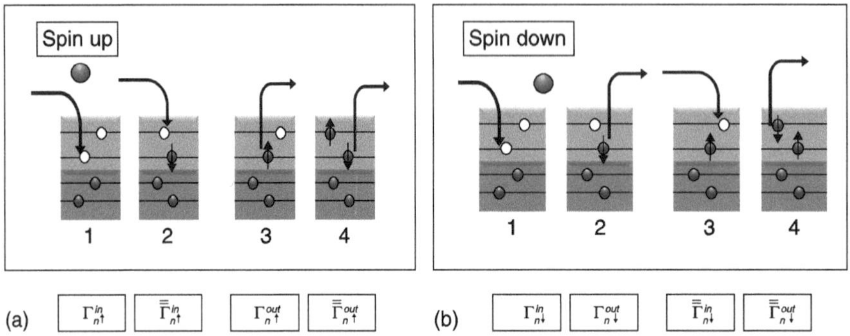


*2.14* Schematic illustrations of sequential tunneling events in SET devices with quantized QD levels for (a) spin up and (b) spin down states.

where $\alpha = L, R$ denotes the left or right electrodes, $R_t$ is the tunneling resistance, which is inversely proportional to the strength of coupling between the contact and island, $M$, $N$ are the initial and final number of island charges which characterize the initial and final states of the SET system, $E_c$ is the electrostatic energy sum due to the total number of or $M$ or $N$ island charges, and $\mu$ is the electrochemical potential of the island or the contact electrode. For SET systems with a semiconductor quantum dot as the island electrode, the presence of quantized energy levels results in a different rate equation given by:

$$\Gamma^{\alpha}_{k_{N+1}jN} = \gamma^{\alpha} f(E_{k_{N+1}} - E_{jN} - \mu_{\alpha}), \qquad [2.13]$$

where $j_N$ ($k_{N+1}$) is the initial (final) quantum dot states with $N(N + 1)$ charges on the quantum dot. $E_{k_{N+1}}/E_{jN}$ is the energy sum of $(N + 1)/N$ charges on the quantum dot. This would translate to $(E_{k_{N+1}} - E_{jN})$ being the highest energy level taken by the last electron in the dot. The electrochemical potential of electrode is given by $\mu_{\alpha}$. Notation $f$ is the Fermi–Dirac distribution function and $\gamma^{\alpha\sigma} = 2\pi|t|^2 \rho^{\alpha\sigma}$ denotes the strength of the tunnel coupling constant, assuming that $t$ is the same for all electrodes. The spin-dependent density of states is given by $\rho^{\alpha\sigma}$ in electrode $\alpha$. In the case of FM electrodes with spin polarization $p^{\alpha}$, one has $\rho^{\alpha\sigma} = (1 + \sigma m^{\alpha} p^{\alpha})\rho^{\alpha 0}$, where $\rho^{\alpha 0} = (\rho^{\alpha\uparrow} + \rho^{\alpha\downarrow})/2$ is the average spin-independent density of states. The spin polarization $m^{\alpha}p^{\alpha}$ changes sign depending on $m^{\alpha}$, whose sign depends on the magnetization orientation of the FM. One can choose $m^{\alpha}$ to be positive for magnetization pointing up and negative for magnetization pointing down.

## 2.3.4 Molecular electronics

Molecular electronics[38] is implemented in a nanoscale device structure similar to those of single electronic systems with metal and semiconductor central islands, except that the central island here is smaller (of molecular size). The main advantage of molecular electronics is that transport properties at this molecular size may be more predictable compared to those of nanoscale SET. On the other hand, conventional silicon-based MOSFET devices, if shrunk to the putative size of typical molecular electronics, would consist of only a small number of atoms, leading to non-statistical and unpredictable transport behavior. Presently, the minimum feature size of the state-of-the-art MOSFET devices stands at 30–40 nm. Further device miniaturization would lead to problems such as insufficient gate oxide thickness, resulting in large leakage current and thus high power consumption. Molecular electronics may then be a candidate to overcome these problems. Table 2.2 presents a summary of advantages in molecular electronics.

Electron transport through the molecular structure depends on the types of molecules inserted between the two contacts. But there is one crucial aspect of molecular electronics that is different from the metal–semiconductor single-electronic system or MOSFET. The tunneling coupling of molecule to the metal electrodes is strong, which could reduce the effect of charge and size quantization. In the simplest linear chain of molecules between leads, electron transport could be a simple wave tunneling under a barrier. In the more elaborate donor–barrier–acceptor configuration, the superexchange process for both electron and hole takes over.

In the case of a molecular quantum dot, single molecule physics is important, e.g. in a single molecule magnet, the Kondo effects could be prominent. In the case of large molecules, such as DNA or protein molecules, intramolecular transport becomes important because now electrons need to navigate through a large molecule with many functional groups. But in general, charge transfer within DNA is coherent for short distance; when many base pairs are involved, then transport is one of incoherent hopping. More complicated still is electron transport through the DNA–metal junction. The donor–bridge–acceptor

*Table 2.2* Summary of some useful device features in molecular electronics

| | Useful device features in molecular electronics | Remarks |
|---|---|---|
| 1. | Small size with predictable behavior | Central island is a single molecule |
| 2. | Chemical self-assembly | In chemistry and nanotechnology, a great deal is known about the technique of self-assembly |
| 3. | Switchable between stable configurations | Molecule can switch between a few stable isomers, each having its own electrical and optical properties |
| 4. | Tunable optical, electrical and structural property by choice of chemical composition and geometry | Molecule synthesis has been well-established and these techniques can be applied to tune the electrical and optical properties of molecular based devices |

configuration[38] provides more possibilities in terms of controlling electron or hole transport through the molecule. The multiple barriers within the molecule can be engineered to effect quantum pumping or current rectification. The bridge between the donor and the acceptor can also be engineered to switch between hopping and superexchange transport mechanisms.

What distinguishes electron transport through molecules from normal semiconductor or metal-based nanostructures is the strong coupling across the junction separating metal from molecule. Tunnel coupling through molecular junction to metal electrode is usually so strong that the correction due to cotunneling and the Kondo effect can be significant. This raises the need to incorporate additional physics into molecular transport. In this respect, one may note that the NEGF method, which takes proper account of intersite coupling, might be a useful method with which to study electron transport in molecular devices.

Here, we give one example of molecular electronics (in this case magnetic) where the molecular properties when connected to electrodes or substrates might become altered significantly with manipulation.[39] The molecular central channel is connected to a substrate and demonstrates a high Kondo effect after manipulation is carried out to the molecular structure. This is the first step toward molecular electronics or spintronics in which molecular structures are to be connected to contacts for the injection of charge current, spin current or even supercurrent. In a magnetic molecule, intramolecular exchange and spin orbit coupling can give rise to high anisotropy energy at zero magnetic field.

By and large, the outlook for molecular electronics is still fraught with fundamental challenges. Controlled fabrication of molecular systems suitable for electronic applications has still not been fully mastered. A self-assembly technique is one of the favorite methods that is being intensively pursued at present. As we have learnt earlier, electron transport within the molecule is easier to understand than the transport across the junction between metal and molecule. In fact accurate measurement of the junction conductance is itself a major experimental challenge. There is also a whole host of other practical issues such as reliability, temperature stability and so forth that have not been fully addressed.

### 2.3.5 Carbon nanotube transistors

Carbon-based electronics[40] can be regarded as a form of molecular electronics. It falls under the linear chain type if it is in the form of nanotube sandwiched between the contacts, or the quantum dot type if it is in the form of C60 fullerene that plays the role of the central island. C60 (fullerene) was first discovered by Harold W. Kroto, Richard E. Smalley, Robert F. Curl in 1985,[41] while the carbon nanotube was discovered by Sumio Ijima in 1991.[42] These carbon-based nanostructures have markedly different electronic, mechanical, optical and chemical properties than their film or bulk counterparts. There has been a large body of work dedicated to studying the fabrication of the CNTs. Generally, they can be divided into:

- Arc discharge, which yields single-walled (anode doped with metals) and multiwalled CNT (both electrodes are pure graphite).
- Laser ablation,[43] which yields single-walled nanotubes (SWNT) mixed with carbon nanoparticles (e.g. fullerenes). The diameter of the tube is controllable by the reaction temperature, but this process is costly.
- Chemical vapor deposition (CVD),[44] which yields both SWNT and multiwalled nanotubes (MWNT). This method also enables selective and directional growth, but tends to produce more defects.

A field-effect transistor (FET) with CNT elements has electrical properties which promise superior FET performance to that of silicon-based MOSFET. To fully appreciate the superiority of CNT FET as opposed to MOSFET, one needs to understand the conventional difficulty faced by the silicon-based MOSFET community. The electronic properties of CNT include high conductance, variable bandgap and ohmic contact problems. Its electronic properties are similar to graphene after considering additional boundary conditions. But unlike graphene, the semiconducting CNT shows a large variable bandgap, lending itself to generating a large on/off current ratio. This property is especially useful for potential transistor application.

With more superior electronics features, photonic and chemical properties (the photonic and chemical properties of these devices are not discussed here), there is the favorable prospect of CNT-based electronics replacing MOSFET electronics. Like the problem faced by molecular electronics, the main hurdle lies in the successful controlled fabrication of a large quantity of CNTs with the desired and repeatable contact and channel qualities.

A semiconducting CNT channel is normally connected to metal contacts. The alignment of Fermi levels depends on the work functions and the bandgaps of the CNT and metal, resulting in the formation of a Schottky barrier. Carrier transport through the CNT–metal interface is mainly via quantum tunneling through the Schottky barrier. Large quantity synthesis of CNT is making headway, and the eventual emergence of CNT electronic products in the marketplace – similar to the success achieved by the spintronic spin valve and MRAM – depends on how fast and far the fabrication techniques can improve.

Metal spintronics have been most successful in terms of commercialization, but semiconductor spintronics still have fundamental challenges to overcome. Does this paradox suggest an opportunity for CNT spintronics? CNT already has the essential ingredients for spintronics (e.g. long spin diffusion length, large current carrying capacity). Its electroluminescence property might even promise some sort of unification which involves charge, spin and light – namely electrophoto-spintronics. This all might seem speculative, but the possibility does exist.

### 2.3.6 Graphene electronics

More advances in 2D nanoscale electronics and physics have been achieved in recent years with the advent of graphene-based electronic devices[45] consisting of monolayer and bilayer carbon-based 2D systems. Graphene has superb or extreme electronic properties, which explains why it might be an ideal candidate for possibly replacing silicon in the ever shrinking world of nanoscale electronics. Fundamentally, the transport in graphene is governed by the Dirac equation rather than the Schrödinger equation, reflecting the effective relativistic behavior of electrons in graphene.

The extremely high mobility of graphene of up to 100 000 $cm^2V^{-1}S^{-1}$ at room temperature is its most striking electronic feature. This property makes graphene one of the leading candidates in the post-silicon era. Besides ambient ballisticity, the large value of Fermi velocity ($v_F$) and low Schottky barrier of graphene could reduce switching time. However, graphene possesses a lower bound to its conductivity, i.e. it has a finite minimal conductivity even when carrier density goes to zero. This poses an obstacle to achieving a good on/off current ratio, which is a key figure-of-merit of transistor devices. In fact, one of the main hurdles that graphene faces in terms of serving as a suitable material for electronic devices is the absence of an energy gap. There have been various proposals (substrate choice, and lateral or magnetic confinement) to induce a gap in graphene. It has been demonstrated that a systematic modulation of the graphene gap can be accomplished by varying the sample thickness. This work has paved the way for possible bandgap engineering in graphene. Some other useful properties of graphene are listed in Table 2.3.

*Graphene spintronics*

Graphene material can be used as the spacer layer between two ferromagnetic electrodes in a current-perpendicular-to-plane spin valve. What is important here

*Table 2.3* Useful transport properties of graphene-based electronics

| Topic | Electronic property | Experimental value | Remark |
|---|---|---|---|
| 1. | Mobility | 15 000 – 100 000 $cm^2V^{-1}S^{-1}$ | 300 K<br>Remains high even with increasing $n > 10^{12}cm^{-2}$ |
| 2. | Mean free path | 300 nm | 300 K<br>Room temperature ballistic is already within reach for today's device sizes |
| 3. | Screening length | 5 *A* | |
| 4. | Minimal conductivity | $4e^2/h$ | Theoretical prediction is $4e^2/h\pi$ |

is the attempt to measure CPP MR using graphene. There have been some experimental indications that spin current can probably travel though graphene without much spin flip, indicating a long spin diffusion length. It was also found in recent experiments that graphene is enough to reduce the exchange coupling between the FM electrodes.

One may also consider that the lack of spin orbit coupling and the absence of a hyperfine effect might render graphene suitable for quantum computation where stability of the spin orbit is crucial for the proper functioning of such devices. Besides, it has been suggested[46] that the valley degree of freedom in graphene can be compared to the spin degree of freedom in spintronics. This is due to the independence and the degeneracy of the valley degree of freedom. Indeed the valley polarization can be achieved by engineering means. Intervalley scattering is very well-suppressed, which makes the valley DOF robust. The concept of valley electronics might lead to new devices utilizing this degree of freedom. In fact, two valley filters in a series may function as an electrostatically controlled valley valve.

*Graphene single electronics*

Graphene is particularly suitable for single electronic devices if one considers the fact that graphene nanostructure is stable down to atomic monolayer thickness. A possible means of introducing CB and single electronic effects in graphene is via nanopatterning of the entire SET circuitry (including the contacts, the central quantum dot and the interconnects) out of a single graphene sheet. As mentioned previously, SET circuits constitute a well-established class of nanodevices with novel transport properties due to correlated tunneling and Coulomb charging effects. However, these distinct transport characteristics are usually erased or washed out at room temperature. The thermal stability of graphene makes it a good material to exhibit SET effects.

It has been conceived that the absence of an energy gap and the effect of Klein tunneling would render it difficult to achieve one of key requirements of single electronics in graphene, i.e. electron confinement in graphene. However, this obstacle can be overcome by making use of lateral confinement. It is well known that a graphene nanoribbon exhibits an energy gap due to confinement. By introducing two constrictions sandwiching the central island, one may be able to generate energy gaps of the order of about 6 meV for electron confinement.

## 2.4 Electronic background

Electronics refers to the transport of electrons in device systems. Recent progress achieved in nanotechnologies has enabled devices to be made very small down to the nanometer scale, giving rise to the emergence of various nanoscale electronics. Electron transport is the underlying physics of conductance properties or I–V characteristics of devices. Much has been spoken about the fact that, in nanoelectronic devices, one has to take into account both boundary effects and neighboring perturbations. For this purpose, the NEGF formalism is the most complete and systematic method to derive the transport equations formally.

However, the general electronic background in the central channel under which the transport equations are to be constructed still needs to be studied separately. Electron transport is often closely connected with its energy properties. The environment is thus best characterized by the energy–density ($E - r$) and the energy–dispersion ($E - k$) relations. For simplicity, we will confine ourselves here to pure condensed matter systems without boundaries or discrete parts, although it is well known that boundaries give rise to sub-bands which alter the ($E - k$) relations.

How does energy affect the transport properties of electron, or in the language of engineering, the I-V characteristics? First of all, knowledge of the $E - r$ relation allows one to determine the correlation energy between electrons and understand when a specific theory of electron propagation breaks down (for instance, in highly correlated systems, the perturbative field theory cannot be applied). An electron propagator is a function of the energy and its wavevector, i.e. $G(E, k)$. The energy band or dispersion profile can thus provide very physical information directly related to the extent of k-space over which one sums the propagator. In other words, energy band information allows one to appreciate the physics of transport over different energy ranges. The energy band profile also determines the carrier's effective mass, as well as the carrier type (e.g. the conduction band of a semiconductor defines an electron carrier), while the valence band defines the hole carrier, and the Rashba sub-band defines a spin particle with a non-scalar, special unitary group of degree 2 (SU(2)) charge.

## 2.5 Non-interacting electron gas

For metals like Na and Al, the ionic charges are taken to be uniformly distributed to form a static positive background which ensures that the total system is neutral. Electrons move in this uniform positive background, which is also known as jellium (see Fig. 2.15). In the absence of Coulomb interaction, one has a free electron gas system. In the presence of Coulomb interaction, no particle is a totally independent particle. Every particle is defined by its intrinsic property as well as interaction with the other particles.

Determination of an accurate expression for $E - r$ is difficult if the crystalline arrangement of the ions is taken into account. For simplicity, the background ions are normally smudged or averaged out to yield the uniform background of positive charges, which is the aforementioned jellium. Thus, under this approximation, the same $E - r$ expression applies for material systems with different crystal structure. For example, the same $E - r$ expression is applicable to both metals and semiconductors, although they have different values of $r_s$, which is a measure of the electron correlation energy.

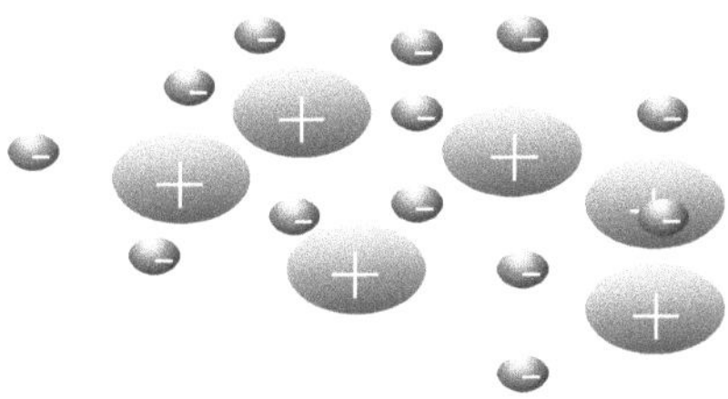

*2.15* Electrons can be treated like gas particles moving randomly in a positively charged background – also known as jellium in many-body physics.

The same cannot be said for the energy band where the metal and semiconductor have distinctly different $E - k$, because the energy band is derived taking into account the crystal arrangement. Thus, different materials such as metals, semiconductors, carbon nanotubes, graphene, insulators, molecules and superconductors give rise to different energy bands. For the non-interacting electron gas system, we will focus on the small $r$ region where the electron system consists primarily of the kinetic energy and behaves like gas. We will apply the non-interacting model known as the jellium model, to derive the energy band.

## 2.5.1 Background energy

In this section, we show that the background energy (or the jellium term) of a material system will be cancelled exactly by an interaction energy term known as the direct energy. Here, we derive the background energy which comprises of the ion ($H_b$) and ion-electron ($H_{be}$) energies of:

$$H_b = \frac{1}{2}e^2\left(\frac{N}{V}\right)^2 \lim_{\mu\to 0}\int\int \frac{e^{-\mu|x-x'|}}{|\boldsymbol{x}-\boldsymbol{x}'|}d^3x\,d^3x' = \lim_{\mu\to 0}\frac{1}{2}e^2\frac{N^2}{V}\frac{4\pi}{\mu^2} \qquad [2.14a]$$

$$H_{be} = -e^2\sum_i^N\left(\frac{N}{V}\right)\lim_{\mu\to 0}\int\frac{e^{-\mu|x-r'|}}{|\boldsymbol{x}-\boldsymbol{r}'|}d^3x = -\lim_{\mu\to 0}e^2\frac{N^2}{V}\frac{4\pi}{\mu^2} \qquad [2.14b]$$

where $N$ is the total number of electron and $V$ volume of the material bulk. Note that due to the long-range nature of the Coulomb interaction, the jellium term is individually divergent. However, it is shown later that the direct energy term cancels this divergent term exactly. The divergent nature gives rise to difficulty in mathematical representation. The Yukawa term $e^{-\mu|x-x'|}$ is therefore introduced to produce a well-defined expression prior to its eventual cancellation. With proper substitution, the integral becomes:

$$I = \int\int\frac{e^{-\mu|x-x'|}}{|\boldsymbol{x}-\boldsymbol{x}'|}d^3x\,d^3x' = \int\int\frac{e^{-\mu y}}{y}d^3y\,d^3x$$

$$= V\int_0^{2\pi}d\phi\int_0^{\pi}\sin\theta\,d\theta\int_0^{\infty}\frac{y^2e^{-\mu y}}{y}dy. \qquad [2.15]$$

Straightforward integration of the above leads to:

$$I = 4\pi V \int_0^\infty y e^{-\mu y}\, dy = 4\pi V \int_0^\infty y d\left(\frac{e^{-\mu y}}{-\mu}\right)$$

$$= 4\pi V \left[\left(y \frac{e^{-\mu y}}{-\mu}\right)_0^\infty + \int_0^\infty dy \left(\frac{e^{-\mu y}}{\mu}\right)\right] = \frac{4\pi V}{\mu^2}. \qquad [2.16]$$

Having evaluated the integral, the total background energy can be found to be as follows:

$$H_b + H_{be} = -\lim_{\mu \to 0} \frac{e^2}{2} \frac{N^2}{V} \frac{4\pi}{\mu^2}. \qquad [2.17]$$

## 2.5.2 Electron kinetic and direct energy

The background energy has been derived based on semiclassical means. The electron energy will now be derived more formally based on quantum methods. We learnt earlier that the electron energy of the system consists of kinetic energy and potential energy. Note that in the following, the scalar amplitude of electron–electron interaction (matrix element) has been explicitly written down. Since the Fermi gas has been written in the field theoretic form, second quantization is required in this section. In second quantization, an operator can be written in the form of its matrix elements summed over all possible states, and multiplied by the creation and annihilation operators of those states. In the following, we show the derivation of these terms as well as the scalar amplitude of electron–electron interaction:

$$\widehat{H}_e = \widehat{H}_k + \widehat{H}_{ee} = \sum_{k_1 k_2 \lambda_1 \lambda_2} \left\langle \boldsymbol{k}_1 \lambda_1 \right| T^v \left| \boldsymbol{k}_2 \lambda_2 \right\rangle a^\dagger_{k_1\lambda_1} a_{k_2\lambda_2}$$

$$+ \frac{e^2}{2V} \sum_{\boldsymbol{kpq}} \sum_{\lambda_1 \lambda_2} \left\langle \boldsymbol{k}_1 \lambda_1, \boldsymbol{k}_2 \lambda_2 \right| \widehat{V} \left| \boldsymbol{k}_3 \lambda_3, \boldsymbol{k}_4 \lambda_4 \right\rangle a^\dagger_{k_1\lambda_1} a^\dagger_{k_2\lambda_2} a_{k_4\lambda_4} a_{k_3\lambda_3} \qquad [2.18]$$

Factor of 2 to account for double accounting

Scalar amplitude of electron–electron interaction

where $T^v$ is the first-quantized kinetic energy operator. The derivation of the scalar amplitude of electron–electron interaction, upon its completion, imposes an important condition that can be interpreted straightforwardly as the conservation of momentum, which dictates:

$$\boldsymbol{k}_1 = \boldsymbol{k} + \boldsymbol{q}$$
$$\boldsymbol{k}_2 = \boldsymbol{p} - \boldsymbol{q}$$
$$\boldsymbol{k}_3 = \boldsymbol{k}$$
$$\boldsymbol{k}_4 = \boldsymbol{p}. \qquad [2.19]$$

With the above, the Hamiltonian can be written in a more intuitive form of:

$$\begin{aligned}\widehat{H}_e = \widehat{H}_k + \widehat{H}_{ee} & \sum_{k\lambda}\left(\frac{\hbar^2 k^2}{2m}\right) a^{\dagger}_{k\lambda} a_{k\lambda} \\ & + \lim_{\mu\to 0}\frac{e^2}{2V}\sum_{\substack{kpq \\ q\neq 0}}\sum_{\lambda_1\lambda_2}\left(\frac{4\pi}{q^2+\mu^2}\right) \underbrace{a^{\dagger}_{k+q,\lambda_1} a^{\dagger}_{p-q,\lambda_2} a_{p,\lambda_2} a_{k,\lambda_1}}_{(\widehat{H}_{ee})_{q\neq 0}} \\ & + \lim_{\mu\to 0}\frac{e^2}{2V}\sum_{kp}\sum_{\lambda_1\lambda_2}\left(\frac{4\pi}{\mu^2}\right) \underbrace{a^{\dagger}_{k,\lambda_1} a^{\dagger}_{p,\lambda_2} a_{p,\lambda_2} a_{k,\lambda_1}}_{(\widehat{H}_{ee})_{q=0}} \end{aligned} \qquad [2.20]$$

Note that the potential energy contains two terms, i.e. the finite momentum transfer and the zero momentum transfer terms. The former is also known as the exchange energy, while the latter is called the direct energy. Recalling that electrons are fermionic particles, the following anti-commutation rules apply:

$$\left\{a_k, a^{\dagger}_p\right\} = \delta_{kp};\quad \left\{a_k, a_p\right\} = 0;\quad \left\{a^{\dagger}_k, a^{\dagger}_p\right\} = 0. \qquad [2.21]$$

Note that $a_{k\lambda}$ is an operator which destroys a particle with momentum $k$ and spin $\lambda$, whereas $a^{\dagger}_{k\lambda}$ is an operator that creates a particle with momentum $k$ and spin $\lambda$. When the two operators are paired together as follows, $\hat{n}_{k\lambda} = a^{\dagger}_{k\lambda} a_{k\lambda}$, they represent the number of particles with momentum $k$ and spin $\lambda$. Since in condensed matter physics we are only dealing with electrons that are fermions, the average of $\hat{n}_{k\lambda}$ in the ground state or $\langle G|\hat{n}_{k\lambda}|G\rangle$ is either 1 or 0, depending on whether momentum $k$ is above or below the Fermi level. Therefore $\langle G|\hat{n}_{k\lambda}|G\rangle$ can also be represented by the Heaviside function of $\theta(k_F - k)$. Note that:

$$\theta(k_F - k) = \begin{cases} 1 \ \mathit{for}\ k < k_F \\ 0 \ \mathit{for}\ k > k_F \end{cases}. \qquad [2.22]$$

We will now study the direct energy, which is in fact the second term of the potential energy.

$$\begin{aligned}\widehat{H}_D &= \frac{e^2}{2V}\sum_{kp}\sum_{\lambda_1\lambda_2}\frac{4\pi}{\mu^2} a^{\dagger}_{k,\lambda_1} a^{\dagger}_{p,\lambda_2} a_{p,\lambda_2} a_{k,\lambda_1} \\ &= \frac{e^2}{2V}\sum_{kp}\sum_{\lambda_1\lambda_2}\frac{4\pi}{\mu^2} a^{\dagger}_{k,\lambda_1} a^{\dagger}_{p,\lambda_2} a_{p,\lambda_2} a_{k,\lambda_1}(1 - \delta_{kp}\delta_{\lambda_1\lambda_2}). \end{aligned} \qquad [2.23]$$

Because destroying a particle with a particular momentum twice yields a zero, i.e. $a_k a_k = 0$, the momenta $k$ and $p$ must be different. Any operation on the system in the ground state should return the system back to the ground state. Since the above Hamiltonian involves zero momentum transfer (i.e. $\Delta k = 0$), the only possibility is that an electron with momentum $k$ is destroyed and replaced by the creation of an electron with the same momentum $k$. The same goes for an electron with momentum $p$; this leads to the pairing as shown below:

$$\widehat{H}_D = \frac{e^2}{2V}\sum_{kp}\sum_{\lambda_1\lambda_2}\frac{4\pi}{\mu^2}(a^\dagger_{\boldsymbol{k},\lambda_1}a_{\boldsymbol{k},\lambda_1})(a^\dagger_{p,\lambda_2}a_{p,\lambda_2})(1-\delta_{kp}\delta_{\lambda_1\lambda_2})$$
$$= \frac{e^2}{2V}\sum_{kp\lambda_1\lambda_2}^{N}\frac{4\pi}{\mu^2}\hat{n}_{k\lambda_2}\hat{n}_{p\lambda_2}. \qquad [2.24]$$

Here, one recognizes that $a^\dagger_{\boldsymbol{k},\lambda_1}a_{\boldsymbol{k},\lambda_1} = \hat{n}_{k,\lambda_1}$ which is the number operator that gives the number of particles with quantum state $k\,\lambda_1$. What can be observed or measured is the expectation or average value of this operator, which is just the number of particles with quantum state $k\,\lambda_1$. In the above, we have made use of the anti-commutation relations, $\{a_k, a_p\} = 0$; $\{a^\dagger_k, a^\dagger_p\} = 0$. The average direct energy is thus:

$$E_D = \left\langle G\middle|\widehat{H}_D\middle|G\right\rangle = \lim_{\mu\to 0}\frac{e^2 4\pi}{2V\mu^2}(N^2 - N) = \lim_{\mu\to 0}\left(\frac{e^2N^2 4\pi}{2V\mu^2} - \frac{e^2 N 4\pi}{2V\mu^2}\right). \qquad [2.25]$$

It can be seen that the first term of the direct energy cancels the average of the background energy. The second term of the direct energy goes to zero with proper limit taking. We will ignore the process of limit taking which makes the second term vanish. The system is thus left with the kinetic energy if the exchange can be neglected for the time being. We discuss the exchange energy in Section 2.6 on interaction. The kinetic energy operator is:

$$\widehat{H}_k = \sum_{k_1k_2\lambda_1\lambda_2}\left\langle \boldsymbol{k}_1\lambda_1\,|\,T^v\,|\,\boldsymbol{k}_2\lambda_2\right\rangle a^\dagger_{k_1\lambda_1}a_{k_2\lambda_2} \qquad [2.26]$$

where $T^v$ is the first-quantized kinetic energy operator. Note that the vector quantity has been written in bold.

---

**Exercise 2.1**

Evaluate the matrix element $\langle \boldsymbol{k}_1\,\lambda_1|T^v|\boldsymbol{k}_2\lambda_2\rangle$ of the kinetic energy.

***Solution***

**Hint:** The matrix element is:

$$\left\langle \boldsymbol{k}_1\lambda_1\,|T^v|\,\boldsymbol{k}_2\lambda_2\right\rangle = \int -(2mV)^{-1}\left\langle \boldsymbol{k}_1\lambda_1\,|x'\right\rangle\left\langle x'|\hbar^2\nabla^2|x''\right\rangle\left\langle x''|\boldsymbol{k}_2\lambda_2\right\rangle d^3x'd^3x''$$
$$= \int e^{-i\boldsymbol{k}_1.x'}\eta^\dagger_{\lambda_1}\frac{-\hbar^2\nabla^2}{2m\,V}e^{i\boldsymbol{k}_2.x''}\delta(x'-x'')\eta_{\lambda_2}\,d^3x'\,d^3x''$$
$$= \frac{\hbar^2k_2^2}{2m\,V}\delta_{\lambda_1\lambda_2}\int d^3x''\,e^{i(\boldsymbol{k}_2-\boldsymbol{k}_1).x''} = \frac{\hbar^2k_2^2}{2m\,V}\delta_{\lambda_1\lambda_2}V\delta_{k_1k_2}$$

---

Thus, the full expression for the kinetic operator is:

$$\hat{H}_k = \sum_{k_1k_2\lambda_1\lambda_2} \frac{\hbar^2 k_2^2}{2m\,V}\delta_{\lambda_1\lambda_2} V\delta_{k_1k_2} a^{\dagger}_{k_1\lambda_1} a_{k_2\lambda_2} = \sum_{k\lambda}\left(\frac{\hbar^2 k^2}{2m}\right) a^{\dagger}_{k\lambda} a_{k\lambda}$$
$$= \sum_{k\lambda}\left(\frac{\hbar^2 k^2}{2m}\right)\hat{n}_{k\lambda}. \qquad [2.27]$$

To find the kinetic energy per particle of the system, one needs to consider the Fermi wavevector. The number of particles in the system can be found by the following expression:

$$N = \sum_{k\lambda}\left\langle G\middle|\hat{n}_{k\lambda}\middle|G\right\rangle = \sum_{k\lambda}\theta(k_F - k) = \frac{V}{(2\pi)^3}\int_0^{kF} d^3k \sum_{\lambda}\theta(k_F - k). \qquad [2.28]$$

Note that the summation over spin above leads to a factor of 2. It is not hard to deduce from Eq. 2.28 the Fermi wavevector expressed in terms of electron density as below:

$$\Rightarrow k_F = \left(3\pi^2\frac{N}{V}\right)^{1/3} = \left(\frac{9\pi}{4}\right)^{1/3}\frac{1}{r_0}. \qquad [2.29]$$

In Eq. 2.29, use has been made of the important relation of:

$$\sum_{k} \rightarrow \frac{V}{(2\pi)^3}\int d\mathbf{k} = \frac{V}{(2\pi)^3}\int dk_x\, dk_y\, dk_z. \qquad [2.30]$$

Equation 2.30 simply follows from the fact that one state fills the k-space volume of $\left(\frac{2\pi}{L_x}\right)\left(\frac{2\pi}{L_y}\right)\left(\frac{2\pi}{L_z}\right) = \frac{(2\pi)^3}{V}$. Averaging the kinetic energy over the ground state, we have:

$$E_k = \left\langle G\,|\,\hat{H}_k\,|\,G\right\rangle = \frac{\hbar^2}{2m}\sum_{k\lambda}k^2\left\langle G\,|\,\hat{n}_{k\lambda}\,|\,G\right\rangle = \frac{\hbar^2}{2m}\sum_{k\lambda}k^2\theta(k_F - k)$$
$$= \frac{\hbar^2}{2m}\frac{V}{(2\pi)^3}\int_0^{\infty} d^3k\sum_{\lambda}k^2\theta(k_F - k) = \frac{3}{5}\frac{\hbar^2 k_F^2}{2m}N = \frac{e^2}{2a_0}N\frac{2.21}{r_S^2}$$
$$\Rightarrow \frac{E_k}{N} = \frac{2.21}{r_S^2} \text{ in Rydberg energy per particle.} \qquad [2.31]$$

Note that to convert SI units to the Rydberg unit, one needs the following relationships:

$$V = \frac{4}{3}\pi r_0^3 N,\ r_S = \frac{r_0}{a_0} \text{ where } a_0 = \frac{\hbar^2}{me^2} \qquad [2.32]$$

where $r_0$ is the radius defining a volume surrounding an electron. Therefore $r_0$ can be interpreted as the average inter-particle spacing. With the earlier expressions found for the Fermi wavevector, the kinetic energy can now be expressed in terms of the Fermi wavevector and electron density. It follows that the density of states per bulk volume can be deduced by proper substitution to be:

$$D(E) = \frac{dN}{dE} = \frac{V}{2\pi^2}\left(\frac{2m}{\hbar^2}\right)^{\frac{3}{2}} E^{\frac{1}{2}}\,\theta(E). \qquad [2.33]$$

The number of particles as well as the total energy of the system will be given by:

$$N = \int D(E)\ dE;\ \ E_T = \int E\,D(E)\ dE. \qquad [2.34]$$

Although we have found the density of states, it does not necessarily imply the density of electrons. Density of states below the Fermi level implies the density of electrons only at zero temperature. At non-zero temperature, the probability of an electron occupying a state is given by the probability function also known as the Fermi–Dirac function:

$$f(E) = \frac{1}{e^{(E-E_F)/kT}+1} \to \lim_{T\to 0} f(E) = \theta(E_F - E). \qquad [2.35]$$

This function is a Heaviside function only at zero temperature, which decrees strict occupation of all states below the Fermi energy and the vacation from any state above the Fermi energy. But at non-zero temperature, the system becomes partially degenerate and electron occupation obeys the smooth Fermi–Dirac distribution. In fact the Fermi–Dirac function is very robust against temperature effect in that its deviation from the Heaviside profile is only affected minimally even at very high temperature. Similarly its derivative is also a strict delta function at zero temperature with minimal deviation as temperature increases.

$$\frac{-\partial f}{\partial E} = \frac{kT/4}{\cosh^2\left[\frac{kT}{2}(E-E_F)\right]} \to \lim_{T\to 0}\frac{-\partial f}{\partial E} = \delta(E_F - E). \qquad [2.36]$$

It is, however, interesting to note that, in semiconductors, electrons above the conduction exceeds the Fermi energy by $(E - E_F) > E_g/2 \gg kT$. Electron distribution can be approximated by the Maxwell Boltzmann distribution instead.

## 2.5.3 Energy band

Besides the ground state, there is one important approximation in the field theoretic description of electron gas, i.e. electrons are free and not bound to the ion cores. In reality, the ion cores are not uniformly distributed but extend periodically in three dimensions. This effect can be fully appreciated by studying the energy dispersion (or $E - k$) relation. The $E - k$ relation takes fully into account the core periodic potential which, unlike the random electron correlation potential, gives rise to the quantization of the electron wavevectors, and hence energy bands and bandgaps. It is fortunate that the effect of periodic potential is not hard to incorporate and the benefit of such modification is tremendous. The emergence of energy band and bandgaps provides an explanation for the important transport concept of conductance and resistance modulation.

Most importantly, energy dispersion studies allow one to distinguish metals from semiconductors and insulators. Under the free electron assumption, all material systems are regarded as consisting of a collection of electrons (forming an electron gas or liquid), and thus cannot be differentiated from one another. But experiments have shown the presence of different classes of materials that vary significantly in terms of their electrical conductance (namely, metals, semiconductors, semimetals, insulators) with different periodic ion arrangements. It is thus reasonable to expect that ionic arrangements can exert a strong influence on the electron dynamics, energy and so forth. One would also suspect that under this influence, the electrons might not travel as freely as was assumed in the free electron model and might exist as some kind of bound waves. When electron motion is constrained, quantization of the wavevector is to be expected.

Bloch's theory of non-interacting electron traveling in periodic cores explains the existence of metal, semiconductor and insulator in 3D electron gas systems. The important point here is the Fermi energy relative to the band. For example, in metals, the Fermi energy is in the middle of the band, i.e. there will be no energy gap between the first unoccupied level and the last occupied level. In insulators, the Fermi energy is within an energy gap (bandgap) above the valence band or the filled band for electrons bound to atomic core. In semiconductors, the situation is similar, except that the bandgap is relatively small, i.e. usually less than 2 eV. Thus, semiconductors like GaAs and Si are insulators at zero temperature. But at room temperature, a sufficient number of electrons can be excited into the conduction band, where they become charge carriers.

Let us consider a spatially periodic potential which gives rise to the energy bands and bandgaps discussed above. For simplicity, we assume periodicity in one dimension only. We know from wave mechanics that when the Bragg condition of $k = \pm\frac{1}{2}G = \pm\frac{n\pi}{a}$ is satisfied, wave reflection can occur. Here, $a$ is the lattice constant of the 1D system and $G = \frac{2\pi n}{a}$ can be viewed as a property of the system related to its periodicity. Thus the first reflection occurs in the range of $-\frac{\pi}{a} < k < \frac{\pi}{a}$, which is also known as the first Brillouin zone of the lattice system. The points of reflection at $k = \pm\frac{n\pi}{a}$ in fact correspond to the points where energy gaps emerge.

We will now explain this in terms of the electron waves. Electron waves with long wavelengths (small wavevectors) can be regarded as mostly free waves since they are bound only by the material edges. The lowest wavevector is $\frac{2\pi}{L}$. These wavevectors increase in multiples of $n\frac{2\pi}{L}$. The energy of these electron waves increases parabolically with the wavevectors since the quantization effect due to material edge is too small to be noticeable. As the wavevector reaches a value that satisfies the Bragg condition, the electron wave is no longer free. In fact, electron waves at these wavevectors exist in the linear superposition consistent with quantum mechanics formalism of single particle wave function:

$$\psi(+) = \exp\left(\frac{i\pi x}{a}\right) + \exp\left(-\frac{i\pi x}{a}\right) = 2\cos\left(\frac{\pi x}{a}\right)$$

$$\psi(-) = \exp\left(\frac{i\pi x}{a}\right) - \exp\left(-\frac{i\pi x}{a}\right) = 2i\sin\left(\frac{\pi x}{a}\right). \qquad [2.37]$$

The above are standing waves and the negative sign is due to the change of phase upon reflection. For a free traveling wave, the probability density is $\rho = \psi^*\psi = |\psi|^2 = 1$. But for the standing waves of the above, we have:

$$\rho(+) \propto \cos^2\left(\frac{\pi x}{a}\right);\ \rho(-) \propto \sin^2\left(\frac{\pi x}{a}\right). \qquad [2.38]$$

The above electron densities give rise to a difference of potential energy between $\psi(+)$ and $\psi(-)$. This difference is the energy gap. The wavefunctions at the first Brillouin zone are, after normalization:

$$\psi(+) = \sqrt{2}\cos\left(\frac{\pi x}{a}\right);\ \psi(-) = \sqrt{2}\sin\left(\frac{\pi x}{a}\right). \qquad [2.39]$$

Writing the potential energy of an electron in the crystal as $U(x) = U\cos 2\pi x/a$, the first energy gap is:

$$E_g = \int_0^1 dx\, U(x)(|\psi(+)|^2 - |\psi(-)|^2)$$

$$= 2\int dx\, U\cos\left(\frac{2\pi x}{a}\right)\left[\cos^2\left(\frac{\pi x}{a}\right) - \sin^2\left(\frac{\pi x}{a}\right)\right] = U. \qquad [2.40]$$

In the following, we present a model of the periodic potential, known as the Kronig–Penny model (Fig. 2.16).

The electron waves in region, II, I and III are, respectively, $\psi_{II} = Ae^{iKx} + Be^{-iKx}$ $\psi_I = Ce^{Qx} + De^{-Qx}$, and $\psi_{III} = \psi_I e^{-ik(a+b)}$, consistent with the Bloch form. Matching the wavefunctions and their derivatives at $x = 0$, one has $A + B = C + D$

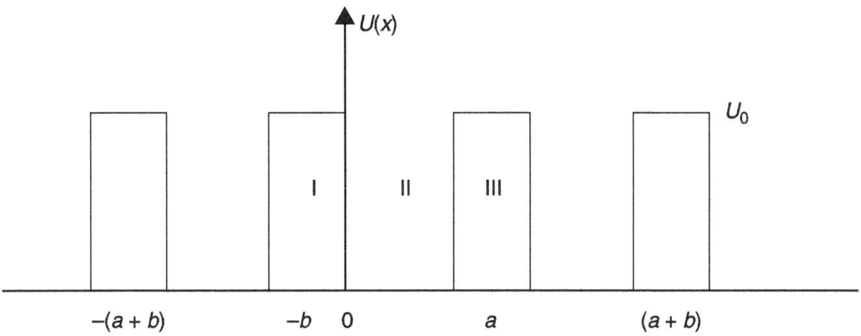


*2.16* Idealized square potential distribution mimic of crystal core potential for illustrating the core crystal effect on energy band.

as well as $iK(A - B) = Q(C - D)$. Matching wavefunctions and derivatives at $x = \alpha$, one has:

$$Ae^{iKa} + Be^{-iKa} = (Ce^{-Qb} + De^{Qb})e^{ik(a+b)}$$
$$iK(Ae^{iKa} + Be^{-iKa}) = Q(Ce^{-Qb} - De^{Qb})e^{ik(a+b)}. \qquad [2.41]$$

The wavefunctions and their derivative equations will only have a solution at $x = 0$ and $x = \alpha$ if the determinant of the coefficients of A, B, C, D vanishes, or:

$$\left[\frac{Q^2 - K^2}{2QK}\right]\sinh Qb \sin Ka + \cosh Qb \cos Ka = \cos k(a+b). \qquad [2.42]$$

One can simplify Eq. 2.42 by setting $b \to 0$, $U_0 \to \infty$, such that $\frac{Q^2ba}{2} = P$. In this limit, $Q >> K$ and $Qb << 1$, and Eq. 2.42 reduces to:

$$\frac{P}{Ka}\sin Ka + \cos Ka = \cos ka. \qquad [2.43]$$

The allowed values of the energy $E$ are given by ranges of $Ka = \left(\frac{2mE}{h^2}\right)^{1/2}$ for which the function lies between ±1. For other values of the energy there are no travelling waves or Bloch-like solutions to the wave equation. Forbidden gaps are thus formed. The energy dispersion relation has been derived above under the non-interacting electron system where electron waves are solutions to the non-interacting Hamiltonian.

One immediate benefit of this analysis is the realization that metal, semiconductor, elemental carbon, single molecules and superconductors can all be induced to conduct electrons, providing five main classes of material choice for the design of new nanoscale electronics. The electron charge and spin provide two DOFs which can be represented by the coupling constant of the gauge potential.

Another DOF is the device dimensionality, which can range from 3D (bulk), to 2D, 1D and 0D. In a crude reasoning, there are at least 40 permutations in terms of possible nanoelectronic systems, thus promising a fertile area for further development. Table 2.4 summarizes the different DOFs that can be exploited in nanoelectronic systems.

*Effective mass*

The dispersion relation ($E - k$) arises due to boundary conditions and, under these conditions, the electron waves are no longer plane waves. To maintain the simple plane wave forms for electron, the effective mass approximation is followed. Refer to references 3 and 4 for the elucidation of the concept of effective mass. The effective mass has to be used in the derivation of the kinetic energy, particle density, average electron energy and so forth.

In summary, we have learnt that, for small particle spacing, the electrons can be regarded collectively as a gaseous system where the kinetic energy is the only relevant energy. Many important quantities like the Fermi wavevector, particle density and average electron energy can subsequently be derived under this approximation. But since these derivations are based on the free electron approximation, where electrons are regarded as plane waves, modification would be required to model Bloch electron waves in crystals, which are more complicated than plane waves.

*Table 2.4* Requisite background conditions for the construction of nanoelectronic transport equations

| | | | $E - k$ | $E - r$ |
|---|---|---|---|---|
| 1. | Device channel material | Metal, semiconductor, carbon, molecule, superconductor | Energy band Crystal-core dependent | Energy correlation (electron gas, liquid, Wigner crystal) |
| 2. | Carrier type (coupling constant) | Electron charge, hole charge | Conduction band Valence band | Exchange energy Kinetic energy |
| | | Angular spin Pseudospin Isospin | Zeeman sub-bands | |
| 3. | Device channel dimension (nanoscale) | 3D bulk 2D sheet 1D nanowire | Geometrical sub-bands | Contact electrode perturbation |
| | | 0D quantum dot | Geometrical/ Coulomb sub-bands | |

## 2.6 Interacting electron gas

The physics of interaction is rather complicated and takes one into the realm of many-body effects. As described earlier, the propagators and Green's functions are used extensively to study electron transport in an interacting system. In this book, we do not delve fully into interacting electron transport. In this section we discuss the effect of electron–electron interaction on the energy and particle density relation in order to give readers an idea of the environment in which the electrons reside or travel.

### 2.6.1 Electron exchange energy

After the cancellation of the background energy by the direct energy, the total energy of the system now consists of just the kinetic energy and the correlation energy with finite momentum transfer, i.e. the exchange energy.

$$\begin{aligned}\widehat{H} &= \widehat{H}_k + (\widehat{H}_b + \widehat{H}_{be} + (\widehat{H}_{ee})_{q=0}) + (\widehat{H}_{ee})_{q\neq 0} \\ &= \sum_{k\lambda} \frac{\hbar^2 k^2}{2m} a^{\dagger}_{k\lambda} a_{k\lambda} + (\widehat{H}_{ee})_{q\neq 0}.\end{aligned} \qquad [2.44]$$

Recalling that the direct energy is $\widehat{H}_D = (\widehat{H}_{ee})_{q=0}$, we hereby label the exchange energy according to $\widehat{H}_{Ex} = (\widehat{H}_{ee})_{q\neq 0}$. The formal expression of the exchange energy is given below:

$$\begin{aligned}\widehat{H}_{Ex} &= \lim_{\mu\to 0} \frac{e^2}{2V} \sum_{\substack{kpq \\ q\neq 0}} \sum_{\lambda_1\lambda_2} \frac{4\pi}{q^2+\mu^2} a^{\dagger}_{k+q,\lambda_1} a^{\dagger}_{p-q,\lambda 2} a_{p,\lambda_2} a_{k,\lambda_1} \\ &= \frac{e^2}{2V} \sum_{\substack{kpq \\ q\neq 0}} \sum_{\lambda_1\lambda_2} \frac{4\pi}{q^2} a^{\dagger}_{k+q,\lambda_1} a^{\dagger}_{p-q,\lambda_2} a_{p,\lambda_2} a_{k,\lambda_1}.\end{aligned} \qquad [2.45]$$

As is the case with the direct energy term, any operation on the ground state of the system that involves an electron being destroyed and replaced should, upon completion of the process, return the system to the ground state. In the case of direct energy, the electron can be replaced by itself which implies the zero momentum transfer picture, as explained earlier. In the case of exchange energy, the momentum transfer is finite, i.e. $q \neq 0$. One way of achieving this is to let an electron of momentum $p$ exchange position with that of $k$ by imparting momentum $q = (p - k)$ to $k$ while maintaining the spin angular momentum of both electrons. This process is mathematically denoted by $\delta_{\lambda_1\lambda_2}\,\delta_{p,k+q}$. We then shift the operator $a_{p,\lambda_2}$ one notch to the left by 'jumping' over $a^{\dagger}_{p-q,\lambda_2}$:

$$E_{Ex} = \langle G \mid \widehat{H}_{Ex} \mid G \rangle = \frac{e^2}{2V} \sum_{\substack{kpq \\ q\neq 0}} \sum_{\lambda_1\lambda_2} \frac{4\pi}{q^2} \langle G | a^{\dagger}_{k+q,\lambda_1} a^{\dagger}_{p-q,\lambda_2} a_{p,\lambda_2} a_{k,\lambda_1} | G \rangle. \qquad [2.46]$$

Since these two operators are different, swapping them generates a factor of –1.

Thus, the final expression for the exchange energy becomes:

$$E_{Ex} = \left\langle G \mid \widehat{H}_{Ex} \mid G \right\rangle = \sum_{\substack{kq\ \lambda_1 \\ q \neq 0}} \frac{-4\pi e^2}{2Vq^2} \left\langle G \mid \hat{n}_{k+q,\lambda_1}\, \hat{n}_{k,\lambda_1} \mid G \right\rangle$$

$$= 2\sum_{\substack{kq \\ q\neq 0}} \frac{-4\pi e^2}{2Vq^2}\theta\left(k_F - |\boldsymbol{k}+\boldsymbol{q}|\right)\theta(k_F - |\boldsymbol{k}|)$$

$$= \frac{-4\pi e^2}{2Vq^2} 2\int\int_0^\infty \frac{d^3\boldsymbol{k}\, d^3\boldsymbol{q}}{(2\pi)^6}\theta(k_F - |\boldsymbol{k}+\boldsymbol{q}|)\theta(k_F - |\boldsymbol{k}|). \qquad [2.47]$$

In the integral equation above, all operators and state vectors have disappeared. The operators are gone but they have left behind a negative sign and the Heaviside functions. In other words, the second quantized form contains information that has been translated to the energy's sign as well as the limit for integration which determines the interaction's strength.

---

## Exercise 2.2

(1) In an electron system, the total energy comprises only the kinetic energy and the correlation energy with finite momentum transfer, i.e. the exchange energy. By proper reasoning and pairing of the electron operators, the exchange energy can be derived as follows:

$$E_{exch} = \sum_{\substack{kq\ \lambda_1 \\ q\neq 0}} \frac{-4\pi e^2}{2Vq^2}\left\langle G \mid \hat{n}_{k+q,\lambda_1}\, \hat{n}_{k,\lambda_1} \mid G\right\rangle = 2\sum_{\substack{kq \\ q\neq 0}} \frac{-4\pi e^2}{2Vq^2}\theta(k_F - |\boldsymbol{k}+\boldsymbol{q}|)\theta(k_F - |\boldsymbol{k}|).$$

The above can be expressed in the form of an integral function:

$$E_{exch} = \frac{-4\pi V e^2}{2q^2} 2\int\int \frac{d^3\boldsymbol{k}\, d^3\boldsymbol{q}}{(2\pi)^6}\theta(k_F - |\boldsymbol{k}+\boldsymbol{q}|)\theta(k_F - |\boldsymbol{k}|).$$

Show that this integral finally leads to the exchange energy per particle expression of:

$$\frac{E_{exch}}{N} = -\frac{e^2}{2a_0}\frac{0.916}{r_S} = -\frac{0.916}{r_S} \text{ in Ryberg per particle.}$$

***Solution***

**Hints:** Let's start from one value of $q$ in any direction. The strength of $q$ must fall in the range $0 < q < 2k_F$ to ensure that, for any $q$, there exists a $\boldsymbol{k} < k_F$ to ensure that $|\boldsymbol{k} + \boldsymbol{q}|$ is less than $k_F$. With this, the limits for this integration as imposed by $\theta(k_F - |\boldsymbol{k}+\boldsymbol{q}|)\theta(k_F - |\boldsymbol{k}|)$ would have been fulfilled.

Thus integration is essentially over the overlapping region between two circles of radius $k_F$ each. Use polar coordinates.

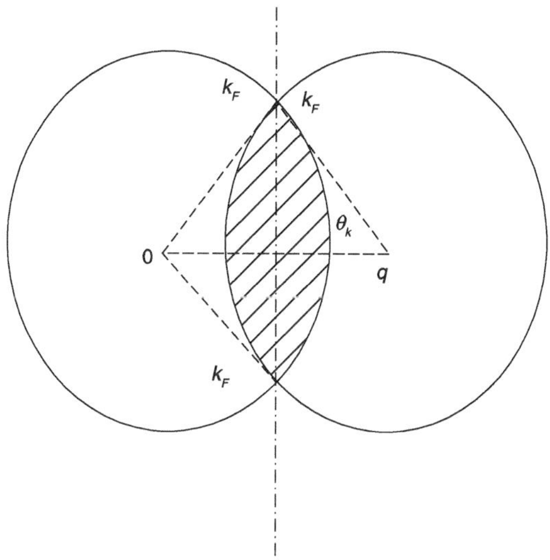


## 2.6.2 Scalar strength of electron–electron interaction

In the above, much has been used of the explicit expression of the matrix element to aid the derivation of the energy terms. Here, we show how the matrix element for the electron–electron interaction can be written down by following the four-field theory commonly used in many-body interaction. The matrix element has the slightly more complicated expression as follows:

$$\langle k_1\lambda_1, k_2\lambda_2 \,|\, V \,|\, k_3\lambda_3, k_4\lambda_4 \rangle$$
$$= \lim_{\mu\to 0}\frac{e^2}{V^2}\int dx_1^3\, dx_2^3 (e^{-ik_1x_1}\eta_{\lambda_1}^{\dagger})(e^{-ik_2x_2}\eta_{\lambda_2}^{\dagger})\frac{e^{-\mu|x_1-x_2|}}{|x_1-x_2|}(e^{ik_3x_1}\eta_{\lambda_3})(e^{ik_4x_2}\eta_{\lambda_4}). \quad [2.48]$$

| Site 1 | Site 2 | Site 1 | Site 2 |
|---|---|---|---|

With the substitution of $x = x_2$; $y = x_1 - x_2$, Eq. 2.48 reduces to:

$$\langle k_1\lambda_1, k_2\lambda_2 | V | k_3\lambda_3, k_4\lambda_4 \rangle = \lim_{\mu\to 0}\frac{e^2}{V^2}\int dx^3\, dy^3 (e^{-i(k_1+k_2-k_3-k_4)\cdot x})(e^{i(k_3-k_1)\cdot y})\frac{e^{-\mu|y|}}{|y|}\delta_{\lambda_1\lambda_3}\delta_{\lambda_2\lambda_4}$$
$$= \lim_{\mu\to 0}\frac{e^2}{V}\int dy^3 (e^{i(k_3-k_1).y})\frac{e^{-\mu|y|}}{|y|}\delta_{\lambda_1\lambda_3}\delta_{\lambda_2\lambda_4}\delta_{k_1+k_2,k_3+k_4}$$
$$= \lim_{\mu\to 0}\frac{e^2}{V}\int dy^3 (e^{iq\cdot y})\frac{e^{-\mu|y|}}{|y|}\delta^{k_1+k_2,k_3+k_4}_{\lambda_1\lambda_3;\lambda_2\lambda_4}. \quad [2.49]$$

Note that in Eq. 2.49, the relation of $\int_{-\infty}^{\infty} e^{-ikx}\, dx = \delta(k)$ or $\int_{-\infty}^{\infty} e^{-i(a-b)x}\, dx = \delta(a-b)$ has been used. In discrete form, $\delta(a-b) = \delta_{ab}$. For the change of variables above where $x_1, x_2 \rightarrow x, y$, the corresponding Jacobian of:

$$J = \begin{vmatrix} \frac{\partial x_1}{\partial x} & \frac{\partial x_1}{\partial y} \\ \frac{\partial x_2}{\partial x} & \frac{\partial x_2}{\partial y} \end{vmatrix} \qquad [2.50]$$

is trivial. Letting $M = \langle \boldsymbol{k}_1\lambda_1, \boldsymbol{k}_2\lambda_2\, |V| \boldsymbol{k}_3\lambda_3, \boldsymbol{k}_4\lambda_4 \rangle$ and integrating over the spherical coordinates:

$$\begin{aligned} M &= \lim_{\mu\to 0} \frac{e^2}{V} y^2\, d\theta\, dy\, sin\, \theta\, d\phi \left( \frac{e^{iqy\cos\theta - \mu|y|}}{|y|} \right) \delta^{k_1+k_2,k_3+k_4}_{\lambda_1\lambda_3;\lambda_2\lambda_4} \\ &= \lim_{\mu\to 0} \frac{e^2}{V} 2\pi \int_0^{\infty} dy \int_{-1}^{1} d\cos\theta \frac{e^{iqy\cos\theta - \mu y}}{y} \delta^{k_1+k_2,k_3+k_4}_{\lambda_1\lambda_3;\lambda_2\lambda_4}. \end{aligned} \qquad [2.51]$$

Note that Eq. 2.51 is equivalent to another change of variables where the Jacobian is simply $y^2 \sin\theta$. To save space, we will now represent $\delta^{k_1+k_2,k_3+k_4}_{\lambda_1\lambda_3;\lambda_2\lambda_4}$ with just a simple notation of $\delta$. It thus follows that:

$$\begin{aligned} M &= \lim_{\mu\to 0} \frac{e^2}{V} \frac{2\pi}{iq} \int_0^{\infty} e^{-\mu y} dy \left[ e^{iqy\cos\theta} \right]_{-1}^{1} \delta = \boxed{\lim_{\mu\to 0} \frac{e^2}{V} \frac{2\pi}{iq} \int_0^{\infty} e^{-\mu y} dy (e^{iqy} - e^{-iqy}) \delta} \\ &= \lim_{\mu\to 0} \frac{e^2}{V} \frac{2\pi}{iq} \int_0^{\infty} e^{-\mu y} dy (2i \sin qy) \delta = \lim_{\mu\to 0} \frac{e^2}{V} \frac{4\pi}{q} \int_0^{\infty} e^{-\mu y} dy (\sin qy) \delta. \end{aligned} \qquad [2.52]$$

Integration by parts yields:

$$\begin{aligned} M &= \lim_{\mu\to 0} \frac{e^2}{V} \frac{-4\pi}{q^2} \int_0^{\infty} \delta e^{-\mu y} d\cos qy \\ &= \lim_{\mu\to 0} \frac{e^2}{V} \left( \frac{-4\pi}{q^2} \right) \left( \left[ e^{-\mu y} \cos qy \right]_0^{\infty} + \int_0^{\infty} \mu \cos qy\, e^{-\mu y} dy \right) \delta \\ &= \lim_{\mu\to 0} \frac{e^2}{V} \left( \frac{-4\pi}{q^2} \right) \left( 1 + \frac{\mu}{q} \int_0^{\infty} e^{-\mu y}\, d\sin qy \right) \delta \\ &= \lim_{\mu\to 0} \frac{e^2}{V} \left( \frac{-4\pi}{q^2} \right) \left( 1 + \left[ \frac{\mu}{q} \sin qy\, e^{-\mu y} \right]_0^{\infty} + \frac{\mu^2}{q} \int_0^{\infty} \sin qy\, e^{-\mu y}\, dy \right) \delta \\ &= \lim_{\mu\to 0} \frac{e^2}{V} \left( \frac{-4\pi}{q^2} \right) \left( 1 + \frac{\mu^2}{q} \int_0^{\infty} \sin qy\, e^{-\mu y} dy \right) \delta. \end{aligned} \qquad [2.53]$$

Recalling the expression in the dashed box in Eq. 2.52 where:

$$M = \lim_{\mu\to 0}\frac{e^2}{V}\frac{4\pi}{q}\left(\int_0^\infty \sin qy\, e^{-\mu y}dy\right)\delta = \lim_{\mu\to 0}\frac{e^2}{V}\left(\frac{-4\pi}{q^2}\right)\left(1+\frac{\mu^2}{q}\int_0^\infty \sin qy\, e^{-\mu y}dy\right)\delta$$

$$\Rightarrow\left(\frac{4\pi}{q}+\frac{4\pi\mu^2}{q^3}\right)\lim_{\mu\to 0}\frac{e^2}{V}\left(\int_0^\infty \sin qy\, e^{-\mu y}dy\right) = \lim_{\mu\to 0}\frac{e^2}{V}\left(\frac{-4\pi}{q^2}\right)$$

$$\Rightarrow\int_0^\infty \sin qy\, e^{-\mu y}dy = \frac{\left(\dfrac{-4\pi}{q^2}\right)}{\left(\dfrac{4\pi}{q}+\dfrac{4\pi\mu^2}{q^3}\right)} = \frac{-1}{q+\dfrac{\mu^2}{q}} \qquad [2.54]$$

one can therefore arrive at:

$$M = \lim_{\mu\to 0}\frac{e^2}{V}\frac{4\pi}{q}\int_0^\infty e^{-\mu y}dy\,(\sin qy)\delta = \lim_{\mu\to 0}\frac{e^2}{V}\frac{4\pi}{q}\left(\frac{-1}{q+\dfrac{\mu^2}{q}}\right)\delta = \lim_{\mu\to 0}\frac{e^2}{V}\left(\frac{-4\pi}{q^2+\mu^2}\right)\delta$$

$$= \frac{e^2}{V}\left(\frac{-4\pi}{q^2}\right)\delta^{k_1+k_2,k_3+k_4}_{\lambda_1\lambda_3;\lambda_2\lambda_4}. \qquad [2.55]$$

In fact $\delta^{k_1+k_2,k_3+k_4}_{\lambda_1\lambda_3;\lambda_2\lambda_4}$ plays the role of ensuring an important physical outcome, that is, momentum conservation and spin conservation. The delta function allows one to simplify the correlation energy expression of electrons, reducing the number of possible interaction or pairing in the four field operators. As we have seen, the matrix element expression above appears frequently in all second-quantized representations of the electron–electron interaction.

Drude's assumption of an independent electron model has been validated by the field theoretic description of the electron system which shows that, indeed at high electron density, electron–electron correlation can be neglected. The field theoretic approach is, however, accurate only for small interparticle spacing where the ground state remains valid.

We have studied the many-body physics of a free electron gas and, indeed, it validates the early conceptualizations of Drude and Somerfeld. We discuss below a few aspects of Drude's conceptualization and compare this with the many-body effects derived earlier. Drude's model assumes independent electron approximation, which says electron–electron interaction can be neglected. This can be validated by derivations that show that, indeed for high electron density, the electron–electron interaction effects scales linearly with density while the kinetic energy scales as the square of the density. At high density, electron interaction can be neglected, validating Drude's assumptions.

Drude's model assumes free electron approximation, that is, in between collisions, the ionic core has no effect on electron motion. But in reality, the ionic core is not evenly distributed, it is periodically arranged. Electrons are not completely free from the ionic core as there will be binding energies associated with its influence on electron dynamics.

The energy for the electron system is:

$$\left(\frac{E}{N}\right)_{r_s \to 0} = \frac{e^2}{2a_0}\left[\frac{2.21}{r_s^2} - \frac{0.916}{r_s} + \cdots\right]. \qquad [2.56]$$

Equation 2.56 shows that, when electron density is very high, the kinetic energy of electrons is very high and thus the potential energy due to interactions can be neglected. Thus, a high-density electron system mimics a gaseous atomic system. At lower electron density, interaction becomes more prominent as opposed to the kinetic energy (although in reality both energies drop with lower density, but the kinetic energy drops more rapidly than the interaction energy). Under this condition, Coulomb interaction or electron–electron interaction can no longer be neglected. A further decrease in density takes the electron system to the liquid phase.

At this stage, the density of an electron system can best be characterized by a characteristic length known as the Wigner–Seitz radius. This radius defines a sphere (3D) or a circle (2D) within which there contains only one electron. Normally when the Wigner–Seitz radius exceeds one, the electron system can be considered as an electron liquid, where interaction is becoming important. Perturbation theory normally does not work well for a Wigner–Seitz radius higher than one. However, due to screening, the Landau Fermi liquid theory works reasonably well in the regime of a Wigner–Seitz radius that slightly exceeds one. For a Wigner–Seitz radius lower than one, the electron system can be treated as a gas, where interaction can be neglected. In a weakly interacting system like this (gas), perturbation theory can be used to study electron–electron interaction. In the electron liquid phase, the Wigner–Seitz radius ranges from 1 to 110 for a 3D system, and from 1 to 35 for a 2D system. Note that, common metallic systems like Na, Mg and Al, have Wigner–Seitz radii of 2–6. Thus, perturbation theory does not apply very well here. For a very low density system, electron–electron correlation is so prominent that the electron liquid transforms to a crystalline phase also known as the Wigner crystal.

Due to the fact that electron correlation increases with interparticle spacing, one can deduce that ground state expectation will be inaccurate when the interparticle spacing is large. This is because we have evaluated the energy based on the expectation value of the ground state. The Rayleigh–Ritz principle states that the exact ground state of a quantum system always has a lower energy than that evaluated by taking the experimental value of the total Hamiltonian in any normalized state. Wigner showed that, in the low density limit, the energy per particle should be given by:

$$\left(\frac{E}{N}\right)_{r_s \to \infty} = \frac{e^2}{2a_0}\left[\frac{-1.79}{r_s} + \frac{2.66}{r_s^{\frac{3}{2}}} + \cdots\right]. \qquad [2.57]$$

A pictorial representation of Eq. 2.57 is shown in Fig. 2.17.

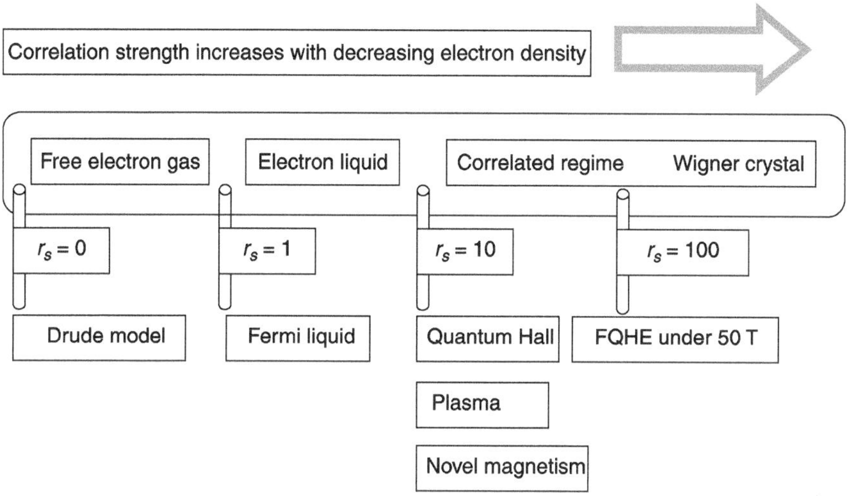


*2.17* Pictorial illustration of electron density effect on the energy of the system. Electron transport expressions are thus closely related to the electron density.

Most of today's devices are based on electron gas systems. Can devices be made from electron liquid systems? The quantum Hall system is one good example. Superconductivity is another interesting system which is not gaseous. Recently, spintronic systems based on the spin orbital effect are treated as some forms of a non-Abelian liquid. We are most familiar with electron gas but we should keep our mind open. There are many possibilities in the world of nanoscale electronics beyond electron gas systems.

Table 2.5 summarizes the energy properties of the various material systems represented by their crystal cores. While the crystal cores are fixed frames, the carriers are mobile and move about randomly. The final column of Table 2.5 describes the carrier types which have direct physical significance with respect to their dynamics and to the types of current one would expect to measure in nanoscale electronic systems.

### Exercise 2.3

(1) In material with a very high electron density or small interparticle spacing (e.g. metals), the electron system can be treated as a non-interacting gas in which only the kinetic energies of individual electrons need to be considered. This is consistent with Drude's assumptions of non-interacting (independent) electron approximation. Explain in terms of energy relation with interparticle

*Table 2.5* Summary of various nanoelectronic systems with respect to their energy systems and carrier types

| | Crystal core | $E-k$ | $E-r$ | Carriers |
|---|---|---|---|---|
| 1. Spin electronics | Metal (FM) | Exchange dispersion | Electron gas | Electron |
| | Semiconductor | Rashba, Dresselhaus, Luttinger, sub-bands | Electron gas | Electron<br>Spin particle |
| | Diluted magnetic semiconductor | | | |
| | Carbon | Cylindrical sub-bands | | Electron |
| 2. Single electronics | Metal | Coulomb sub-bands | Correlated | Electron |
| | Semiconductor | Geometrical/ Coulomb sub-bands | Few electron system | Electron |
| 3. Carbon nanotube (CNT) electronics | Carbon | Cylindrical sub-bands<br><br>(Semiconductor/ metal-like) | Luttinger liquid | Quasiparticle |
| 4. Molecular electronics | Organic/carbon | (Semiconductor/ insulator-like) | Correlated few electron system | Electron |
| 5. Graphene electronics | Carbon | Relativistic neutrino<br>Relativistic massive | Electron gas | Massless/<br><br>Massive Dirac |
| Graphene nanoribbon (GNR) | Carbon | Sub-bands (semiconductor/ metal like) | | Pseudospin<br>Valley-spin |
| 6. Quantum Hall systems and devices | Semiconductor/ graphene | (IQHE) Landau sub-band | Electron correlated | Electron |
| | | (FQHE) Landau sub-band | Electron liquid | Anyon |
| | Topological insulators | Dirac cone | | Spin particle |
| 7. Superconducting | Superconductor | Superconductor bandgap | Electron liquid | Cooper pairs |
| 8. Plasmon and photonics | Metal–insulator | | Electron liquid | Plasmon and polaron |

Note: Carrier types are classified with respect to their minimal coupling to the momentum operator.

spacing why Drude's assumptions are only accurate for high-density electronic systems.

(2) While the non-interacting (independent) electron assumption is generally correct for most metal and semiconductor materials, the free electron approximation which ignores the periodic distribution of the ion cores is problematic. Explain how the development of band theoretic method provides the final explanation for why materials behave like metal, semiconductor, and insulator.

(3) The current density in diffusive transport is given by:

$$j = nev$$

where $e$ is the electron charge, $n$ is the bulk electron density, $v$ is the drift velocity. This expression seems to suggest that electric current is contributed by all electrons below the Fermi energy. But the energy-resolved experiment as well as Boltzmann theory show that the measured electric current is contributed by the dynamic of electron close to the Fermi energy. Explain the apparent contradiction of the current expression to experimental and theoretical consensus.

---

**Exercise 2.4**

The number of particles in a non-interacting system is formally described by the following expression

$$N = \sum_{k\lambda} \left\langle G \middle| \hat{n}_{k\lambda} \middle| G \right\rangle = \frac{V}{(2\pi)^3} \int_{-\infty}^{\infty} d^3k \sum_{\lambda} \theta(k_F - |k|)$$

where $\hat{n}_{k\lambda}$ is the number operator, $V$ is the volume of the sample, $\theta(k_F - |k|)$ is a Heaviside function, $|k| = \sqrt{k_x^2 + k_y^2 + k_x^2}$. Using the spherical coordinates or other methods, deduce that:

(1) the Fermi wavevector of a bulk electron gas is given by $k_F = (3\pi^2 n)^{1/3}$

(2) the kinetic energy per particle in the ground state is $E_{avg} = \frac{3}{5}\frac{\hbar^2 k_F^2}{2m}$

(3) the density of states per bulk volume is $D(E) = \frac{1}{2\pi^2}\left(\frac{2m}{\hbar^2}\right)^{\frac{3}{2}} E^{\frac{1}{2}}\theta(E)$

where $n$ is carrier density.

---

## 2.7 Electron localization

The phase relaxation length ($l_\phi$) is a special quantum mechanical relaxation length which has no analogs in classical physics. Namely, classical motion can be described as evolution of the probability of finding a particle at a given point at a given time. However, in quantum mechanics, the state is characterized by the wavefunction which has a phase. The phase is important in the so-called interference phenomena, where the electron wavefunctions having different pre-history are collected at the same point. If the phases of the waves are not destroyed, a specific quantum interference phenomenon can be observed. The phase relaxation time, $\tau_\phi$ describes the relaxation of this phase memory. It is clear that scattering against any static, spin-independent potential cannot lead to the phase relaxation. Indeed, in any stationary potential the equations of motion are time-reversible. The processes that can be responsible for phase relaxation are the ones which break the symmetry with respect to time reversal. These processes are, among others, inelastic scattering by phonons and spin–flip processes.

One can thus visualize that, within a phase coherence length, an electron experiences many elastic collisions during a typical time $\tau_\phi$. Since the electron is considered to be moving diffusively, i.e. subject to many scattering events, the length can be estimated to be $l_\phi$, where:

$$l_\phi = \sqrt{\tau_\phi D} \qquad [2.58]$$

and where $D = \frac{1}{d} vl$ is the diffusion constant ($d$ is the dimensionality of the electron gas). The relaxation process mentioned above is relevant to the interference of the wavefunctions belonging to a single-electron state. However, interference can also be important for the interaction of two electrons having close energies. Indeed, if the energy difference between the electrons is $\approx kT$ they travel almost coherently during the time $\hbar/kT$. Thus the characteristic length of coherent propagation is $l_T = \sqrt{\hbar D/kT}$ which is known here as the thermal dephasing length ($l_T$).

Today's MOSFET technology, which makes semiconductor as short as 100 nm, can already deliver commercial 2DEG devices shorter than the mean free path (MFP). Electron transport can be considered ballistic. Since $l_\phi$ is normally longer than the mean free path because it is harder for collision process to destroy phase coherence than to scatter momentum, nanoscale devices normally fall within $l_\phi$ (Fig 2.18). Electron phase is randomized (losing phase coherence) if the device is longer than $l_\phi$. Within the $l_\phi$, device, resistance scales with an extra quantity beyond the normal linear scaling effect due to only momentum scattering. Such scaling becomes exponential when the device length approaches a critical value known as the localization length. In other words, electron transport becomes impossible through a conductor whose length extends beyond the localization length.

At first glance, the reasoning above seems to suggest that any conductor longer than the critical localization length will become an insulator. But this is not consistent with our knowledge in electrical engineering. Electrical engineers have been able to lay long wires for data transmission. This is because the localization effect is valid only when the electron phase is coherent. In other words, the device length must be shorter than the phase coherent length but comparable to the localization length in order for this effect to manifest (Fig 2.19).

A conductor which is shorter than its $l_\phi$ is known as the phase coherent conductor. Take the example of a metal. The localization length of a metal is $M \times L$ where $L$ is of the order of the MFP and $M$ is the number of modes. An electrical wire with cross section 200 nm × 200 nm has nearly $10^6$ modes. With the MFP in metal approximately 1 nm, the localization length works out to be 1 mm. An electron in metal would have had its phase randomized before it can experience the localization effect. Nonetheless, weak localization has been observed in very thin metal wires at very low temperature. In semiconductors, the number of modes is much less than metals; weak and strong localization have been observed in 1D/2D semiconductors.

If a conductor's length is comparable to the localization length, it is said to be in a strong localization regime; if the conductor length is much shorter, it is in a weak localization regime. Weak localization in 1D/2D materials entails conductivity correction. As temperature increases, $l_\phi$ decreases and the weak localization effect will disappear. Earlier we mentioned that the localization effect can only manifest for phase-coherent electrons. It is also known that electron–electron scattering could give rise to similar conductivity correction with increasing temperature. Since the two effects are similar, they can be distinguished by applying a low magnetic field. A low magnetic field destroys a weak localization effect and hence the conductivity correction due to it. More descriptions of localization effects can be found in field theoretic textbooks.[8,9]

Table 2.6 provides a summary of the length scales discussed in this chapter with respect to electron transport under the electron gas system. The length scales are compared to the many areas of emerging nanoscale electronics to provide contextual understanding.

*Table 2.6* Summary of physical length scales in the context of sizes relevant to modern nanoelectronic devices and systems

| | |
|---|---|
| 1 mm | |
| | Mean free path in the quantum Hall regime<br>Phase relaxation length of metal |
| 100 μm | |
| | Mean free path/phase relaxation length in high-mobility semiconductor at $T < 4$ K |
| 10 μm | |
| | Spin diffusion length in GaAs at low temperature |
| 1 μm | |
| | MOSFET devices (1990)<br>Spin diffusion length in semiconductor |

*(Continued)*

*Table 2.6* Continued.

| | |
|---|---|
| 100 nm | |
| | Mean free path in 2DEG at room temperature<br>Fermi wavelength in semiconductors<br>Quantum Hall devices<br>Graphene devices<br>MOSFET 22 nm (2010)<br>Flash memory 18 nm (2011)<br>Mean free path in polycrystalline metallic films<br>CNT electronic devices<br>Molecular electronic devices |
| 10 nm | |
| | Optical lithography limit<br>Metal spintronic devices (CPP thickness)<br>Spin diffusion length in metal<br>Single-electronic devices (channel radius)<br>Electron beam lithography<br>Fermi wavelength in metals |
| 1 nm | |
| | Distance between atoms |
| 1 Å | |

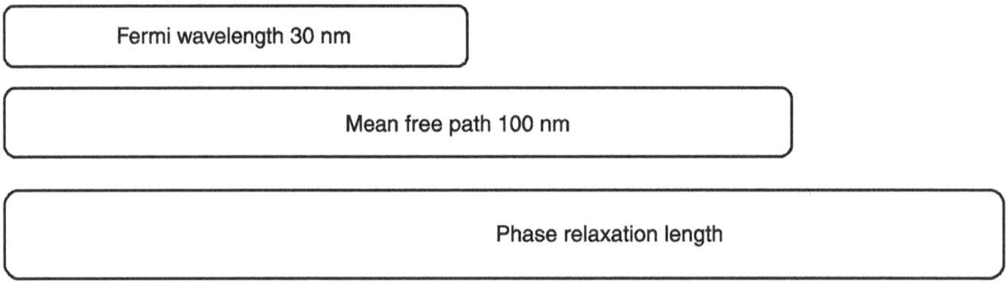


*2.18* Contextual comparison of Fermi wavelength, mean free path and phase relaxation length of a semiconductor channel.

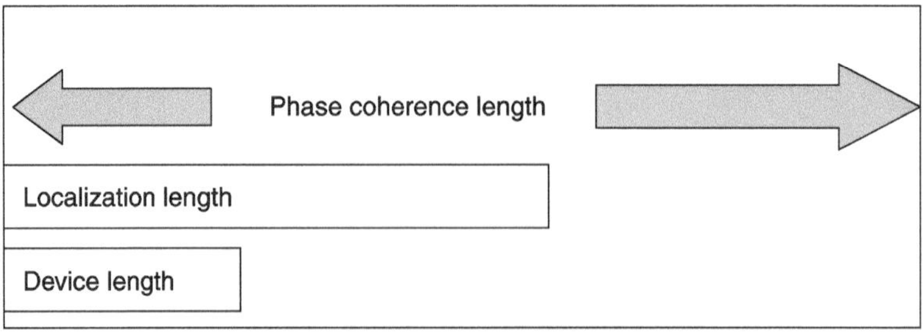


*2.19* Illustration of localization effect occurring in a device with length scale in the above context.

## 2.8 References

[1] Ashcroft N W and Mermin N D (1976), *Solid State Physics*, Brooks/Cole Publishing, Pacific Grove, CA.

[2] Kittel C (1953), *Introduction to Solid State Physics*, John Wiley & Sons, New York; Marder M P (2003), *Condensed Matter Physics*, Wiley InterScience, New York.

[3] Sze S M (1981), *Physics of Semiconductor Devices*, John Wiley & Sons, New York; Sze S M (1985), *Semiconductor Devices Physics and Technology*, John Wiley & Sons, New York.

[4] Chang C Y and Sze S M (1996), *ULSI Technology*, McGraw-Hill, New York.

[5] Schroder D K (1998), *Semiconductor Material and Device Characterization*, John Wiley & Sons, New York.

[6] Merzbacher E (1998), *Quantum Mechanics*, John Wiley & Sons, New York.

[7] Schiff L I (1968), *Quantum Mechanics*, McGraw-Hill International, Singapore.

[8] Bruus H and Flensberg K (2004), *Many-Body Quantum Theory in Condensed Matter Physics*, Oxford University Press, Oxford.

[9] Fetter A L and Walecka J D (2003), *Quantum Theory of Many Particle Systems*, Dover Publications, Mineola, NY.

[10] Datta S (2005), *Quantum Transport – Atom to Transistor*, Cambridge University Press, Cambridge.

[11] Haug H and Jauho A-P (2008), *Quantum Kinetics in Transport and Optics of Semiconductors*, Springer-Verlag, Berlin and Heidelberg.

[12] Wen X G (2004), *Quantum Field Theory of Many-Body Systems*, Oxford University Press, Oxford.

[13] Atland A and Simons B (2006), *Condensed Matter Field Theory*, Cambridge University Press, Cambridge.

[14] Nagaosa N (1999), *Quantum Field Theory in Condensed Matter Physics*, Springer-Verlag, Berlin and Heidelberg.

[15] Binasch G, Grünberg P, Saurenbach F and Zinn W (1989), 'Enhanced magnetoresistance in layered magnetic structures with antiferromagnetic interlayer exchange', *Phys. Rev. B*, 39, 4828.

[16] Baibich M N, Broto J M, Fert A, Nguyen Van Dau F, Petroff F *et al.* (1988), 'Giant magnetoresistance of (001)Fe/(001)Cr magnetic superlattices', *Phys. Rev. Lett.*, 61, 2472.

[17] Yuasa S and Djayaprawira D D (2007), 'Giant tunnel magnetoresistance in magnetic tunnel junctions with a crystalline MgO(0 0 1) barrier', *J. Phys. D: Appl. Phys.*, 40, R337.

[18] Butler W H, Zhang X-G, Schulthess, T C and MacLaren, J M (2001), 'Spin-dependent tunneling conductance of Fe|MgO|Fe sandwiches', *Phys. Rev. B*, 63, 054416.

[19] Hoshiya H and Hoshino K (2004), 'Current-perpendicular-to-the-plane giant magnetoresistance in structures with half-metal materials laminated between CoFe layers', *J. Appl. Phys.*, 95, 6774.

[20] Seneor P, Fert A, Maurice J-L, Montaigne F, Petroff F *et al.* (1999), 'Large magnetoresistance in tunnel junctions with an iron oxide electrode', *Appl. Phys. Lett.*, 74, 4017.

[21] Zheng Y K, Wu Y H, Li K B, Qiu J J, Han G C *et al.* (2007), 'Magnetic random access memory', *J. Nanosci. Nanotechnol.*, 7, 117; Ishigaki T, Kawahara T, Takemura R, Ono K, Ito K *et al.* (2010), 'A multi-level-cell spin-transfer torque memory with

series-stacked magnetotunnel gunctions', *Symposium on VLSI Technology Digest of Technical Papers*, 47, 234.

[22] Berger L (1996), 'Emission of spin waves by a magnetic multilayer traversed by a current', *Phys. Rev. B*, 54, 9353; Slonczewski J C (1996), 'Current-driven excitation of magnetic multilayers', *J. Magn. Magn. Mater.*, 159, L1.

[23] Katine J A, Albert F J, Buhrman R A, Myers E B and Ralph D C (2000), 'Current-driven magnetization reversal and spin-wave excitations in Co/Cu/Co pillars', *Phys. Rev. Lett.*, 84, 3149.

[24] Ikeda S, Miura K, Yamamoto H, Mizunuma K, Gan H D *et al.* (2010), 'A perpendicular-anisotropy CoFeB–MgO magnetic tunnel junction', *Nat. Mater.*, 9, 721.

[25] Houssameddine D, Ebels U, Delaët B, Rodmacq B, Firastrau I *et al.* (2007), 'Spin-torque oscillator using a perpendicular polarizer and a planar free layer', *Nat. Mater.*, 6, 447.

[26] Seneor P, Bernand-Mantel A and Petroff F (2007), 'Nanospintronics: when spintronics meets single electron physics', *J. Phys.: Condens. Matter*, 19, 165222.

[27] Monsma D J, Ladder J C, Popma J A and Dieny B (1995), 'Perpendicular hot electron spin-valve effect in a new magnetic field sensor: The spin-valve transistor', *Phys. Rev. Lett.*, 74, 5260.

[28] Appelbaum I, Huang B and Monsma D J (2007), 'Electronic measurement and control of spin transport in silicon', *Nature*, 447, 295.

[29] Schmidt G, Ferrand D, Molenkamp L W, Filip A T and van Wees B J *et al.* (2000), 'Fundamental obstacle for electrical spin injection from a ferromagnetic metal into a diffusive semiconductor', *Phys. Rev. B*, 62, R4790.

[30] Tan S G, Jalil M B A, Bala Kumar S B, Teo K L and Liew T (2006), 'Utilization of magnetoelectric potential in ballistic nanodevices', *J. Appl. Phys.*, 99, 084305.

[31] Datta S and Das B (1990), 'Electronic analog of the electro-optic modulator', *Appl. Phys. Lett.*, 56, 665.

[32] Meier L, Salis G, Shorubalko I, Gini E, Silke Schön S *et al.* (2007), 'Measurement of Rashba and Dresselhaus spin–orbit magnetic fields', *Nat. Phys.*, 3, 650; Nitta J, Bergsten T, Kunihashi Y and Kohda M (2009), 'Electrical manipulation of spins in the Rashba two dimensional electron gas systems', *J. Appl. Phys.*, 105, 122402.

[33] Likharev K K (1988), 'Correlated discrete transfer of single electrons in ultrasmall tunnel junctions', *IBM J. Res. Dev.*, 32, 144; Likharev K. K. (1999), 'Single-electron devices and their applications', *Proc. IEEE*, 87, 4; Zhuang L, Guo L and Chou S Y (1998), 'Silicon single-electron quantum-dot transistor switch operating at room temperature', *Appl. Phys. Lett.*, 72, 1205.

[34] Krupenin V A, Presnov D E, Zorin A B and Niemeyer J (2000), 'Aluminum single electron transistors with islands isolated from a substrate', *J. Low Temp. Phys.*, 118, 287; Bolotin K I, Kuemmeth F, Pasupathy A N and Ralph D C (2004), 'Metal-nanoparticle single-electron transistors fabricated using electromigration', *Appl. Phys. Lett.*, 84, 3154.

[35] Augke R, Eberhardt W, Single C, Prins F E, Wharam D A *et al.* (2000), 'Doped silicon single electron transistors with single island characteristics', *Appl. Phys. Lett.*, 76, 2065; Foxman E B, McEuen P L, Meirav U, Wingreen N S, Meir Y *et al.* (1993), 'Effects of quantum levels on transport through a Coulomb island', *Phys. Rev. B*, 47, 10020.

[36] Klein D L, Roth R, Lim A K, Alivisatos A P and McEuen P L (1997), 'A single-electron transistor made from a cadmium selenide nanocrystal', *Nature*, 389, 699;

Jung M, Hirakawa K, Kawaguchi Y, Komiyama S, Ishida S *et al.* (2005), 'Lateral electron transport through single self-assembled InAs quantum dots', *Appl. Phys. Lett.*, 86, 033106.

[37] Jalil M B A and Tan S G (2005), 'Spin transfer and current-induced switching in a ferromagnetic single-electron transistor', *Phys. Rev. B*, 72, 214417 ; Braun M, Konig J and Martinek J (2004), 'Theory of transport through quantum-dot spin valves in the weak-coupling regime', *Phys. Rev. B*, 70, 195345; Hamaya K, Masubuchi S, Kawamura M, Machida T, Jung M *et al.* (2007), 'Spin transport through a single self-assembled InAs quantum dot with ferromagnetic leads', *Appl. Phys. Lett.*, 90, 053108; K. Hamaya, Kitabatake M, Shibata K, Jung M, Kawamura M *et al.* (2007), 'Electric-field control of tunneling magnetoresistance effect in a Ni/InAs/Ni quantum-dot spin valve', *Appl. Phys. Lett.*, 91, 022107.

[38] James R H and Mark A R (2003), 'Molecular electronics', *Phys. Today*, 56, 43.

[39] Zhao A, Li Q, Chen L, Xiang H, Wang W *et al.* (2005), 'Controlling the Kondo effect of an adsorbed magnetic ion through its chemical bonding', *Science*, 309, 1542; Crommie M F (2005), 'Manipulating magnetism in a single molecule', *Science*, 309, 1501.

[40] Avouris P (2009), 'Carbon nanotube electronics and photonics', *Phys. Today*, 62, 34.

[41] Curl R F and Smalley R E (1988), 'Probing $C_{60}$', *Science*, 242, 1017; Kroto H (1988), 'Space, stars, $C_{60}$ and soot', *Science*, 242, 1139.

[42] Ijima S (1991), 'Helical microtubules of graphitic carbon', *Nature*, 354, 56.

[43] Guo T, Nikolaev P, Thess A, Colbert D T and Smalley R T (1995), 'Catalytic growth of single-walled nanotubes by laser vaporization', *Chem. Phys. Lett.*, 243, 49; Thess A, Lee R, Nikolaev P, Dai H, Petit P *et al.* (1996), 'Crystalline ropes of metallic carbon nanotubes', *Science*, 273, 483.

[44] Dai H J, Rinzler A G, Nikolaev P, Thess A, Colbert D T *et al.* (1996), 'Single-wall nanotubes produced by metal-catalyzed disproportionation of carbon monoxide', *Chem. Phys. Lett.*, 260, 471; Kong J, Soh H T, Cassell A M, Quate C F and Dai H (1998), 'Synthesis of individual single-walled carbon nanotubes on patterned silicon wafers', *Nature*, 395, 878.

[45] Geim A K and Novoselov K S (2007), 'The rise of graphene', *Nat. Mater.*, 6, 183; Geim A K and MacDonald A H (2007), 'Graphene: Exploring carbon flatland', *Phys. Today*, 60, 35.

[46] Rycerz A, Tworzydo J and Beenakker C W J (2007), 'Valley filter and valley valve in graphene', *Nat. Phys.*, 3, 172.

# 3 Electron dynamics in nanoscale devices

**Abstract:** This chapter introduces the concept of linear response and conductivity, which is by far the most appropriate quantification of electron transport. The concept of non-equilibrium necessary for nanoscale devices is then incorporated, introducing the modern non-equilibrium Green's function (NEGF) formalism, which has become a standard computational method in nanoscale device transport. The equation of motion and the matrix formalisms are discussed. The formula for current is also introduced, taking into account the sizes and shapes of the device.

**Key words:** linear response, conductivity, non-equilibrium Green's function, NEGF, equation of motion, matrix, current.

## 3.1 Introduction to electron transport

The underlying physics which crucially distinguishes the various fields of nanoelectronics is the transport or dynamics of electrons in these systems. The construction of dynamic equations requires the background conditions described in the first two chapters:

- electron spatial density ($E - r$)
- electron dispersion relation ($E - k$)
- carrier types (charge scalar/ spin matrix).

The background conditions determine the forms and types of transport equations, their approximations and assumptions as well as providing the parametric input to these equations. The spatial density of electrons determines if they are to be collectively known as gas, liquid or even solid (solid due to localized electrons, not atoms). Electron gas is by far most useful for nanoelectronic applications because of the ease with which conductance modulation can be effected on these systems. In Chapter 2, we have understood the energy properties and their direct relevance to electron transport. For example, the independent and free electron approximation in Boltzmann transport is only valid in electron gas that can be formally categorized in the $E - r$ studies of electronic energies. The perturbative expansion used in equilibrium or non-equilibrium electron transport is also valid within a certain spectrum of $E - r$, thus allowing one to contemplate when these formalisms might break down or become less accurate.

From the electronic standpoint, different material systems simply means different crystal arrangement (ions) which give rise to the energy bands of metals, semiconductors, insulators, superconductors and semi-metals. Crystal nature determines the types of bonds with which the atoms are bonded and hence the number of free or bound electrons that might, under non-equilibrium conditions (e.g. application of external fields), participate in some processes of collective motions. Thus, electron dynamic equations are normally constructed within certain bands; $E - k$ studies allow one to decide the range over which the energy limit is to be taken. Particularly in semiconductors, which have complicated bandstructures, electron transport can be simplified by understanding the band effects (e.g. effective mass) and restricting studies to within the band of interest.

Carrier type is particularly important in nanoelectronics that involve internal degrees of freedom (DOF), for example, the electron spin in metals, semiconductors, graphene and insulator spintronics, and pseudospin and valley spin in graphene electronics. The transport equations (e.g. Boltzmann or the non-equilibrium Green's function) are modified to incorporate the internal DOF. Since spin states are quantum vectors complete in two dimensions, transport equations become a 2 × 2 matrix as a result of the modification of the coupling constant due to carrier type. We refer readers to Table 2.4 for a clear recapitulation of these descriptions.

## 3.2 Equilibrium Green's function in electron transport

In condensed matter physics, the field theoretic version of Green's function has been used extensively[1] to study the ground state energy, excited lifetime, linear response and so forth of electron particles or holes depending on their energy level relative to the Fermi energy. These methods, similar to those developed for the studies of particle propagation in high energy physics, have been particularly useful for studying the many-body effects of the electron gas, nuclear matter and superconductivity.

In modern electronics where mesoscopic physics is gaining importance, the Green's function methods are useful for studying the equilibrium microscopic effects on the functioning of these devices. For example, in spintronics, the spin flip and spin orbital effects affect the transport of both charge and spin fluxes in these devices; understanding these effects goes a long way towards resolving many problems popular in the area of applied physics such as spin transfer switching, spin oscillations and spin injection.

In the language of quantum field theory, the field operator and its conjugate for particles are given by:

$$\psi(x_1) = \sum_{\lambda} a_{\lambda}(t_1)\phi_{\lambda}(x_1); \quad \psi^{\dagger}(x_2) = \sum_{\lambda} a^{\dagger}_{\lambda}(t_2)\phi^{\dagger}_{\lambda}(x_2) \qquad [3.1]$$

where $a_{\lambda}(t_1) = a_{\lambda}e^{-ie_{\lambda}t_1}$. The annihilation field operator destroys a particle above the Fermi level and creates a hole below the Fermi level as expressed below:

$$\psi(x) = \sum_{k>k_F} a_k\phi_k(x)e^{-iw_k t} + \sum_{k<k_F} b^{\dagger}_{-k}\phi_k(x)e^{-iw_k t}. \qquad [3.2]$$

On the other hand, the creation operator creates a particle above the Fermi level and destroys a hole below the Fermi energy as below:

$$\psi^{\dagger}(x) = \sum_{k>k_F} a_k^{\dagger}\phi_k^{\dagger}(x)e^{iw_k t} + \sum_{k<k_F} b_{-k}\phi_k^{\dagger}(x)e^{iw_k t}. \qquad [3.3]$$

In Keldysh formalism for devices, the lesser and greater Green's functions are crucial for non-equilibrium effects. Below are definitions of the greater, lesser, retarded and advanced Green's functions:

$$G^{>}(x_1,x_2) = -i\langle\psi(x_1)\psi^{\dagger}(x_2)\rangle = -i\sum_{\lambda}(1-n_{\lambda})\phi_{\lambda}(x_1)\phi_{\lambda}^{\dagger}(x_2)e^{-ie_{\lambda}(t_1-t_2)}$$
$$G^{<}(x_1,x_2) = i\langle\psi^{\dagger}(x_2)\psi(x_1)\rangle = i\sum_{\lambda}n_{\lambda}\phi_{\lambda}(x_1)\phi_{\lambda}^{\dagger}(x_2)e^{-ie_{\lambda}(t_1-t_2)}$$
$$G_{\lambda}^{RA}(x_1t_1,x_2t_2) = \mp i\theta(\pm t_1 \mp t_2)$$
$$\langle\psi_{\lambda}(x_1t_1)\psi_{\lambda}^{\dagger}(x_2t_2) + \psi_{\lambda}^{\dagger}(x_2t_2)\psi_{\lambda}(x_1t_1)\rangle. \qquad [3.4]$$

The Green's function or the propagator is normally expressed in the perturbative expansion or the path integral formalism. But before discussing the perturbative expansion, we first show that a propagator can be written in a consecutive series of intermediate propagators. If one likes, each of these propagators can be expanded perturbatively to include the effect of global scattering (e.g. electron–electron or electron–phonon) during that interval of propagation. Using the definition of $\psi_I(t) = e^{iH_0t}\,\psi_S(t)$, one could work out the wavefunction of a single particle in the interaction and the Schrödinger pictures as follows:

$$\psi_I(t) = e^{iH_0t}.e^{-iH_S(t-t_0)}.e^{-iH_0t_0}\,\psi_I(t_0) \equiv U_I(t,t_0)\psi_I(t_0)$$
$$\psi_S(t) = e^{-iH_S(t-t_0)}\psi_S(t_0) \equiv U_S(t,t_0)\psi_S(t_0) \qquad [3.5]$$

where $H_S = H_0 + V$. Equation 3.5 shows how a wavefunction evolves according to the rules of quantum dynamics. One can establish a path integral formalism by choosing either the interaction or the Schrödinger picture. The generator of time evolution would be $H = H_0 + V$ and, in the absence of interaction, the generator is simply $H_0$. One could now express evolution in interaction and Schrödinger pictures, respectively, in the form of:

$$\psi_I(x,t) = \int\langle x|e^{iH_0t}\cdot e^{-iH_S(t-t_0)}.e^{-iH_0t_0}|x_0\rangle\langle x_0|t_0\rangle dx_0$$
$$\equiv \int G(xt,x_0t_0)\,\psi_I(x_0,t_0)dx_0$$
$$\psi_S(x,t) = \int\langle x|e^{-iH_S(t-t_0)t}|x_0\rangle\langle x_0|t_0\rangle dx_0 \equiv \int G(xt,x_0t_0)\,\psi_S(x_0,t_0)dx_0 \qquad [3.6]$$

which essentially describes the evolution of a particle from one spatial point to another, $G(xt, x_0t_0)$ would then be the propagator between these times. Under the path integral formalism, a Green's function between two points can always be expressed as a consecutive series of Green's functions, each correlating a pair of spatial points between the initial and the final points as shown below:

$$G(xt,x_0t_0)=\int\langle x|U(tt_n)|x_n\rangle\langle x_n|U(t_nt_{n-1})|x_{n-1}\rangle \\ \ldots\langle x_1|U(tt_0)|x_0\rangle\, dx_1dx_2\,\ldots.dx_n, \qquad [3.7]$$

where $x$ and $t$ are the $(n+1)^{\text{th}}$ terms of the spatial and the time points, respectively. Here we will focus specifically on the Schrödinger picture, so that:

$$\langle x_n|U(t_nt_{n-1})|x_{n-1}\rangle=\sqrt{\frac{m}{2\pi i\hbar\Delta t}e^{\frac{im}{2\hbar}\left(\frac{x_n-x_{n-1}}{\Delta t}\right)^2\Delta t}}.e^{-\frac{iV}{\hbar}\Delta t} \qquad [3.8]$$

is the propagator between a pair of spatial points of $x_n$ and $x_{n-1}$. Taking the limit, $\Delta t\rightarrow 0$, yields:

$$\psi_S(x,t)$$

$$=\int\left[\left(\frac{m}{2\pi i\hbar\Delta t}\right)^{\frac{n+1}{2}}e^{\left[\frac{m}{2}\left(\frac{x_{n+1}-x_n}{\Delta t}\right)^2-V(x_n)\right]\frac{i\Delta t}{\hbar}}\ldots..e^{\left[\frac{m}{2}\left(\frac{x_1-x_0}{\Delta t}\right)^2-V(x_0)\right]\frac{i\Delta t}{\hbar}}dx_n\ldots dx_1\right]\psi_S(x_0t_0)dx_0$$

$$=\int\left[\left(\frac{m}{2\pi i\hbar\Delta t}\right)^{\frac{n+1}{2}}e^{\frac{iS(t)}{\hbar}}dx_n\ldots dx_1dx_0\right]\psi_S(x_0t_0) \qquad [3.9]$$

where $S(T)=\frac{i}{\hbar}\int_0^T\frac{m}{2}\left(\frac{dx}{dt}\right)^2-V(x_n)dt$ would be the action of the system. Note that $\left(\frac{m}{2\pi i\hbar\Delta t}\right)^{\frac{n+1}{2}}$ has the dimension of $\left(\frac{1}{l}\right)^{n+1}$, which cancels those due to $dx_n\ldots dx_1dx_0$. The propagator between a pair of space–time points in the path integral is:

$$G(xt,x_0t_0)=\langle x|e^{iH_ot}\cdot e^{-iH_S(t-t_0)}\cdot e^{-iH_0t_0}|x_0\rangle;$$

$$G(xt,x_0t_0)=\int\left(\frac{m}{2\pi i\hbar\Delta t}\right)^{\frac{n+1}{2}}e^{\frac{iS(t)}{\hbar}}dx_n\ldots dx_1. \qquad [3.10]$$

One can now look into the perturbative expansion of the individual propagator for any space–time point. Two important theories are crucial for the basic formalism of the perturbative Green's function, namely the generation of time evolution by the interaction expressed in the form of the integral equation, and the four-field theory which describes the interaction itself. In the perturbative expansion for $U_I(t,t_0)$, one takes note that $[H_I(t), H_I(t')] \neq 0$, and $H_I(t_1) = e^{iH_0t}Ve^{-iH_0t}$. Due to the non-commutativity of the Hamiltonian in time, $\psi_I(t)$ cannot be expressed in the simple form of:

$$\psi_I(t) = U_I(t,t_0)\psi_I(t_0) = e^{-i\int_{t_0}^{t} H_I(t')dt'}\psi_I(t_0) \quad [3.11]$$

but instead:

$$\psi_I(t) = U_I(t,t_0)\psi_I(t_0) = \mathcal{T}e^{-i\int_{t_0}^{t} H_I(t')dt'}\psi_I(t_0) \quad [3.12]$$

where $\mathcal{T}$ is the time-ordering operator. Here $\psi_I(t)$ is treated as a field operator instead of a function. The steps leading to the solution for the above are involved and the final results are:

$$\psi_I(t) = \left[1 + (-i)\int_{t_0}^{t} dt_1 H_I(t_1) + \sum_{n=2}^{\infty}(-i)^n \int_{t_0}^{t} dt_1 \ldots \int_{t_0}^{t_{n-1}} dt_n \prod_{j=1}^{n} H_I(t_j)\right]\psi_I(t_0). \quad [3.13]$$

Following standard mathematics, the above can be expressed in the more useful form of:

$$\psi_I(t) = \left[1 + (-i)\int_{t_0}^{t} dt_1 H_I(t_1) + \sum_{n=2}^{\infty}\frac{(-i)^n}{n!}\int_{t_0}^{t} dt_1 \ldots \int_{t_0}^{t} dt_n \mathcal{T} \times\left\{H_I(t_1)H_I(t_2)\ldots H_I(t_n)\right\}\right]\psi_I(t_0). \quad [3.14]$$

It is clear from the above that the generator of the evolution is the interaction Hamiltonian itself. Readers are referred to reference 1 for standard derivations of field theoretic Green's functions.

In non-interacting systems, the Green's function being considered will thus be the propagator with, $H_S = H_0$ i.e. $V$ set to zero. Considering its retarded version, one has the bare propagator for a uniform system as follows:

$$\begin{aligned} G^R(xt,x_0t_0) &= -i\theta(t-t_0)\sum_{\alpha,\gamma}\langle x|\gamma\rangle\langle\gamma|e^{-iH(t-t_0)}|\alpha\rangle\langle\alpha|x_0\rangle \\ &= -i\theta(t-t_0)\sum_{\alpha}\phi_\alpha(x)\phi_\alpha^\dagger(x_0)e^{-i\varepsilon_\alpha(t-t_0)} \end{aligned} \quad [3.15]$$

if $|\alpha\rangle$, $|\gamma\rangle$ are chosen to be eigenstates of $H$. Replacing $t - t_0$ with $\tau$, the time Fourier transform of the above, yields:

$$\int_{-\infty}^{\infty} G^R(xt,x_0t_0)e^{iE\tau}d\tau = \sum_{\alpha}\frac{\phi_\alpha(x)\phi_\alpha^\dagger(x_0)}{E-\varepsilon_\alpha + i\eta}. \quad [3.16]$$

In fact one can check that:

$$(i\partial_t - H^f)\left[-i\theta(t-t_0)\sum_\alpha \phi_\alpha(x)\phi_\alpha^\dagger(x_0)e^{-i(E-\varepsilon_\alpha)\tau}\right] = \delta(x-x_0)\delta(\tau). \quad [3.17]$$

Similarly, the Fourier transform version needs to satisfy the following:

$$(E-H^f)\left[\sum_\alpha \frac{\phi_\alpha(x)\phi_\alpha^\dagger(x_0)}{E-\varepsilon_\alpha+i\eta}\right] = \delta(x-x_0) \quad [3.18]$$

$$G^R(xx_0,E) = \langle x|(E-H^v+i\eta)^{-1}|x_0\rangle \quad [3.19]$$

where we have taken care to distinguish $H^f$ from $H^v$. The significance of such distinction is discussed in Chapter 1.

## 3.3 Electric current under linear response

In Drude's theory, besides the independent and free electron approximation which relates to the study of electron interaction and its effect on the system's energy/particle, the concept of mean free time has been discussed in the context of electron collision and its effect on the electron's propagation under a slightly out-of-equilibrium condition in the Fermi gas. Drude attributed electron collisions to electrons bouncing off the impenetrable ion cores. Under the concept of mean free time or mean free path and the relaxation time approximation, one could conjecture that an electron picked at random at a given moment will, on average, travel for a time $\tau$ before its next collision and have been travelling for time $\tau$ since its last collision. Immediately after each collision, an electron is taken to emerge with a velocity that is not related to its velocity just before the collision, but is randomly directed and with a speed appropriate to the temperature prevailing at where the collision takes place. This approximation is important for deriving the conductivity as well as the electric current density of the out-of-equilibrium electron gas in the diffusive regime based on the Boltzmann method.

In linear response, a weak perturbation generates a small out-of-equilibrium response that is proportional to this perturbation. In the context of nanoelectronics, an external voltage is the applied perturbation. The current response is expected to be proportional to this perturbation, where the response coefficient will be the conductivity, which is independent of the strength of the external voltage applied through the channel. Linear response marks the first small step towards the fully non-equilibrium treatment of carrier transport in nanoscale devices. There are numerous ways to derive the conductivity that fulfill the linear response theory. The most general treatment was prescribed by the Kubo's formula, which describes the linear response conductivity as a retarded correlation function of the internal current operators of the electronic channel. The general Kubo expression could, under the appropriate specialization technique, lead to:

- Boltzmann conductivity in the diffusive treatment
- Landauer conductivity in the ballistic treatment.

Nonetheless, it should be remembered that these two formalisms were derived independently of the Kubo approach. Conceptual details of these methods can be found in solid state and mesoscopic transport reference books.

The semiclassical non-equilibrium transport applies mainly to few-electron systems such as the quantum dot or metallic island channels. For simplicity, carrier type is treated as electron charge with scalar coupling constant. Table 3.1 provides a summary of the different types of electron transport on the nanoscale.

*Table 3.1* Summary of the physics of electron transport in various nanoscale electronic systems

| Type of nanoscale device | Linear response | Semiclassical non-equilibrium | Quantum non-equilibrium |
|---|---|---|---|
| Diffusive current | Boltzmann (macro)<br>Kubo (micro) | Master equation (correlated) | NEGF (micro) |
| Ballistic current | Quantum tunneling<br>Landauer conductance<br>Kubo (ballistic) | Master equation (ballistic) | NEGF (ballistic) |

## 3.4 General Kubo conductivity

The purpose of this section is to introduce readers to the equilibrium conductivity formula which, upon expansion, can incorporate the effects of microscopic scattering. As nanoscale devices or system get smaller, many-body effects cannot be approximated with the statistical averaging that worked well for large systems (e.g. the statistical methods based on Boltzmann derivation for bulk conductivity). On the small scale, quantum mechanics is required to give a description that takes into account many-body interactions.

In quantum mechanics, one can view the state and operators in different pictures. Here, the interaction picture is required. The following is a quick revision:

$$\left|\psi_I(t)\right\rangle = e^{iH_0 t}\left|\psi_S(t)\right\rangle \qquad [3.20]$$

which leads to:

$$\left|\psi_S(t)\right\rangle = e^{-iH_0 t}\left|\psi_I(t)\right\rangle = e^{-iH_0 t}U(t,t_0)\left|\psi_I(t_0)\right\rangle. \qquad [3.21]$$

$U(t,t_0)$ is an integral equation that captures all the interaction effect that affects the time evolution of the state. This explains the need to express operators and state vectors in the interaction picture. It is clear that $U(t,t_0)$ plays the role of evolving a particular state from its initial condition to the final condition. One deduces that:

$$\left|\psi_I(t_0)\right\rangle = e^{iH_0 t_0}\left|\psi_S(t_0)\right\rangle. \qquad [3.22]$$

We have seen how the quantum state of the electron system evolves with time. In the following, we study the Hamiltonian of the electron system in the presence of interaction as given below:

$$H(t) = H_0 + V(t)\theta(t - t_0). \qquad [3.23]$$

The Heaviside function shows that interaction is turned on at $t = t_0$.

Note that the $U$ matrix is an evolution operator which consists of multiple integral functions to describe the time evolution of the state vector in the interaction picture. This operator would thus contain information about the effect of interaction on the time evolution of the state vector. It is, however, important to understand the physical implication of this function. The evolution operator in its explicit form is:

$$\begin{aligned} U(t,t_0) &= 1 + \frac{1}{i}\int_{t_0}^{t} V(t')\,dt' + \frac{1}{i^2}\int_{t_0}^{t} V(t')\,dt' \int_{t_0}^{t'} V(t'')\,dt'' + \ldots\ldots\ldots \\ &= 1 + \frac{1}{i}\int_{t_0}^{t} V(t')\,dt' + \frac{1}{2i^2}\int_{t_0}^{t}\int_{t_0}^{t} \mathcal{T}\left\{V(t')V(t'')\right\}dt'\,dt''. \end{aligned} \qquad [3.24]$$

One can understand the physical outcome of the above expression by examining the individual terms. It is not hard to notice that the $U$ matrix contains only the interaction part of the Hamiltonian (i.e. the $V$). In other words, in the absence of interaction, $U$ is reduced to 1. One can thus imagine that the higher order terms correspond to the effects of electron interaction. This interaction could affect the observables or the propagators of a many-body system depending on how it is used.

In the following we apply this to the study of the effect of microscopic interaction on the average value of a particular observable, namely $\langle A\rangle$. It is understood that $A$ is an operator; it can only be measured when its average or expectation is given by $\langle G|A|G\rangle$. In the language of nanoscale engineering, an 'observable' refers to something that one can measure or detect. Examples of observables are current, momentum, etc. Since temperature has an effect on the average of an observable or measureable, one can deduce from statistical quantum mechanics that an observable under the effect of finite temperature is to be given by:

$$\left\langle A(t)\right\rangle = \frac{1}{Z_0}\sum_n \left\langle \psi_S^n(t)\right|A_S\left|\psi_S^n(t)\right\rangle e^{-\beta E_n}. \qquad [3.25]$$

We are thus led to:

$$\begin{aligned} \left\langle A(t)\right\rangle &= \frac{1}{Z_0}\sum_n \left\langle \psi_I^n(t_0)\right|U^\dagger(t,t_0)e^{iH_0 t}A_S\,e^{-iH_0 t}\,U(t,t_0)\left|\psi_I^n(t_0)\right\rangle e^{-\beta E_n} \\ &= \frac{1}{Z_0}\sum_n \left\langle \psi_I^n(t_0)\right|\left(1 + i\int_{t_0}^{t} V(t')dt'\right)A_I(t)\left(1 - i\int_{t_0}^{t} V(t')dt'\right)\left|\psi_I^n(t_0)\right\rangle e^{-\beta E_n} \end{aligned} \qquad [3.26]$$

where the symbol † means transpose conjugate. In the above, use has been made of $(ab)^\dagger = b^\dagger a^\dagger$ and $|\psi_S(t)\rangle = e^{-iH_0t}|\psi_I(t)\rangle = e^{-iH_0t}U(t,t_0)|\psi_I(t_0)\rangle$. The expectation of an observable should not depend on the picture used, i.e. Schrödinger, Heisenberg or interaction pictures should all yield the same expectation, consistent with the fact that one only measures one type of average. Thus one can write:

$$\langle A(t)\rangle = \frac{1}{Z_0}\sum_n \langle \psi_I^n(t_0)|A_I(t)|\psi_I^n(t_0)\rangle\, e^{-\beta E_n}$$

$$-i\int_{t_0}^{t} dt' \frac{1}{Z_0}\sum_n e^{-\beta E_n}\langle \psi_I^n(t_0)|A_I(t)V(t') - V(t')A_I(t)|\psi_i^n(t_0)\rangle. \qquad [3.27]$$

One thus has:

$$\langle A\rangle_t = \langle A\rangle_0 - i\int_{t_0}^{t} dt' \langle [A_I(t), V(t')]\rangle_0. \qquad [3.28]$$

The final expression is the remarkable Kubo formula, which states that the deviation of an observable from its unperturbed state due to the application of external field (e.g. $V$) can be expressed as a correlation function of the observable and the external effect $V$:

$$\langle A\rangle_t - \langle A\rangle_0 = -i\int_{t_0}^{t} dt' \langle [A_I(t), V(t')]\rangle_0 = \delta\langle A\rangle. \qquad [3.29]$$

Since the $U$ matrix is a time-ordered function, the correlation is naturally retarded (i.e. the correlation function can be described with a Heaviside function) and is now known as the retarded correlation function:

$$\delta\langle A\rangle = -i\theta(t-t')\int_{t_0}^{\infty} dt' \langle [A_I(t), V(t')]\rangle_0 = \int_{t_0}^{\infty} dt'\, C_{AV}^R(t,t'), \qquad [3.30]$$

where $C_{AV}^R(t, t') = -i\theta(t-t')\langle[A_I(t), V(t')]\rangle_0$. One can express the interaction as $V(t') = Bf(t')$, so that $C_{AV}^R(t, t') = C_{AB}^R(t-t')f(t')$. One could then deduce that the non-equilibrium deviation is a convolution function. Note that $C_{AV} = [A,V]$ and $C_{ABf} = [A,B]f = C_{AB}f$. It is thus clear that $C_{AB}^R(t-t')f(t')$ is a convolution of the response function $C_{AB}^R(t-t')$. Let, $t_0 \to -\infty$, one arrives at:

$$\delta\langle A\rangle = \int_{-\infty}^{\infty} dt'\, C_{AB}^R(t-t')f(t'), \qquad [3.31a]$$

$$\delta\langle A(\omega)\rangle = C_{AB}^R(\omega)f(\omega). \qquad [3.31b]$$

Now, we will take it a step further to consider the external potential to be slightly more complicated. Instead of $V(t') = Bf(t')$, we take the interaction potential to be:

$$V(t') = \int dr'\, B^\beta(r')f^\beta(r', t') \qquad [3.32a]$$

$$V\langle\omega\rangle = \int dr'\, B^{\beta}(r') f^{\beta}(\omega,r'). \quad [3.32b]$$

One would now obtain:

$$\delta\langle A(\omega)\rangle = \int C^{R}_{AB^{\beta}}(\omega) f^{\beta}(\omega,r')\, dr'. \quad [3.33]$$

Therefore, one can write in complete analogy to Eq. 3.33 that:

$$\delta\langle J_0^{\alpha}(r,\omega)\rangle = \int C^{R}_{J_0^{\alpha}(r)\,B^{\beta}}(\omega) f^{\beta}(\omega,r')\, dr' \quad [3.34]$$

where $J_0 = J_0^{para} + \frac{e}{m} A_0\, \rho$ contains the paramagnetic and the diamagnetic components. Let us now compare $V(\omega)$ to the external perturbation due to an electric field where:

$$V(\omega) = \frac{e}{i\omega}\int dr'\, J_0^{\beta}(r')\, E^{\beta}_{ext}(\omega,r'). \quad [3.35]$$

It is clear that $B^{\beta}(r') \equiv J_0^{\beta}(r')$; $f^{\beta}(\omega,r') \equiv \frac{e}{i\omega} E^{\beta}_{ext}(\omega,r')$. With this, one can now write:

$$\delta\langle J_0^{\alpha}(r,\omega)\rangle = \int C^{R}_{J_0^{\alpha}(r)\,J_0^{\beta}(r')}(\omega) \frac{e}{i\omega} E^{\beta}_{ext}(\omega,r')\, dr'. \quad [3.36]$$

Since $\delta\langle J_0^{\alpha}(r,\omega)\rangle = \langle J_0^{\alpha}(r,\omega)\rangle$:

$$\langle J_0^{\alpha}(r,\omega)\rangle = \int C^{R}_{J_0^{\alpha}(r)\,J_0^{\beta}(r')}(\omega) \frac{e}{i\omega} E^{\beta}_{ext}(\omega,r')\, dr'. \quad [3.37]$$

Since $\langle J^{\alpha}(r,\omega)\rangle - \langle J_0^{\alpha}(r,\omega)\rangle + \frac{e}{i\omega m} n(r) E^{\alpha}_{ext}(r,\omega)$ and noting that $J_e^{\alpha}(r,\omega) = -e\langle J^{\alpha}(r,\omega)\rangle$, we are now ready to compare current expressions in the following manner:

$$J_e^{\alpha}(r,\omega) = \int \sigma^{\alpha\beta}(r,r',\omega)\, E^{\beta}_{ext}(\omega,r')\, dr' \quad [3.38a]$$

$$J_e^{\alpha}(r,\omega) = -e\langle J_0^{\alpha}(r,\omega)\rangle - \frac{e^2}{i\omega m} n(r) E^{\alpha}_{ext}(r,\omega). \quad [3.38b]$$

---

**Exercise 3.1**

Show by comparing the above current expressions that the conductivity is given by:

$$\sigma^{\alpha\beta}(r,r',\omega) = \frac{ie^2}{\omega} C^{R}_{J_0^{\alpha}(r)\,J_0^{\beta}(r')}(\omega) - \frac{e^2 n(r)}{i\omega m}\delta(r-r')\delta_{\alpha\beta}$$

***Solution***

$$J_e^{\alpha}(r,\omega) = -e\left\langle J_0^{\alpha}(r,\omega)\right\rangle - \frac{e^2}{i\omega m} n(r) E_{ext}^{a}(r,\omega)$$

$$= -\int C^R_{J_0^{\alpha}(r)\,J_0^{\beta}(r')}(\omega)\frac{ie^2}{\omega} E_{ext}^{\beta}(\omega,r')\,dr' - \frac{e^2}{i\omega m} n(r) E_{ext}^{\alpha}(r,\omega)$$

$$= -\int C^R_{J_0^{\alpha}(r)\,J_0^{\beta}(r')}(\omega)\frac{e^2}{i\omega} E_{ext}^{\beta}(\omega,r')\,dr'$$

$$-\frac{e^2 n(r)}{i\omega m}\int \delta(r-r')\delta_{\alpha\beta} E_{ext}^{\beta}(r',\omega)\,dr'$$

$$= \int\left( C^R_{J_0^{\alpha}(r)\,J_0^{\beta}(r')}(\omega)\frac{ie^2}{\omega} - \frac{e^2 n(r)}{i\omega m}\delta(r-r')\delta_{\alpha\beta}\right)$$

$$\times E_{ext}^{\beta}(r',\omega)\,dr'$$

---

The equilibrium studies of electron transport will not yield any net current and conclude with the derivation of an equilibrium property of the system (e.g. conductivity). But in the above, the non-equilibrium effect arises via the linear response method, which is an approximation to study the non-equilibrium effect using basically equilibrium methods. Note that the entire derivation is based on the equilibrium time-ordered but not the non-equilibrium contour-ordered Keldysh Green's function. However, the linear response approximation is useful because it allows one to derive a current expression which is essentially an equilibrium conductivity multiplied by a low applied voltage, or a conductivity expression which is essentially a current–voltage differential.

Explicitly, the dissipative part of the general Kubo conductivity is:

$$Re\left[\sigma^{\alpha\beta}(r,r',\omega)\right] = Re\left[\frac{ie^2}{\omega} C^R_{J_0^{\alpha}(r)\,J_0^{\beta}(r')}(\omega) - \frac{e^2 n(r)}{i\omega m}\delta(r-r')\delta_{\alpha\beta}\right]$$

$$= -\frac{e^2}{\omega} Im\, C^R_{J_0^{\alpha}(r)\,J_0^{\beta}(r')}(\omega). \qquad [3.39]$$

After Fourier transformation, the DC Kubo conductivity is:

$$Re\left[\sigma^{\alpha\beta}(q,\omega)\right] = -Im \lim_{\substack{q\to 0\\ \omega\to 0}} \frac{e^2}{\omega} C^R_{J_0^{\alpha}\,J_0^{\beta}}(q,\omega). \qquad [3.40]$$

Based on the general Kubo conductivity derived earlier, one can decide to use this formula to study interacting (diffusive) or non-interacting electron transport. For diffusive transport, a perturbative expansion of the general expression is required. In this book, we will merely show that the general Kubo expression can be written for perturbative expansion but we will not delve into the details of such expansion.

The zeroth order of this expansion will be the conductivity for a non-interacting system. The general Kubo formula is a retarded correlation function.

Rewriting the current correlation function to give greater clarity using more simplified notations gives:

$$C^R(r\,r',t\,t') = -i\theta(t-t')\left\langle\left[J^\alpha(rt), J^\beta(r't')\right]\right\rangle_0. \qquad [3.41]$$

Note that $J_0^\alpha(rt)$ has been replaced with $J^\alpha(rt)$ and $C^R_{J_0^\alpha(rt)\,J_0^\beta(r't')}$ with $C^R\,(r\,r',\,t\,t')$.

The Fourier transform of Eq. 3.41 with respect to space yields:

$$C^R(q,t-t') = -i\theta(t-t')\frac{1}{V}\left\langle\left[J^\alpha(q,t), J^\beta(-q,t')\right]\right\rangle_0. \qquad [3.42]$$

In the frequency domain, the correlation function is $C^R(q,t-t') \rightarrow C^R(q,\omega)$. We will look particularly at the free electron expression where in the absence of interaction:

$$J^\alpha(q,t) = \sum_{k\sigma}\left(k+\frac{1}{2q}\right)a^\dagger_{k\sigma}a_{k+q\,\sigma}e^{i(E_k-E_{k+q})t} = \sum_{k\sigma}J^\alpha a^\dagger_{k\sigma}a_{k+q\,\sigma}e^{i(E_k-E_{k+q})t}. \qquad [3.43]$$

One can obtain:

$$C^R_{free}(q,\omega) = -\frac{1}{V}\sum_{k\sigma}\left[\frac{n(E_k)-n(E_{k+q})}{E_k-E_{k+q}+\omega+i\eta}\right]J^\alpha J^\beta. \qquad [3.44]$$

On the other hand, one can also arrive at the above using the Matsubara approach. The Matsubara version of the current correlator $C^R(q,t-t')$ is:

$$C^R(q,\tau-\tau') = -\frac{1}{V}\left\langle\mathcal{T}_\tau\, J^\alpha(q,\tau)\,J^\beta(-q,\tau')\right\rangle_0 \qquad [3.45]$$

where $\mathcal{T}_\tau$ is the Matsubara version of the time ordering operator. In the frequency domain:

$$C^R(q,iq_n) = -\frac{1}{\beta V}\left\langle J^\alpha\,(q,iq_n)\,J^\beta(-q,-iq_n)\right\rangle_0 \qquad [3.46]$$

where $iq_n$ is the Bosonic frequency. By analytic continuation, one can convert a Matsubara Bosonic frequency function into the retarded correlation function, i.e.

$$C^R(q,iq_n\rightarrow\omega+i\eta) = C^R(q,\omega) \qquad [3.47]$$

where $\eta$ is a small number. Alternatively, one can proceed with the Matsubara version and derive that:

$$C^R(q,iq_n) = -\frac{1}{\beta V}\sum_{ik_n}\sum_{k\sigma}J^\alpha G_0(k+q\sigma,ik_n+iq_n)\,J^\beta\,G_0(k\sigma,ik_n). \qquad [3.48]$$

Equation 3.48 can be derived by applying Wick's theorem to $C^R(q,\tau-\tau') = -\frac{1}{V}\left\langle\mathcal{T}_\tau\,J^\alpha(q,\tau)\,J^\beta(-q,\tau')\right\rangle_0$ to first obtain:

$$C^R(q,\tau-\tau')=\frac{1}{\mathcal{V}}\sum_{k\sigma}J^\alpha G_0(k+q\sigma,\tau-\tau')\,J^\beta G_0(k\sigma,\tau'-\tau). \quad [3.49]$$

The two Green's functions are a product function in the time domain and the Fourier transform of the product function will be a convolution function in the frequency domain. Thus performing a Fourier transform now yields the important equation, Eq. 3.48. Alternatively, one performs Fourier transform directly on $-\frac{1}{\mathcal{V}}\left\langle T_\tau\, J^\alpha(q,\tau)\,J^\beta(-q,\tau')\right\rangle_0$ to obtain $-\frac{1}{\beta\mathcal{V}}\left\langle J^\alpha(q,iq_n)\,J^\beta(-q,-iq_n)\right\rangle_0$. Wick's theorem is then performed to obtain Eq. 3.48.

$$\begin{aligned}C^R(q,iq_n)&=\frac{1}{\beta\mathcal{V}}\sum_{ik_n}\sum_{k\sigma}J^\alpha G_0(k+q\sigma,ik_n+iq_n)J^\beta\, G_0(k\sigma,ik_n)\\&=\frac{1}{\mathcal{V}}\sum_{k\sigma}J^\alpha J^\beta\left[n(E_k)G_0(k+q\sigma,E_k+iq_n)\right.\\&\left.+n(E_{k+q}-iq_n)G_0(k\sigma,E_{k+q}-i\,q_n)\right)\\&=\frac{1}{\mathcal{V}}\sum_{k\sigma}\left[\frac{n(E_k)-n(E_{k+q})}{E_k-E_{k+q}+iq_n}\right]J^\alpha\, J^\beta. \end{aligned} \quad [3.50]$$

Again, analytic continuation can be applied, and one obtains the expression identical to $C^R_{Free}(q,\omega)$.

---

### Exercise 3.2

The retarded correlator is one of the most important functions in nanoscale electronic transport. For example, in the linear response, the conductivity has been found to be a current–current correlator $\chi^G(rr',tt')$. Falling short of a full-scale non-equilibrium treatment, the general Kubo conductivity provides the best approximation. In a similar fashion, the polarization function is a charge–charge correlator $\chi^P(rr',tt')$. Below, the correlators are shown explicitly:

$$\chi^G(rr',tt')=-i\theta(t-t')\left\langle\left[j(r,t),\,j(r',t')\right]\right\rangle$$

$$\chi^P(rr',tt')=-i\theta(t-t')\left\langle\left[\rho(r,t),\,\rho(r',\,t')\right]\right\rangle.$$

(1) Show that in momentum space, electron polarization in the device channel is given by:

$$\chi^P(q,tt')=-i\theta(t-t')\frac{1}{\mathcal{V}}\left\langle\left[\rho(q,t),\rho(-q,t')\right]\right\rangle$$

where $\mathcal{V}$ is the volume of the channel and $q$ is the transfer of momentum in the interaction.

(2) In the absence of interaction, charge density of the device in momentum space is:

$$\rho(q,t) = \sum_{k\sigma} c^{\dagger}_{k\sigma} c_{k+q,\sigma} e^{i(E_k - E_{k+q})t};$$

$$\rho(-q,t') = \sum_{k'\sigma'} c^{\dagger}_{k'\sigma'} c_{k'-q,\sigma'} e^{i(E_{k'} - E_{k'-q})t'}.$$

Noting that $\langle c^{\dagger}_k c_k \rangle = n_k = n(E_k)$, show that:

$$\chi^P(q,tt') = -i\theta(t-t')\frac{1}{V}\sum_{k\sigma}\left[n(E_k) - n(E_{k+q})\right]e^{i(E_k - E_{k+q})(t-t')}.$$

(3) Show that in the frequency domain, one has:

$$\chi^P(q,\omega) = \frac{1}{V}\sum_{k\sigma}\left[\frac{n(E_k) - n(E_{k+q})}{E_k - E_{k+q} + \omega + i\eta}\right].$$

---

## 3.5 Non-equilibrium electron transport

Mesoscopic physics became important in the 1980s with the advances in nanotechnologies that enabled quantum-sized nanostructures to be fabricated and characterized. This has led to discoveries in modern phenomena (e.g. quantum Hall, giant magnetoresistance) and advances in single-electronics, quantum dot electronics, spintronics, carbon-based electronics and other novel forms of devices. Each of these electronic systems normally comprises a few discrete structures with dimensions on the nanometer scale. Of key interest to researchers here is the physics of electron transport across these nanostructures, the apparent departure of which from classical circuit and semiclassical transport theory is a result of quantum effects.

In traditional equilibrium condensed matter physics, the Green's function method has played an important role in the studies of electron transport and many-body effects such as screening, conductivity, localization and others. In modern day nanoscale electronic systems, the traditional Green's function description of electron transport in an equilibrium condensed matter system needs new modifications. The new Green's function methods need to take into account:

1. Interaction with electrodes or device parts
2. Non-equilibrium electron distribution in device channel.

### 3.5.1 Interaction with electrodes or device parts

Nanoelectronics demands a form of Green's function that conveniently reflects the discrete effects due to spatial separation between device parts as shown in Fig. 3.1.

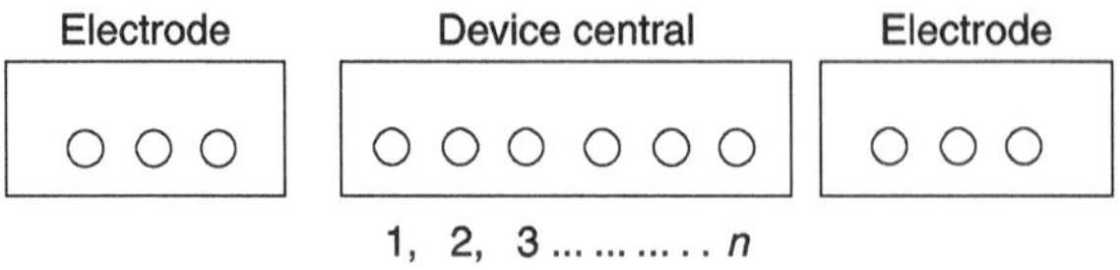


*3.1* Three-region device representing the left electrode, right electrode and central regions. The central region is discretized into points labeled from left from 1 to *n*. This one-dimensional horizontal structure can adequately represent even a three-dimensional device if the other two dimensions consists of homogeneous material.

A Hamiltonian discretized in a real space lattice, which spans the entire device, is particularly suitable for describing electronic energy and propagation in periodic ion crystal or devices with discrete parts. The Hamiltonian can be derived by second quantization or the finite difference method. A Green's function based upon these Hamiltonians suits the discontinuous nature of these devices, and could thus provide an intuitive and physical description of electron propagation. For second-quantized Hamiltonian, equation of motion or Feynman diagram methods can be used to derive the Green's function. For finite difference Hamiltonian, a matrix-based Green's function method can be used. Alternatively, the finite difference Hamiltonian can be converted to the second-quantized Hamiltonian. Therefore in this formalism, electron transport can be studied under the combined mesoscopic boundary effects of device components and the microscopic effects of electron–electron, electron–phonon scattering.

### 3.5.2 Non-equilibrium electron distribution in device channel

Under the application of external electric potential across the device (e.g. by connecting a battery or current source to the two electrodes), electron distribution in the central device is driven out of equilibrium. The current considered here is the net device current, which conserves across the entire device. This is unlike the equilibrium system where local currents cancel, resulting in zero net flow of device current. It is thus important for readers to keep in mind the difference between the internal, local current of an equilibrium condensed matter system and the net device current of a non-equilibrium device system. Net electron current flows through the device from high electron potential to low. In conventional electrical engineering, electric current flows in the opposite direction to electron current. To avoid confusion, device current here refers to electron current which flows from the negative battery terminal to the positive.

The expression for device current requires the knowledge of the lesser Green's function, whose non-equilibrium expression departs markedly from its equilibrium version. The Keldysh formalism provides a systematic approach to derive the non-equilibrium Green's function, which includes the time-ordered, retarded, advanced, lesser and greater versions. The non-equilibrium lesser Green's function can be expressed in terms of the retarded and advanced Green's function of the central region and the self-energy due to the electrodes, i.e. $G^< = G^R \Sigma^< G^A$.

The application of non-equilibrium Green's function (NEGF) methods to the transport in discrete electronic systems began in earnest in the early 1990s,[2–4] providing a quantum transport approach in mesoscopic physics that is consistent with the semiclassical Landauer–Buttiker[5] method. The discrete or lattice Green's function method has since been refined with more application oriented studies.[6–9] Its adaptation to different nanoscale applications results in the construction of a simulation platform for the entire range of modern nanoelectronics, e.g. current imaging in nanostructures,[10,11] electron transport in single-electronic and quantum dot,[12–14] spin transport in spintronic systems,[15–18] and carbon-based or graphene systems.[19]

## 3.6 Electron propagation – physics of Green's function

### 3.6.1 Equation of motion

In the following, we derive the retarded (advanced) Green's function with the equation of motion (EOM). We also present a derivation of the lesser Green's function based on deriving the Keldysh equation using the Langreth theorem. The device shown in Fig. 3.1 can be described by the second-quantized Hamiltonian shown in Eq. 3.51, where $H_{EL/ER}$ is the Hamiltonian for the left/right electrodes, $H_D$ is the Hamiltonian for the central region and $H_T$ is the tunneling Hamiltonian which describes electron propagation from electrodes to leads and vice versa.

$$
\begin{aligned}
H_{EL} &= \sum_{l=(0\ldots-\infty)} \varepsilon_l a_l^\dagger a_l - t_l a_l^\dagger a_{l-1} - t_l a_{l-1}^\dagger a_l \\
H_{ER} &= \sum_{r=(n+1\ldots+\infty)} \varepsilon_r a_r^\dagger a_r - t_r a_r^\dagger a_{r-1} - t_r a_{r-1}^\dagger a_r \\
H_D &= \sum_{j=1}^{n} \varepsilon_j d_j^\dagger d_j + \sum_{j=1}^{n-1} t d_j^\dagger d_{j+1} + h.c. \\
H_T &= t_L a_0^\dagger d_1 + t_R a_{n+1}^\dagger d_n + h.c.
\end{aligned}
\qquad [3.51]
$$

Let's begin with the equilibrium time-ordered Green's function for the central region: $G_{jj'}^t(t-t') = -i < T\{d_j(t) d_{j'}^\dagger(t')\} >$, in which $T$ is the time-ordering operator. One can then obtain the kinetic equation of the time-ordered Green's function for a point in the central region of the device, denoted by $j$:

$$
i\partial_t G_{jj'}^t(t-t') = \delta(t-t')\delta_{jj'} - i\left\langle T\left\{-\left[H(t), d_j(t)\right] d_{j'}^\dagger(t')\right\}\right\rangle, \qquad [3.52]
$$

in which $H = H_{EL/ER} + H_c + H_T$. It is clear that $-[H_{EL/ER}, d_j] = 0$, since the electrode's intrinsic Hamiltonian is unrelated to electron existence in the central region. The non-vanishing terms are:

$$-\left[H_D, d_j\right] = \varepsilon_j d_j + t(1-\delta_{jn})d_{j+1} + t(1-\delta_{j1})d_{j-1}$$
$$-\left[H_T, d_j\right] = t_L \delta_{1j} a_0 + t_R \delta_{nj} a_{n+1} \quad [3.53]$$

where operators above are understood to be in time $t$.

---

**Exercise 3.3**

Prove, where $d_j^\dagger$, $d_j$, $n_j$ are fermionic operators, that:

(1) $[d_j, n_j] = d_j$
(2) $[d_j^\dagger, n_j] = -d_j^\dagger$

---

For simplicity, consider a discretized device system where, $H_D = \varepsilon_1 d_1^\dagger d_1$, $H_T = t_L a_0^\dagger d_1 + t_R a_2^\dagger d_1 + h.c.$. The particular case of $G_{11}^t$ is:

$$(i\partial_t - \varepsilon_1)G_{11}^t(\tau) = \delta(\tau) + t_R G_{21}^t(\tau) + t_L G_{01}^t(\tau) \quad [3.54]$$

where $\tau = t - t'$ Consider the central region is being isolated, one would have:

$$(i\partial_t - \varepsilon_1)g_{11}^t(\tau) = \delta(\tau). \quad [3.55]$$

For convenience, we use $R_{11}(t-t'), = \int dt_1\, \delta(t-t_1)R_{11}(t_1 - t')$ to represent the right-hand side of Eq. 3.54. It can then be expressed as:

$$R_{11}(\tau) = \int dt_1 (i\partial_t - \varepsilon_1) g_{11}^t(t-t_1) R_{11}(t_1 - t'), \quad [3.56]$$

One then arrives at:

$$G_{11}^t(\tau) = \int dt_1 g_{11}^t(t-t_1)\left\{\delta(t_1 - t') + t_R G_{21}^t(t_1 - t') + t_L G_{01}^t(t_1 - t')\right\}. \quad [3.57]$$

According to the theory of Keldysh NEGF, the equilibrium expression above has precisely the same form as its non-equilibrium version, except that in the latter the intermediate time integration has to be performed on the Keldysh loop in the complex contour, i.e.

$$G_{11}^c(\mathcal{T}, \mathcal{T}') = \int d\tau_1 g_{11}^c(\mathcal{T}, \mathcal{T}_1)\left\{\delta(\mathcal{T}_1, \mathcal{T}')\right.$$
$$\left. + t_R G_{21}^c(\mathcal{T}_1, \mathcal{T}') + t_L G_{01}^C(\mathcal{T}_1, \mathcal{T}')\right\} \quad [3.58]$$

where $\mathcal{T}$ is the time stamp in the Keldysh contour. Applying the Langreth theory[2] to the contour integral yields the following retarded Green's function in the normal time loop:

$$G_{11}^R(t, t') = \int dt_1 g_{11}^R(t, t_1)\left\{\delta(t_1 - t') + t_R G_{21}^R(t_1 - t') + t_L G_{01}^R(t_1 - t')\right\}. \quad [3.59]$$

If we assume the Hamiltonian to be time independent, the Green's function is only dependent on the time difference ($t - t'$). In this case, it is more convenient to perform Fourier transform in the energy space:

$$G_{11}^{R}(\omega) = g_{11}^{R}(\omega)\left\{1 + t_R G_{21}^{R}(\omega) + t_L G_{01}^{R}(\omega)\right\}. \qquad [3.60]$$

To evaluate the last term on the right-hand side of Eq. 3.60, we start from $G_{01}^{t}(t-t') = \int dt_1 (g_{0-1}^{t}(t-t_1)t_{-11} + g_0^{t}(t-t_1)t_{01})G_{11}^{t}(t_1 - t')$, which leads to:

$$G_{01}^{t}(t-t') = \int dt_1\, t_L g_0^{t}(t-t_1)G_{11}^{t}(t_1 - t') \qquad [3.61]$$

due to the fact that $t_{-11} = 0$. Thus, its retarded counterpart in energy space is:

$$G_{01}^{R}(\omega) = t_L g_0^{R}(\omega)G_{11}^{R}(\omega). \qquad [3.62]$$

Substituting the above into Eq. 3.60, one obtains:

$$1 = \left\{(g_{11}^{R})^{-1} - t_L g_0^{R} t_L\right\}G_{11}^{R} - tG_{21}^{R}. \qquad [3.63]$$

Similarly, one can deal with other lattice points of Fig. 3.1 and obtain $1 = \{(g_{jj}^{R})^{-1} - \Sigma_{jj}^{R}\}G_{jj}^{R} - t(1-\delta_{jn})G_{j+1,j}^{R} - t(1-\delta_{j1})G_{j-1,j}^{R}$, in which $\Sigma_{jj}^{R} = t_L\delta_{j0}g_j^{R}t_L + t_R\delta_{jn}g_j^{R}t_R$ is the self-energy due to the leads. If now one carries out the above for $n = 3$ instead of $n = 1$, one obtains:

$$\left[G^{R}\right]_{\mu\nu} = \begin{bmatrix} (g_{11}^{R})^{-1} - t_L g_0^{R} t_L & -t & 0 \\ -t & (g_{jj}^{R})^{-1} & -t \\ 0 & -t & (g_{nn}^{R})^{-1} - t_R g_4^{R} t_R \end{bmatrix}_{\mu\nu}^{-1} \qquad [3.64]$$

in which $(g_{jj}^{r})^{-1} = E - \varepsilon_j + i\eta$; $g_{0(4)}^{r} = (E - \varepsilon_{0(4)} + i\eta)^{-1}$. Readers should keep the result of Eq. 3.64 in mind and compare it with results to be obtained using the finite difference method.

Before we proceed to the matrix Green's function and the finite difference method, we complete this section with the derivation of the lesser Green's function, which is the most crucial aspect of the NEGF. We derive the expression for the lesser Green's function, i.e. the Keldysh equation. Let us consider the system in Fig. 3.1 again where $n = 1$ is considered:

$$(i\partial_t - \varepsilon_1)G_{11}^{t}(\tau) = \delta(\tau) + t_R G_{21}^{t}(\tau) + \int dt_1\, t_L\, g_0^{t}(t-t_1)t_L G_{11}^{t}(t_1 - t'). \qquad [3.65]$$

one can then obtain:

$$(i\partial_t - \varepsilon_j)G_{jj}^{t}(\tau) = \delta(\tau) + t(1-\delta_{jn})G_{j+1,j}^{t}(\tau) + (1-\delta_{j0})G_{j-1,j}^{t}(\tau). \qquad [3.66]$$

The matrix representation for an $n = 3$ isolated device (i.e. without lead effects) would then be:

$$[M_0][G_0]^t(\tau) = \delta(\tau)I \qquad [3.67]$$

where explicitly:

$$[M_0] = \begin{bmatrix} i\partial_t - \varepsilon_1 & -t & 0 \\ -t & i\partial_t - \varepsilon_2 & -t \\ 0 & -t & i\partial_t - \varepsilon_3 \end{bmatrix}. \qquad [3.68]$$

Similarly, one can deal with the $G^t_{jj'}(\tau)$, and rewrite them as:

$$[M_0][G]^t(\tau) = \delta(\tau)I + \int dt_1[\Sigma]^t(t-t_1)[G]^t(t_1 - t'). \qquad [3.69]$$

Following similar procedures, one finds that the time-ordered and contour-ordered Green's functions satisfy the Dyson's equation as follows:

$$[G]^t(t-t') = \int dt_1[G_0]^t(t-t_1)\delta(t_1 - t')I + \iint dt_1\, dt_2[G_0]^t(t-t_1)[\Sigma]^t(t_1 - t_2)[G]^t(t_2 - t') \qquad [3.70a]$$

$$[G]^c(\mathcal{T} - \mathcal{T}') = \int d\mathcal{T}_1[G_0]^c(\mathcal{T} - \mathcal{T}_1)\delta(\mathcal{T}_1 - \mathcal{T}')I + \iint d\mathcal{T}_1\, d\mathcal{T}_2[G_0]^c(\mathcal{T} - \mathcal{T}_1)[\Sigma]^c(\mathcal{T}_1 - \mathcal{T}_2)[G]^c(\mathcal{T}_2 - \mathcal{T}') \qquad [3.70b]$$

where:

$$[\Sigma]^t(t_1 - t_2)[G]^t(t_2 - t') = \begin{bmatrix} \Sigma^t_{11} & 0 & 0 \\ 0 & 0 & 0 \\ 0 & 0 & \Sigma^t_{33} \end{bmatrix} \begin{bmatrix} G^t_{11} & G^t_{12} & G^t_{13} \\ G^t_{21} & G^t_{22} & G^t_{23} \\ G^t_{31} & G^t_{32} & G^t_{33} \end{bmatrix}. \qquad [3.71]$$

Applying Langreth theorem, the retarded Green's function can also be found to satisfy the Dyson's equation, as would be expected:

$$[G]^R = [G_0]^R + [G_0]^R[\Sigma]^R[G]^R. \qquad [3.72]$$

In the above, there has been no difference between the equilibrium and non-equilibrium versions of the retarded and advanced Green's functions. Of particular interest here is the lesser Green's function, which by Langreth theory, yields a non-equilibrium version which differs markedly from its equilibrium expression. The non-equilibrium lesser Green's function is:

$$[G]^< = [G_0]^< + [G_0]^R[\Sigma]^R[G]^< + [G_0]^R[\Sigma]^<[G_0]^A + [G_0]^<[\Sigma]^A[G]^A. \quad [3.73]$$

By recursive substitution, the above can be simplified to:

$$[G]^< = [G]^R[\Sigma]^<[G]^A \quad [3.74]$$

in which the matrix component $[\Sigma]^<_{ij} = t_L\delta_{i0}g^<_{ij}t_L\delta_{j0} + t_R\delta_{in}g^<_{ij}t_R\delta_{jn}$.

---

**Exercise 3.4**

(1) Prove by using the Dyson equations of $G^R = G_0^R + G_0^R\Sigma^R G^R$ and $G^A = G_0^A + G_0^A\Sigma^A G^A$, and by iteration of $[G]^< = [G_0]^< + [G_0]^R [\Sigma]^R [G]^< + [G_0]^R [\Sigma]^< [G]^A + [G_0]^< [\Sigma]^A [G]^A$, that the following is true:

$[G]^< = (1 + [G]^R [\Sigma]^R) [G^0]^< (1 + [\Sigma]^A [G]^A) + [G]^R [\Sigma]^< [G]^A$.

(2) In the fluctuation dissipation theorem, where $f$ is the equilibrium Fermi–Dirac distribution of the central region:

$[G_0]^< = ([G_0]^A - [G_0]^R) f.$

Relate this expression to the electron density, density of states and occupation probability.

(3) Prove, using the above identities, that:

$[G]^< = [G]^R [\Sigma]^< [G]^A$

---

## 3.6.2 Matrix Green's function

*Finite difference Hamiltonian in a single-band device*

In a nanoscale electronic device, one needs a formalism to study the propagation of electron from one discrete part of the device to another. In device technology, a basic electronic device is commonly designed into three regions[3,4] with distinct parametric differences in each region. It becomes clear that a Hamiltonian which reflects neighboring site perturbation might be a suitable one to use here since one can perceive the device system to be, in the crudest form, a central region subject to the perturbation of the neighboring source and drain electrodes. One conjectures a Hamiltonian that takes a site eigenstate instead of a momentum, and yields the energy related to the site as well as its nearest neighbor perturbations. Since the perturbation effect repeats for each spatial site, it will manifest in the propagator as a self-energy. Since what is key to reflecting the perturbation effects of discrete parts lies in the Hamiltonian, we perform a simple mathematical manipulation for the Hamiltonian in function space $H^f = -\frac{\hbar}{2m}\frac{d^2}{dx^2} + U(x)$. One first applies $H^f$ to a particular lattice point discretized with an intersite distance $a$ as shown in Fig. 3.1:

$$(H^f \psi)_j = \left( -\frac{\hbar}{2m}\frac{d^2\psi}{dx^2} \right)_j + U_j \psi_j \qquad [3.75]$$

where, for compactness, $\psi_j$ represents $\psi(x = ja)$. Using the finite difference approximation where $a$ is small:

$$\left( \frac{d\psi}{dx} \right)_{j+\frac{1}{2}} \to \frac{1}{a}(\psi_{j+1} - \psi_j); \left( \frac{d^2\psi}{dx^2} \right)_j \to \frac{1}{a^2}(\psi_{j+1} - 2\psi_j + \psi_{j-1}). \qquad [3.76]$$

For a more vivid illustration, we introduce the basic three-part metal–oxide–semiconductor field-effect transistor (MOSFET) device, which has also been widely employed in new areas such as spintronics, carbon and molecular electronics, and even quantum dot style single-electronics. Each part of the device can be regarded as an isolated system containing the unperturbed states, e.g. the spatial orbital states.

To start we choose a spatial site within the central region denoted generally by $j$ as shown in Fig. 3.1 and write the Hamiltonian about this point. This is repeated for all other spatial sites. The central region has thus been discretized into a finite number ($n$) of discrete spacings with intersite distance $a$, while the source part of the device extends semi-infinitely to the left with the same discretization. The above approximation leads, with straightforward substitution, to:

$$(H^f \psi)_j = (U_j + 2t)\psi_j - t\psi_{j+1} - t\psi_{j-1} \qquad [3.77]$$

where $t = \hbar^2/2ma^2$. Normally $(H^f \psi)_j \propto k^2\psi_j$ but with the finite difference method, one has $(H^f \psi)_j \propto g(\psi_j, \psi_{j+1}, \psi_{j-1})$, which clearly reflects the effects of local perturbation from neighboring sites. In fact, one could now define a $H^r$, which acts upon an orbital site $\psi_j$, to yield the energy associated with the site and its nearest neighbors, i.e.

$$H^r \psi_j = (U_j + 2t)\psi_j - t\psi_{j+1} - t\psi_{j-1}. \qquad [3.78]$$

In fact Eq. 3.78 can be expressed in the more general form of $H^r \psi_j = \alpha \psi_j - \beta \psi_{j+1} - \beta^\dagger \psi_{j-1}$ where under specific cases, $\alpha = 2t, \beta = t$. If one represents a lattice site using the Hilbert space vector $|x_j\rangle$ instead of the scalar $\psi_j$, and summing over all lattice sites, the Hamiltonian has to be represented in vector space as follows:

$$H^v = \sum_i 2t|x_i\rangle\langle x_i| - t|x_i\rangle\langle x_{i-1}| - t|x_i\rangle\langle x_{i+1}|. \qquad [3.79]$$

$H^v$ applies generally to any discrete site represented by $|x_j\rangle$. One can substitute Eq. 3.79) into the following $\langle x_j| H^v | \psi\rangle = \langle x_j |2t| \psi\rangle - \langle x_{j-1} |t|\psi\rangle - \langle x_{j+1} |t|\psi\rangle$ to recover Eq. 3.78. Similarly, one also sees that $\langle x_i | H^v | x_j\rangle = \alpha\, \delta_{ij} - \beta\, \delta_{ij+1} - \beta^\dagger\, \delta_{ij-1}$ in a more general form. In the second-quantized form, $H^v$ will be written as below:

$$H^v = \sum_i 2ta_i^\dagger a_i - ta_i^\dagger a_{i-1} - t\, a_i^\dagger a_{i+1} \qquad [3.80]$$

where $a_i^\dagger$ and $a_i$ replace $|x_i\rangle$ and $\langle x_i\,|$, respectively. The Green's function can be derived with the second-quantized Hamiltonian by noting that:

$$\begin{aligned}\sum_i 2t\left\langle x_j \left| a_i^\dagger a_i (E-H^v)^{-1} \right| x_j \right\rangle &= \sum_i 2t\left(a_i^\dagger a_i \left| x_j \right\rangle\right)^\dagger (E-H^v)^{-1}\left| x_j \right\rangle \\ &= \sum_i 2t\left(a_i^\dagger \delta_{ij} \left| 0 \right\rangle\right)^\dagger (E-H^v)^{-1}\left| x_j \right\rangle \\ &= \sum_i 2t\, \delta_{ij} \left\langle x_i \left| (E-H^v)^{-1} \right| x_j \right\rangle \end{aligned} \qquad [3.81a]$$

and:

$$\begin{aligned}\sum_i t\left\langle x_j \left| a_i^\dagger a_{i-1} (E-H^v)^{-1} \right| x_j \right\rangle &= \sum_i t\left(a_{i-1}^\dagger a_i \left| x_j \right\rangle\right)^\dagger (E-H^v)^{-1}\left| x_j \right\rangle \\ &= \sum_i t\left(a_{i-1}^\dagger \delta_{ij} \left| 0 \right\rangle\right)^\dagger (E-H^v)^{-1}\left| x_j \right\rangle \\ &= \sum_i t\, \delta_{ij} \left\langle x_{i-1} \left| (E-H^v)^{-1} \right| x_j \right\rangle. \end{aligned} \qquad [3.81b]$$

To derive the second-quantized Hamiltonian, the following substitution is carried out with the Hamiltonian in vector space:

$$\begin{aligned}&\left\langle x_j \left| (E-H^v)\frac{1}{E-H^v} \right| x_j \right\rangle = \\ &\quad \left\langle x_j \left| \sum_i \left( E - 2t|x_i\rangle\langle x_i| + t|x_i\rangle\langle x_{i-1}| + t|x_i\rangle\langle x_{i+1}| \right) \right| \frac{1}{E-H^v} \left| x_j \right\rangle. \end{aligned} \qquad [3.82]$$

Denoting $\left\langle x_j \left| \frac{1}{E-H^v} \right| x_j \right\rangle$ with $G_{jj}$, we have the following:

$$\begin{aligned}(E-H^f)G_{jj} = \sum_i \Bigg[ &\mathrm{E}\left\langle x_j \left| \frac{1}{E-H^v} \right| x_j \right\rangle - 2t\,\delta_{ij}\left\langle x_i \left| \frac{1}{E-H^v} \right| x_j \right\rangle \\ &+ t\delta_{ij}\left\langle x_{i-1} \left| \frac{1}{E-H^v} \right| x_j \right\rangle + t\,\delta_{ij}\left\langle x_{i+1} \left| \frac{1}{E-H^v} \right| x_j \right\rangle \Bigg] \end{aligned} \qquad [3.83]$$

where $\langle x_j | x_i \rangle = \delta_{ij}$. Note that $H^f$ is the Hamiltonian in function space whose relation to the Hamiltonian in vector space is described in Chapter 1. Since we are only dealing with the retarded Green's function here, $G$ implies $G^R$. It thus follows that:

$$(E-H^f)\, G_{jj} = \left[(E-2t)G_{jj} + tG_{j-1,j} + tG_{j+1,j}\right]. \qquad [3.84]$$

To give a more explicit illustration, one can repeat this through the longitudinal points of the central region for $j = 1,2,3$ and obtain a series of equations:

$$\begin{aligned}(E-H^f)G_{11} &= (E-2t)G_{11} + tG_{01} + tG_{21}\\(E-H^f)G_{22} &= (E-2t)G_{22} + tG_{12} + tG_{32}\\(E-H^f)G_{33} &= (E-2t)G_{33} + tG_{23} + tG_{43}.\end{aligned} \quad [3.85]$$

This series of equations are the discrete single-electron Green's function equations for the different spatial sites in the central region. Solving these equations simultaneously allows one to find the self-energy of the Green's function discretized at a particular site. Generalizing the above to a three-region device (see Fig. 3.1), where the labeling and notations follow the standard format used throughout this chapter, we arrive at a block matrix representation of all discrete spatial points which now compactly but clearly depicts a three-region device system:

$$\begin{bmatrix} E-H_L & \tau^{LC} & 0 \\ \tau^{CL} & E-H_C & \tau^{CR} \\ 0 & \tau^{RC} & E-H_R \end{bmatrix}\begin{bmatrix} G^L & G^{LC} & 0 \\ G^{CL} & G^C & G^{CR} \\ 0 & G^{RC} & G^R \end{bmatrix} = I. \quad [3.86]$$

$E - H_{L/R}$ are $m \times m$ infinite matrices that describe the left (right) contacts from site 0 $(n + 1)$ onwards to the left (right), $\tau^{LC}$ ($\tau^{RC}$) are the coupling matrices between the left (right) contacts to the central region and have dimensions of $m \times n$, and $\tau^{CL}(\tau^{CR})$ are the transpose of the coupling matrix of dimensions $n \times m$. The central region is described by $E - H_C$, a $n \times n$ matrix representing the discrete sites 1 *to* $n$ of the central region. Note also that $H_{L/R}$ are scalar terms related to the coupling between neighboring discrete sites in different parts of the device. Matrix multiplication results in:

$$\begin{aligned}(E+i\eta-H_L)G^{LC} - \tau^{LC}G^C &= 0\\(E+i\eta-H_R)G^{RC} - \tau^{RC}G^C &= 0.\end{aligned} \quad [3.87]$$

The equations above can now be expressed as semi-infinite matrices of $m \times n$:

$$G^{LC} = g^L\tau^{LC}G^C \; ; \; G^{RC} = g^R\tau^{RC}G^C \quad [3.88]$$

and $g^L = (E + i\eta - H_L)^{-1}$ and $g^R = (E + i\eta - H_R)^{-1}$ are the $m \times m$ Green's function matrices of the isolated left and right leads, respectively. With a few straightforward derivations, one obtains for the central region the well-established and intuitively compact expression of the Green's function correlating the lattice sites of the central region:

$$\begin{aligned}&(E-H_C-\tau^{CL}g^L\tau^{LC}-\tau^{CR}g^R\tau^{RC})G^C = I\\&\Rightarrow G^C = (E-H_C-\Sigma^L-\Sigma^R)^{-1}\end{aligned} \quad [3.89]$$

where $\Sigma^L$, $\Sigma^R$ are the $n \times n$ self-energy matrices, which can now be interpreted as perturbation effects due to adjacent sites on the contacts to the left and right of the central region on the propagation of electrons in the central region of the device. Taking the specific example of $n = 3$, the Green's function matrix for central region is explicitly expressed as:

$$G^C = \left( \begin{bmatrix} E-H_{C1} & t & 0 \\ t & E-H_{C2} & t \\ 0 & t & E-H_{C3} \end{bmatrix} - \begin{bmatrix} 0 & 0 & t_{01} \\ 0 & 0 & 0 \\ 0 & 0 & 0 \end{bmatrix} g_0 \begin{bmatrix} 0 & 0 & 0 \\ 0 & 0 & 0 \\ t_{10} & 0 & 0 \end{bmatrix} - \begin{bmatrix} 0 & 0 & 0 \\ 0 & 0 & 0 \\ t_{34} & 0 & 0 \end{bmatrix} g_4 \begin{bmatrix} 0 & 0 & t_{43} \\ 0 & 0 & 0 \\ 0 & 0 & 0 \end{bmatrix} \right)^{-1} \qquad [3.90]$$

where $g_0$ is the Green's function for site $-1$ on the left lead and $g_4$ is the Green's function for site $n+1$ on the right lead. It is clear from the above that the central region Green's function correlating point $\mu$ to $v$ is given by the matrix component:

$$[G^C]_{\mu v} = \begin{bmatrix} E-H_{C1}-t_L g_0 t_L & t & 0 \\ t & E-H_{C2} & t \\ 0 & t & E-H_{C3}-t_R g_4 t_R \end{bmatrix}^{-1}_{\mu v}. \qquad [3.91]$$

For physical clarity, we show below that the Green's function at the central site (2,2) is given by an expression which shows the effect of lead perturbation appearing as self-energies:

$$G^{22} = \frac{1}{E_x - 2t - \dfrac{t^2}{E_x - 2t - t_{34} g^4 t_{43}} - \dfrac{t^2}{E_x - 2t - t_{01} g^0 t_{10}}}. \qquad [3.92]$$

Although the above analysis allows a microscopic analysis of electron propagation in one dimension only, it can actually be used to study a three-dimensional nanoelectronic system in which periodicity is assumed for the other two dimensions. The discretization is carried out only for the longitudinal and not the transverse directions. The full matrix for such a system (e.g. $n = 3$ for ease of comparison with the above) would then, with proper rearrangement and substitution, be:

$$G^C = \left( \begin{bmatrix} E-H_{C1}-E_{yz}-t_{01} g^0 t_{10} & t & 0 \\ t & E-H_{C2}-E_{yz} & t \\ 0 & t & E-H_{C3}-E_{yz}-t_{34} g^4 t_{43} \end{bmatrix} \right)^{-1}. \qquad [3.93]$$

It is worth noting that in the above, $E$ has been used to denote the total energy, i.e. $E = E_x + E_{yz} = E_x + \frac{\hbar^2 k_p^2}{2m_p^*}$ where $m^*$ is the effective mass in the parallel ($p$) or $y$, $z$ directions, assuming they are the same. If $G^C$ represents transmission of one mode, summation has to be carried out over $k_p$ and $E_x$ where the Green's function matrix would now be:

$$[G^C]_{\mu\nu} = \begin{bmatrix} E_x - 2t - t_{01}g^0 t_{10} & t & 0 \\ t & E_x - 2t & t \\ 0 & t & E_x - 2t - t_{34}g^4 t_{43} \end{bmatrix}^{-1}_{\mu\nu}. \qquad [3.94]$$

---

**Exercise 3.5**

Given that $\langle a| (E - H^\nu) (E - H^\nu)^{-1} |b\rangle = \delta_{ab}$, show that:

$\boldsymbol{I} = (E\boldsymbol{I} - \boldsymbol{H})\,\boldsymbol{G}$

where $\boldsymbol{H}$ is the matrix whose components are $[\boldsymbol{H}]_{ij} = \langle i| H^\nu |j\rangle$.

***Solution***

Since $\langle a| (E - H^\nu) (E - H^\nu)^{-1}|b\rangle = \delta_{ab}$:

$$\sum_\nu \langle a|(E - H^\nu)|\nu\rangle\langle \nu|(E - H^\nu)^{-1}|b\rangle = \sum_\nu \left(\delta_{a\nu}E - \langle a|H^\nu|\nu\rangle\right)G_{\nu b}$$

$$\rightarrow \sum_\nu (\delta_{a\nu}E - [\boldsymbol{H}]_{a\nu})[\boldsymbol{G}]_{\nu b} = \delta_{ab}$$

In matrix form, one can represent the above as:

$\boldsymbol{I} = (E\boldsymbol{I} - \boldsymbol{H})\,\boldsymbol{G}.$

---

*Discrete two-dimensional planar structure*

The one-dimensional discretization approach might be adequate for most nanoelectronic systems if microscopic analysis of the other two dimensions is not needed. If, however, one is interested in the spatial current distribution within the bulk nanostructure, one might need to extend the present formalism to the other dimensions. For illustration of how this can be carried out, we will expand the central device into a two-dimensional geometry as shown schematically in Fig. 3.2. A similar approach to that used in the previous sections is followed to derive the Green's function matrix. Specifically for illustration, we have chosen

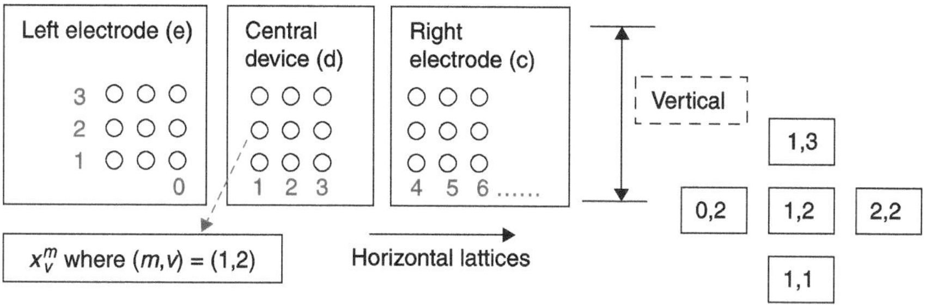


*3.2* Expanded central device with right and left leads in two-dimension discretization. Arrow shows the horizontal or longitudinal direction. The other dimension is known as transverse or vertical.

the site of $x_v^m$ where $(m,v) = (1,2)$, where one could write the Schrödinger equation at site 1,2 as follows:

$$\langle x_2^1|H^v|\psi\rangle = \langle x_2^1|4t|\psi\rangle - \langle x_2^0|t|\psi\rangle - \langle x_2^2|t|\psi\rangle - \langle x_1^1|t|\psi\rangle - \langle x_3^1|t|\psi\rangle. \quad [3.95]$$

Note that Eq. 3.95 has been deduced from Eq. 3.79 with the assumption that $\beta_x = \hbar^2 / 2m_x a_x^2 = \beta_y = \hbar^2 / 2m_y a_y^2$. The labeling convention for the longitudinal direction follows the one-dimensional system in the previous section. The vertical dimension begins with 1 denoting the lowest vertical site. Superscript denotes the longitudinal site while subscript denotes the vertical site.

Following the approach of the previous section, the Green's function equation for a two-dimensional equation can be derived as follows:

$$\begin{aligned}(E - H^v)\langle x_2^1|(E - H^v)^{-1}|x_2^1\rangle \\ &= E\langle x_2^1|(E - H^v)^{-1}|x_2^1\rangle \\ &\quad - \langle x_2^1|[\,|x_2^1\rangle\langle x_2^1|4t - |x_2^1\rangle\langle x_2^0|t - |x_2^1\rangle\langle x_1^1|t - |x_2^1\rangle\langle x_3^1|t](E - H^v)^{-1}|x_2^1\rangle \\ &= (E - 4t)\langle x_2^1|(E - H^v)^{-1}|x_2^1\rangle + t\langle x_2^0|(E - H^v)^{-1}|x_2^1\rangle + t\langle x_1^1|(E - H^v)^{-1}|x_2^1\rangle \\ &\quad + t\langle x_3^1|(E - H^v)^{-1}|x_2^1\rangle = 1.\end{aligned} \quad [3.96]$$

This leads to the Green's function equation of:

$$(E - H^f)G_{22}^{11} = (E - 4t)G_{22}^{11} + t_{22}^{12}G_{22}^{21} + t_{22}^{10}G_{22}^{01} + tG_{12}^{11} + tG_{32}^{11} = I. \quad [3.97]$$

Grouping these equations for different sites, one obtains a matrix which contains the sub-matrices as below:

$$\left[\begin{array}{ccc|ccc} E-4t & t & 0 & t & 0 & 0 \\ t & E-4t & t & 0 & t & 0 \\ 0 & t & E-4t & 0 & 0 & t \\ \hline t & 0 & 0 & E-4t & t & 0 \\ 0 & t & 0 & t & E-4t & t \\ 0 & 0 & t & 0 & t & E-4t \end{array}\right]\left[\begin{array}{ccc|ccc} G_{11}^{00} & G_{12}^{00} & G_{13}^{00} & G_{11}^{01} & G_{12}^{01} & G_{13}^{01} \\ G_{21}^{00} & G_{22}^{00} & G_{23}^{00} & G_{21}^{01} & G_{22}^{01} & G_{23}^{01} \\ G_{31}^{00} & G_{32}^{00} & G_{33}^{00} & G_{31}^{01} & G_{32}^{01} & G_{33}^{01} \\ \hline G_{11}^{10} & G_{12}^{10} & G_{13}^{10} & G_{11}^{11} & G_{12}^{11} & G_{13}^{11} \\ G_{21}^{10} & G_{22}^{10} & G_{23}^{10} & G_{21}^{11} & G_{22}^{11} & G_{23}^{11} \\ G_{31}^{10} & G_{32}^{10} & G_{33}^{10} & G_{31}^{11} & G_{32}^{11} & G_{33}^{11} \end{array}\right]$$

$$= I \qquad [3.98]$$

Following the approach of the previous section, the above matrix can be partitioned into matrix blocks based on device parts, e.g. leads and central region. The following algebra is performed:

$$[\boldsymbol{EI} - \boldsymbol{H}_L]_{av}^{00}[\boldsymbol{G}]_{vb}^{01} + [\boldsymbol{\tau}]_{av}^{01}[\boldsymbol{G}]_{vb}^{11} = 0 \Rightarrow [\boldsymbol{G}]_{ab}^{01} = -[\boldsymbol{g}]_{av}^{00}[\boldsymbol{\tau}]_{vr}^{01}[\boldsymbol{G}]_{rb}^{11} = 0. \quad [3.99]$$

Note that the convention of summation over double index has been followed in Eq. 3.99 and that bold letters denote matrices. This is to avoid confusion as the presence of subscripts normally implies matrix components. With proper substitution and rearrangement, one obtains the Green's function equation similar to those expressed for the one-dimensional system in the previous section:

$$[\boldsymbol{EI} - \boldsymbol{H}_L]_{av}^{11}[\boldsymbol{G}]_{vb}^{11} + [\boldsymbol{\tau}]_{av}^{10}[\boldsymbol{G}]_{vb}^{01} = [\boldsymbol{EI} - \boldsymbol{H}_L]_{av}^{11}[\boldsymbol{G}]_{vb}^{11} - [\boldsymbol{\tau}]_{av}^{10}[\boldsymbol{g}]_{vr}^{00}[\boldsymbol{\tau}]_{rf}^{01}[\boldsymbol{G}]_{fb}^{11} \qquad [3.100]$$

which leads to the well-established form of:

$$\left[\boldsymbol{EI} - \boldsymbol{H}_C - [\boldsymbol{\tau}]_{ar}^{10}[\boldsymbol{g}]_{rf}^{00}[\boldsymbol{\tau}]_{fv}^{01}\right][\boldsymbol{G}]_{vb}^{11} = 0 \qquad \rightarrow \left[\boldsymbol{EI} - \boldsymbol{H}_C - [\boldsymbol{\Sigma}]_{av}^{00}\right][\boldsymbol{G}]_{vb}^{11} = 0. \quad [3.101]$$

One has to be aware that here $[\boldsymbol{\Sigma}]_{av}^{00} = [\boldsymbol{\tau}]_{ar}^{10}[\boldsymbol{g}]_{rf}^{00}[\boldsymbol{\tau}]_{fv}^{01}$ but, in some other representations, it is sometimes written as $[\boldsymbol{\Sigma}]_{av}^{11} = [\boldsymbol{\tau}]_{ar}^{10}[\boldsymbol{g}]_{rf}^{00}[\boldsymbol{\tau}]_{fv}^{01}$.

---

**Exercise 3.6**

Prove that:

$$[\boldsymbol{EI} - \boldsymbol{H}_L]_{av}^{00}[\boldsymbol{G}]_{vb}^{01} + [\boldsymbol{\tau}]_{av}^{01}[\boldsymbol{G}]_{vb}^{11} = 0$$

and:

$$[\boldsymbol{G}]_{ab}^{01} = -[\boldsymbol{g}]_{av}^{00}[\boldsymbol{\tau}]_{vr}^{01}[\boldsymbol{G}]_{rb}^{11} = 0.$$

***Solution***

$$(E-\boldsymbol{H})\boldsymbol{G}=\boldsymbol{I}$$

$$\Rightarrow\left\langle {}^{0}_{a}\middle|(E-H)G\middle|{}^{1}_{b}\right\rangle=0\Rightarrow\sum_{x,y}\left\langle {}^{0}_{a}\middle|E-H\middle|{}^{x}_{y}\right\rangle\left\langle {}^{x}_{y}\middle|G\middle|{}^{1}_{b}\right\rangle=0$$

$$\Rightarrow\sum_{x,y}\left\langle {}^{0}_{a}\middle|E-H_L+\tau_L\middle|{}^{x}_{y}\right\rangle\left\langle {}^{x}_{y}\middle|G\middle|{}^{1}_{b}\right\rangle=0$$

$$\Rightarrow\sum_{x,y}\left\langle {}^{0}_{a}\middle|E-H_L\middle|{}^{x}_{y}\right\rangle\left\langle {}^{x}_{y}\middle|G\middle|{}^{1}_{b}\right\rangle+\left\langle {}^{0}_{a}\middle|\tau_L\middle|{}^{x}_{y}\right\rangle\left\langle {}^{x}_{y}\middle|G\middle|{}^{1}_{b}\right\rangle=0$$

$$\Rightarrow\sum_{y}[E\boldsymbol{I}-\boldsymbol{H}_L]^{00}_{ay}\boldsymbol{G}^{01}_{yb}+[\boldsymbol{\tau}_L]^{01}_{ay}[\boldsymbol{G}]^{11}_{yb}=0.$$

By definition $g_L(E-H_L)=1$

$$\sum_{a}[\boldsymbol{g}_L]^{00}_{ca}[\boldsymbol{E}-\boldsymbol{H}_L]^{00}_{ay}=\delta_{cy}.$$

Thus, multiplying the last line of the above by $[\boldsymbol{G}_L]$ gives:

$$\sum_{y,a}[\boldsymbol{g}_L]^{00}_{ca}[E\boldsymbol{I}-\boldsymbol{H}_L]^{00}_{ay}\boldsymbol{G}^{01}_{yb}+[\boldsymbol{g}_L]^{00}_{ca}[\boldsymbol{\tau}_L]^{01}_{ay}\boldsymbol{G}^{11}_{yb}=0$$

$$\Rightarrow\sum_{y}\delta_{cy}[\boldsymbol{G}]^{01}_{yb}+\sum_{y,a}[\boldsymbol{g}_L]^{00}_{ca}[\boldsymbol{\tau}_L]^{01}_{ay}[\boldsymbol{G}]^{11}_{yb}=0$$

$$\Rightarrow[\boldsymbol{G}]^{01}_{cb}=-\sum_{y,a}[\boldsymbol{g}_L]^{00}_{ca}[\boldsymbol{\tau}_L]^{01}_{ay}[\boldsymbol{G}]^{11}_{yb}.$$

---

## 3.7 Device current formalism

In this section we study the current formalism suitable for device systems with discrete parts. We start with the current expression[20] for a single longitudinal bond, joining two neighboring lattice sites, crossing from the lead to the central region of the device. One can thus apply Dyson's equation, which prescribes an expression of Green's function that suits the physical interpretation of electron propagation across a nanoelectronic system that consists of discrete parts like the emitter (e), central region (d) and collector (c). The expression for a longitudinal bond current crossing over from the left lead (emitter) to the central region is:

$$I_L=\frac{2e}{\hbar}\int\frac{dE}{2\pi}\sum_{e,d}^{\substack{e\in(emitter)\\ d\in(central)}}\left[t_{ed}G^{<}_{de}-t^{*}_{ed}G^{<}_{ed}\right]. \qquad [3.102]$$

The Dyson's equation provides a Green's function expression which decouples the emitter from the device as follows:

$$G^{<}_{ed}=g^{R}_{ee'}(-t_{e'd'})G^{<}_{d'd}+g^{<}_{ee'}(-t_{e'd'})G^{A}_{d'd} \qquad [3.103]$$

where $e$, $d$ can be regarded as some general lattice states belonging to the lead and the central region, respectively.

Under the general framework of the above one can, by proper assignment of lead or central region states to the $e$, $d$ lattice points, define various current expressions. We will show how a longitudinal bond current with vertical/longitudinal site perturbations can be developed. Subsequently, momentum states are assigned to $e$, $d$, and we show the generation of current in momentum representation from the same general current and Dyson's equation which generate the spatial current. Before we proceed, it is worth recalling that the correlation functions $G^<_{de}$ or $g^<_{de}$ are by definition:

$$g^<_{de}(t,t') = i\left\langle a^\dagger_e(t')a_d(t)\right\rangle. \qquad [3.104]$$

In momentum space, where $d$ and $e$ are eigenstates, Eq. 3.104 would be $g^<_{de}(t, t') = in_d\delta_{ed}e^{-iE_d(t-t')}$. Since $g^<_{ee}(t, t') = in_e e^{-iE_e(t-t')}$, it is not hard to see that $g^<_{de}(t, t') = g^<_{ee}(t, t')\delta_{ed}$. In the above, use has been made of the interaction representation where:

$$a^\dagger_e(t) = a^\dagger_e e^{iE_e t}; \qquad a_e(t) = a_e e^{-iE_e t}. \qquad [3.105]$$

### 3.7.1 Device transverse length effect

To study the longitudinal bond current subject to the vertical site perturbation effect, Eq. 3.40 is adapted to follow the mapping rule below:

$$e = (m_e, v_e), \qquad d = (m_d, v_d), \qquad e' = (m'_e, v'_e), \qquad d' = (m'_d, v'_d) \qquad [3.106]$$

where the horizontal site of the central region is denoted by only one lattice point, i.e. $m_d = 1$. The bond current between the left lead (emitter) and the central region would then be that between $m_e = 0$ and $m_d = 1$. The longitudinal branch current would then be given by:

$$I^{01}_{v_e v_d} = \frac{2e}{\hbar}\sum_{k_z}\int\left[t^{01}_{v_e v_d}\left(G^{10}_{v_d v_e}\right)^< - t^{10}_{v_d v_e}\left(G^{01}_{v_e v_d}\right)^<\right]\frac{dE}{2\pi} \qquad [3.107]$$

which, together with its transverse counterpart, can be used to image the spatial distribution of bond current over the entire nanostructure. In Eq. 3.107, summation over $k_z$ appears because the energy integral is only carried out for energy due to the two discrete dimension of $x$ and $y$. Thus for a fixed sub-band due to the vertical ($z$) confinement, the above bond current and its transverse version image the spatial distribution of bond current in a two-dimensional ($x$, $y$) system.

In fact, the above has also been further adapted to image spin current in two-dimensional spintronic or graphene systems. The summation over momentum state is a simple theoretical concept but has a numerical implication. We hereby give an example where discretization is carried out for one dimension, say $x$. In this case, summation over $k_z$ and $k_y$ would be required, leading to:

$$I=\sum_{k_y,k_z}\int T(E_x)(f_L-f_R)dE_x=\int\int T(E_x)(f_L-f_R)\,dE_x\frac{dk_z dk_y}{4\pi^2}L^2 \quad [3.108]$$

where $L^2$ is the transverse dimension of the device and $T\,(E_x)$ is the trace of a Green's function matrix multiplication which corresponds to electron transmission through the device; $f_L$,$f_R$ are the Fermi–Dirac distribution. In this case integration over $k_z$ and $k_y$ or $k_p$ (parallel wave-vector) applies to $f_L$ and $f_R$ only, which yields expressions that can be solved analytically. In short, a homogeneous material system (take the example of a three-layer central region) allows one to express the transmission function $T(E_x)$ and the Fermi–Dirac function in terms of $E_x$; the Green's function would be:

$$[G^C]_{\mu\nu}=\begin{bmatrix} E_x-2t-\Sigma^L & t & 0 \\ t & E_x-2t & t \\ 0 & t & E_x-2t-\Sigma^R \end{bmatrix}^{-1}_{\mu\nu};\quad f=\left[1+e^{-\beta\left(Ex+\frac{\hbar^2k_p^2}{2m_1}-E_F\right)}\right]^{-1} \quad [3.109]$$

where $\beta=\frac{1}{kT}$, and integration with respect to $k_p$ leads to:

$$I=\int T(E_x)(f_L-f_R)\,dE_x dk_p=\int T(E_x)(F_L-F_R)\,dE_x \quad [3.110]$$

where $F=\frac{2mkT}{\hbar^2}\ln[1+e^{\beta(E_F-E_x)}]$. However, in an inhomogeneous device where there are different materials within the device, the central region matrix will contain different effective mass at different diagonal points, which would then result in:

$$[G^C]_{\mu\nu}=\begin{bmatrix} E_{x1}-2t_1-\Sigma^L & t_{12} & 0 \\ t_{21} & E_{x2}-2t_2 & t_{23} \\ 0 & t_{32} & E_{x3}-2t_3-\Sigma^R \end{bmatrix}^{-1}_{\mu\nu} \quad [3.111]$$

as $E=E_x+E_{yz}=E_x+\frac{\hbar^2k_p^2}{2m*}$ would now have to be $E=E_{xn}+\frac{\hbar^2k_p^2}{2m_n^*}$ where $n$ indicates the material type. For a three-layer central region made from material of $m_1$, $m_2$ and $m_3$, respectively, $E_{xn}$ will take on $E_{x1}$, $E_{x2}$ and $E_{x3}$, which will now appear in the matrix above. Thus one now has to re-express the Green's function and $f$ in terms of total energy $E$ as follows:

$$[G^c]^{-1}_{\mu\nu} = \begin{bmatrix} E-2t-\hbar^2k_p^2/2m_1^*-\Sigma^L & t & 0 \\ t & E-2t-\hbar^2k_p^2/2m_2^* & t \\ 0 & t & E-2t-\hbar^2k_p^2/2m_3^*-\Sigma^R \end{bmatrix}_{\mu\nu}. \quad [3.112]$$

This matrix contains $k_p$, while the Fermi–Dirac function is $f = \left[1+e^{\frac{-(E-E_F)}{kT}}\right]^{-1}$. In this case, calculation is normally carried out by integrating over $E$ and summing over $k_p$ by numerical methods as opposed to analytical integration. For illustration, we show this below for a homogenous material system, and compare the current obtained with analytical integration and numerical integration methods.

It is clear from Fig. 3.3 that the numerical method requires a large number of $k_y$ points in order to give an accurate estimate of the current. As shown in previous sections, $G^C$ can be found by inverting the Hamiltonian. It is therefore important to convert $\left(G^{10}_{v_d v_e}\right)^<$ and $\left(G^{01}_{v_e v_d}\right)^<$, which straddle between the left lead and the device's central region in terms of the Green's function for the isolated lead and isolated central region. Figure 3.4 shows a three-part device discretized vertically into multiple sites to consider the effect of vertical perturbation on current.

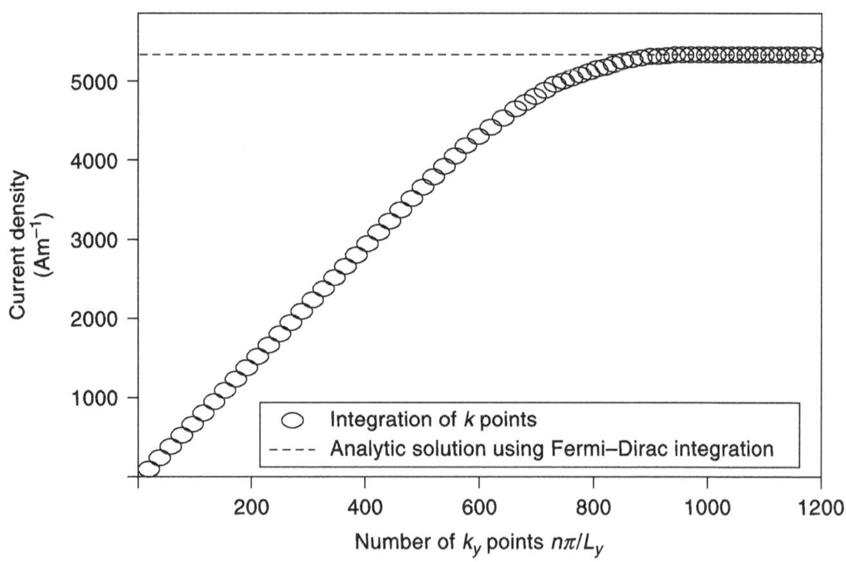


*3.3* Estimate of current using analytical integration and numerical integration methods. For large number of points, the numerical integration method shows results approaching that of analytical integration method.

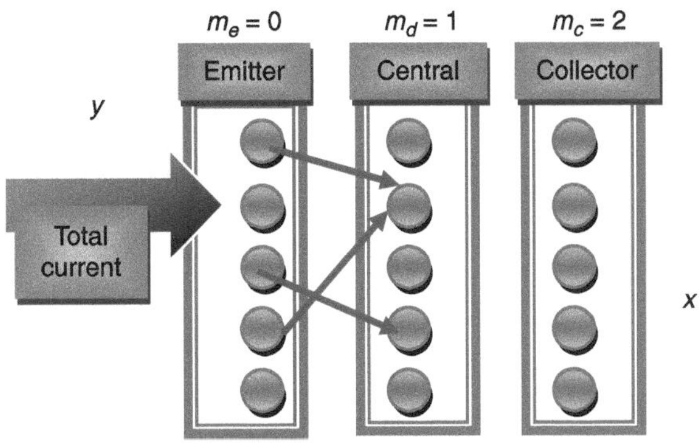


*3.4* Schematic of a three-part device discretized vertically to incorporate the vertical perturbation effect on current. Arrow shows the bond current between a pair of spatial sites across the emitter–device junction. This device has been discretized in *x* and *y* directions but not *z*. Total current would thus require summation over $k_z$.

One now uses the Dyson's equation and applies the mapping to arrive at:

$$\left(G^{01}_{v_e v_d}\right)^< = \left(g^{00}_{v_e v'_e}\right)^R \left(t^{01}_{v'_e v'_d}\right)\left(G^{11}_{v'_d v_d}\right)^< + \left(g^{00}_{v_e v'_e}\right)^< \left(t^{01}_{v'_e v'_d}\right)\left(G^{11}_{v'_d v_d}\right)^A \qquad [3.113]$$

$$\left(G^{10}_{v_d v_e}\right)^< = \left(G^{11}_{v_d v'_d}\right)^R \left(t^{10}_{v'_d v'_e}\right)\left(g^{00}_{v'_e v_e}\right)^< + \left(G^{11}_{v_d v'_d}\right)^< \left(t^{10}_{v'_d v'_e}\right)\left(g^{00}_{v'_e v_e}\right)^A. \qquad [3.114]$$

It is clear that the dummy indices $v_{e'}$ and $v_{d'}$ run through the vertical sites, thus incorporating the effects of vertical site perturbations on bond current. In tight binding approximation, $t^{01}_{v'_e v'_d}$ and $t^{10}_{v'_d v'_e}$ are diagonal, i.e. $t^{01}_{v'_e v'_d} = t^{01}_{v' v'}$ and $t^{10}_{v'_d v'_e} = t^{10}_{v' v'}$. With these, one could now show the longitudinal bond current between horizontal site 0 and site 1 to be:

$$J^{01}_{v_e v_d} = \frac{2e}{A\hbar}\sum_{k_z}\int \frac{dE}{2\pi} t^{01}_{v_e v_d}\left[\left(G^{11}_{v_d v'}\right)^R \left(t^{10}_{v' v'}\right)\left(g^{00}_{v' v_e}\right)^< + \left(G^{11}_{v_d v'}\right)^< \left(t^{10}_{v' v'}\right)\left(g^{00}_{v' v_e}\right)^A\right]$$
$$- t^{10}_{v_d v_e}\left[\left(g^{00}_{v_e v'}\right)^R \left(t^{01}_{v' v'}\right)\left(G^{11}_{v' v_d}\right)^< + \left(g^{00}_{v_e v'}\right)^< \left(t^{01}_{v' v'}\right)\left(G^{11}_{v' v_d}\right)^A\right]. \qquad [3.115]$$

Equation 3.115 is the expression for a longitudinal bond current with one set of $v_e$ and $v_d$. But if one could recall, the total current indeed requires a summation over *e* and *d*. In this case, it simply means summing over $v_e$ and $v_d$ as $m_e$ and $m_d$ have been fixed to 0 and 1, respectively. The summation process physically means summation over all bonds in all manners of vertical correlation. Under the tight binding approximation, one notes again that $t^{01}_{v_e v_d} = t^{01}_{vv}$ and $t^{10}_{v_d v_e} = t^{10}_{vv}$, which simply discards all correlation pairs beyond nearest neighbors. Thus, the total current expression is greatly simplified to an intuitive form of:

$$I_{total} = \sum_{v} I_{vv}^{01} = I_{11}^{01} + I_{22}^{01} + \ldots\ldots \qquad [3.116]$$

It is apparent that the total current is simply the summation over all the parallel branches. This expression is both simple and physically intuitive for it simply means the total current is the vertical sum of all the longitudinal bond current in parallel, consistent with the Kirchoff current law in electrical circuit. This leads to the useful insight that the Kirchoff law applied to most MOSFET circuit calculations is not to be assumed; however, it is a valid approximation. Using the identities of:

$$\left(g_{v_e v'_e}^{00}\right)^{<} = -f_L\left[\left(g_{v_e v'_e}^{00}\right)^{R} - \left(g_{v_e v'_e}^{00}\right)^{A}\right] \qquad [3.117]$$

$$\left(g_{v_d v'_d}^{11}\right)^{<} = -f_R\left[\left(g_{v_d v'_d}^{11}\right)^{R} - \left(g_{v_d v'_d}^{11}\right)^{A}\right] \qquad [3.118]$$

one now obtains a current expression in terms of the leads' distribution functions:

$$\begin{aligned} I_{vv}^{01} = \frac{2e}{\hbar}\sum_{k_Z}\int\frac{dE}{2\pi}&\left[-\left(G_{vv'}^{11}\right)^{R}\left(t_{v'v'}^{10}\right)f_L\left[\left(g_{v'v}^{00}\right)^{R} - \left(g_{v'v}^{00}\right)^{A}\right]t_{vv}^{01} + \left(G_{vv'}^{11}\right)^{<}\left(t_{v'v'}^{10}\right)\left(g_{v'v}^{00}\right)^{A}t_{vv}^{01}\right] \\ &+\left[-t_{vv}^{10}\left(g_{vv'}^{00}\right)^{R}\left(t_{v'v'}^{01}\right)\left(G_{v'v}^{11}\right)^{<} + f_L t_{vv}^{10}\left[\left(g_{vv'}^{00}\right)^{R} - \left(g_{vv'}^{00}\right)^{A}\right]\left(t_{v'v'}^{01}\right)\left(G_{v'v}^{11}\right)^{A}\right]. \end{aligned}$$

[3.119]

It is common in Green's function formalism to define self-energy and coupling energy constants in terms of the Green's function as follows:

$$\Sigma_{dd'}^{R} = t_{de}\, g_{ee'}^{R}\, t_{e'd'}; \quad \Gamma_{dd'}^{ee'} = -i\left(t_{de}\, g_{ee'}^{R}\, t_{e'd'} - t_{de} g_{ee'}^{A}\, t_{e'd'}\right). \qquad [3.120]$$

We have noted earlier that $[\boldsymbol{\tau}]_{v_d v_e}^{10}\,[\boldsymbol{g}]_{v_e v'_e}^{00}\,[\boldsymbol{\tau}]_{v'_e v'_d}^{01}$ can be written as either $[\boldsymbol{\Sigma}]_{v_d v'_d}^{00}$ or $[\boldsymbol{\Sigma}]_{v_d v'_d}^{11}$; it is only a matter of choice. In the following we will use the format of $[\boldsymbol{\Sigma}]_{v_d v'_d}^{11}$. Applying the following mapping for the left and right leads:

$$\begin{aligned} &e = (m_e, v_e), \quad d = (m_d, v_d), \quad e' = (m'_e, v'_e), \quad d' = (m'_d, v'_d) \ \textit{left lead} \\ &c = (m_c, v_c), \quad d = (m_d, v_d), \quad c' = (m'_c, v'_c), \quad d' = (m'_d, v'_d) \ \textit{right lead} \end{aligned}$$

[3.121]

and with $m_e = m_d - 1$, $m_c = m_d + 1$, $m_d = 1$, $m_e = 0$, $m_c = 2$, one has:

$$\begin{aligned} \Sigma_{(R<)}^{1/r} &= t_{v_d, v_e/v_c}^{m_d, m_e/m_c}\, g_{v_e/v_c, v'_e/v'_c(R<)}^{m_e/m_c, m_e/m_c}\, t_{v'_e/v'_c, v'_d}^{m_e/m_c, m_d}; \\ \Gamma_{vv'}^{1/r} &= -i\Big(t_{v_d, v_e/v_c}^{m_d, m_e/m_c}\, g_{v_e/v_c, v'_e/v'_c(R)}^{m_e/m_c, m_e/m_c}\, t_{v'_e/v'_c, v'_d}^{m_e/m_c, m_d} \\ &\qquad - t_{v_d, v_e/v_c}^{m_d, m_e/m_c}\, g_{v_e/v_c, v'_e/v'_c(A)}^{m_e/m_c, m_e/m_c}\, t_{v'_e/v'_c, v'_d}^{m_e/m_c, m_d}\Big). \end{aligned} \qquad [3.122]$$

With proper substitutions and rearrangements, one obtains the current expression:

$$I_{vv}^{01} = \frac{2e}{\hbar}\sum_{k}\int\frac{dE}{2\pi}\left[i\left(G_{vv'}^{11}\right)^{R} f_L\left(\Gamma_{v'v}^{L}\right)+\left(G_{vv'}^{11}\right)^{<}\left(\Sigma_{v'v}^{L}\right)^{A}\right]$$
$$-\left[\left(\Sigma_{vv'}^{L}\right)^{R}\left(G_{v'v}^{11}\right)^{<} + if_L\left(\Gamma_{vv'}^{L}\right)\left(G_{v'v}^{11}\right)^{A}\right]. \quad [3.123]$$

In matrix form the above is written as follows:

$$I_{vv}^{01} = \frac{2e}{\hbar}\sum_{k_z}\int\frac{dE}{2\pi}\left(if_L[\mathbf{G}^R\mathbf{\Gamma}^L - \mathbf{\Gamma}^L\mathbf{G}^A]+\left[\mathbf{G}^<(\mathbf{\Sigma}^L)^A-(\mathbf{\Sigma}^L)^R\mathbf{G}^<\right]\right)_{vv}. \quad [3.124]$$

It is worth noting that all the Green function matrices above correspond to the particular longitudinal site $m_d$, while the superscript $L$ for the $\mathbf{\Gamma}$ and $\mathbf{\Sigma}$ matrices indicates the correspondence to longitudinal site $m_e$. The total longitudinal current would be the sum of all parallel longitudinal bond currents (refer to Fig. 3.4):

$$I_{total} = \sum_{v} I_{vv}^{01} = \frac{2e}{\hbar}\sum_{k_z}\int\frac{dE}{2\pi}\,Trace\left(if_L[\mathbf{G}^R\mathbf{\Gamma}^L - \mathbf{\Gamma}^L\mathbf{G}^A]\right.$$
$$\left.+\left[\mathbf{G}^<(\mathbf{\Sigma}^L)^A-(\mathbf{\Sigma}^L)^R\mathbf{G}^<\right]\right). \quad [3.125]$$

Noting that $Trace(\mathbf{AB}) = Trace(\mathbf{BA})$, and $\mathbf{\Gamma}^L = i(\mathbf{\Sigma}^L)^R - i(\mathbf{\Sigma}^L)^A$, one has:

$$I_{total} = \frac{2e}{\hbar}\sum_{k_z}\int\frac{dE}{2\pi}\,Trace\; i\mathbf{\Gamma}^L\left(f_L[\mathbf{G}^R - \mathbf{G}^A]+\mathbf{G}^<\right). \quad [3.126]$$

To obtain the standard current form for computation, a process known as symmetrization is used. We will not elaborate on the standard symmetrization process as the current expression above is adequate for illustrating the physics of NEGF transport.

## 3.7.2 Device channel length effect

The previous section showed the longitudinal bond current crossing over to the central region is subject to the perturbative effect of the vertical sites. However, perturbation due to the horizontal sites was ignored because discretization along this direction was kept to the minimum of three lattice sites, representing the respective emitter, central and collector parts. To be more accurate, current crossing any central structure is also subject to perturbation due to the horizontal sites. Here we consider a horizontal structure with multiple horizontal sites.

For simplicity, we begin our studies of the longitudinal site perturbation effect by restricting the vertical dimension to a single vertical site. What was applied to account for the vertical site perturbation has a direct analogy to what we can apply here to account for the longitudinal site perturbation. Since discrete points in the emitter, central and collector are denoted by $ee'$, $dd'$ and $cc'$, respectively, one simply needs to remember that now they run through the horizontal sites in much the same way that they ran through the vertical sites in the previous section. The vertical indices can be considered to have been set according to $v_e = v_d = 1$, which applies to every term and is therefore not shown above. For ease of visualization in both longitudinal and vertical perturbations $e$, $d$ can be first be regarded to have been fixed to $e = (m_e = 0, v_e = 1)$ and $d = (m_d = 1, v_d = 1)$, respectively. But the actual current expression shows summation over $e$, $d$ (as shown earlier). Since summation over $e$ and $d$ in the vertical effect means summing over $v_e$ and $v_d$ under the tight binding approximation, it is equivalent to the summation of all parallel longitudinal bond currents, similar to Kirchoff's law. But in the longitudinal effect, summation over $e$, $d$ means summing over $m_e$ and $m_d$. The physical interpretation for this process is not apparent with respect to circuit law in MOSFET. Here, the running indices of $e' = (m'_e$ and $v'_e$ and $d' = (m'_d$ and $v'_d)$ through horizontal sites in the lead and the central region means summing over $m'_e$ and $m'_d$. In the vertical effect, they ran through vertical sites, i.e. summing over $v'_e$ and $v'_d$. We also note in the above that use has been made of $G^<_{de} = -(G^<_{ed})^\dagger$, $t_{ed} = t_{de}$ and $t = t^*$. Recalling the Dyson equation one arrives at:

$$I_L = \frac{2e}{\hbar}\sum_{k_y k_z}\int\frac{dE}{2\pi}\sum_{ee',dd'}^{\substack{e\in(emitter)\\ d:(central)}}\Big[\left(t_{de}g^R_{ee'}t_{e'd'}G^<_{d'd} + t_{de}g^<_{ee'}t_{e'd'}G^A_{d'd}\right)^\dagger + \left(t_{de}g^R_{ee'}t_{e'd'}G^<_{d'd} + t_{de}\,g^<_{ee'}\,t_{e'd'}G^A_{d'd}\right)\Big]. \qquad [3.127]$$

Note that here discretization is in the $x$ direction and that summing over $k_y$ and $k_z$ would be required to find the total current. Similar mapping can be carried out for the horizontal site perturbation, but to keep the notation simple, we will not carry out explicit mapping. One only needs to keep in mind that the running indices of $e$, $d$ or $e'$ $d'$ denote the horizontal components only. The subscripts $e$, $d$, $e'd'$, which now represent the horizontal sites only, sum from $0 \rightarrow -\infty$, $1 \rightarrow n$, respectively, consistent with the labeling format for the leads and central regions of Fig. 3.4. Taking note that $t_{de}g^R_{ee'}\,t_{e'd'} = \Sigma^R_{dd'}$ and $t_{de}g^<_{ee'}t_{e'd'} = \Sigma^<_{dd'}$, and that $e$, $e'$ are discrete points, we now have:

$$I_L = \frac{2e}{\hbar}\sum_{k_y k_z}\int\frac{dE}{2\pi}\sum_{dd'}^{N}\Big[\left(\Sigma^R_{dd'}\,G^<_{d'd} + \Sigma^<_{dd'}\,G^A_{d'd}\right)^\dagger + \left(\Sigma^R_{dd'}\,G^<_{d'd} + \Sigma^<_{dd'}\,G^A_{d'd}\right)\Big]. \qquad [3.128]$$

Note that $d'$ is the dummy index across the longitudinal sites within the central region. There is also an implicit summation over $e$, $e'$ that has been absorbed into $t_{de}g^R_{ee'}\,t_{e'd'} = \Sigma^R_{dd'}$. It follows that:

$$I_L = \frac{2e}{\hbar}\sum_{k_y k_z}\int\frac{dE}{2\pi}\sum_{ee',\,dd'}^{N}\Big[-G^<_{dd'}\,\Sigma^A_{d'd} + \Sigma^R_{dd'}\,G^<_{d'd} + if_L\Gamma_{dd'}G^A_{d'd} - if_L G^R_{dd'}\Gamma_{d'd}\Big]. \qquad [3.129]$$

Noting that $Trace(\boldsymbol{AB}) = Trace(\boldsymbol{BA})$, one obtains:

$$\begin{aligned} I_L &= \frac{2e}{\hbar}\sum_{k_y k_z}\int \frac{dE}{2\pi}\, Trace\left((\boldsymbol{\Sigma}^R - \boldsymbol{\Sigma}^A)\boldsymbol{G}^< - if_L(\boldsymbol{G}^R - \boldsymbol{G}^A)\right) \\ &= \frac{2e}{\hbar}\sum_{k_y k_z}\int -\frac{dE}{2\pi}\, Trace\, i\boldsymbol{\Gamma}\left(\boldsymbol{G}^< + f_L(\boldsymbol{G}^R - \boldsymbol{G}^A)\right). \end{aligned} \quad [3.130]$$

The final current expression for the one-dimensional system contains a double index sum or Trace over $d$ which, as is the case of $d'$, essentially means the running of $m_d$ or $m_{d'}$ from longitudinal sites 1 to $n$ within the central region. This compares to the vertical perturbation where double index sum or Trace over $d$ or $d'$ is that of $v_d$ or $v_{d'}$ running over vertical sites 1 to $n$, and which has the physical meaning of summing of all parallel branches of the longitudinal bond current. Although the physical meaning for the summation over $m_d$ or $m_{d'}$ is not very apparent, it would be interesting to examine $\Sigma^R_{dd'} = t_{de} g^R_{ee'} t_{e'd'}$ to appreciate the simple results that can be deduced from the horizontal summation. In the horizontal effect, since only $t_{10}$ and $t_{01}$ are considered, $\Sigma^R_{11} = t_{10} g^R_{00} t_{01}$. This is because site 1 in the central region is closest to the emitter. The only point in the emitter that has perturbation effect would be site 0. It is also worth noting that $\Sigma^R_{11} = t_{10}\, g^R_{00} t_{01}$ is defined by $g^R_{00}$, while in the case of vertical perturbation, self-energy based on $g^R_{00}$ is sometimes labeled $\Sigma^R_{00}$ in some literature. A quick examination of $\boldsymbol{\Gamma}$ shows that only the diagonal components are non-vanishing.

### 3.7.3 Longitudinal current in momentum space

We now make a direct comparison of the above real space lattice to the momentum lattice. For ease of comparison, we will use $e$ to represent all momentum states in the left lead and $d$ to represent all momentum states in the central device, i.e., $e \rightarrow (k\sigma)_e$, and $\rightarrow (k\sigma)_d$. Figure 3.5 is an illustration of the momentum states in the electrode and the device channel.

The Dyson's equations of $G^<_{de} = \left[g^t_{ee} t_{ed'} G^<_{d'd} - g^<_{ee} t_{ed'} G^{\bar{t}}_{d'd}\right]$ and $G^<_{de} = \left[g^<_{ee} t^*_{ed'} G^t_{dd'} - g^{\bar{t}}_{ee} t^*_{ed'} G^<_{dd'}\right]$ used here are somewhat simpler, i.e. no iteration by $e'$ is required. It can be shown later that, in momentum representation, $e' = e$. It thus makes sense to start with a simple expression. In previous sections, $e'$ represents

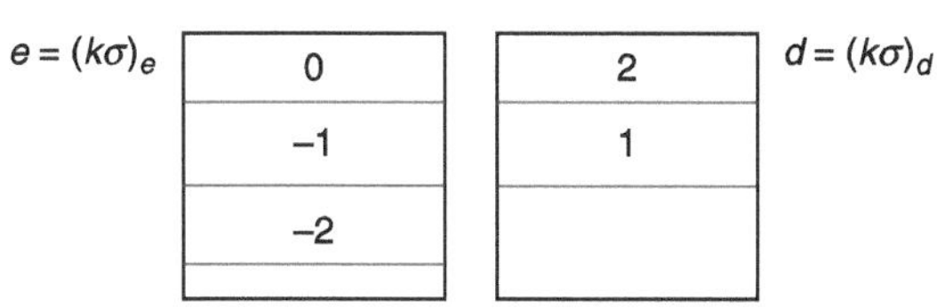


*3.5* Schematic structure of a device with discrete momentum states in the device channel ($d$) and emitter($e$). Summation over discrete states in the emitter would later be converted to summation over continuum states.

vertical or horizontal site perturbation in the leads; however, the index of $e'$ in previous sections was also inconsequential as eventually $m_e' = m_e = 0$. The current would now be:

$$I_L = \frac{e}{\hbar}\int \frac{dE}{2\pi} \sum_{e,d}^{\substack{e:0\to-\infty \\ d:1\to N}} \left[ t_{ed} G_{de}^{<} - t_{ed}^{*} G_{ed}^{<} \right]$$

$$= \frac{e}{\hbar}\int \frac{dE}{2\pi} \sum_{e,d,d'}^{\substack{e:0\to-\infty \\ d:1\to N}} \Big[ t_{ed'}^{*} \left[ g_{ee}^{<} t_{ed} G_{dd'}^{t} - g_{ee}^{\bar{t}} t_{ed} G_{dd'}^{<} \right]$$

$$- t_{ed}^{*} \left[ g_{ee}^{t} t_{ed'} G_{d'd}^{<} - g_{ee}^{<} t_{ed'} G_{d'd}^{\bar{t}} \right] \Big]. \qquad [3.131]$$

---

## Exercise 3.7

Show using the identity $G^{>} - G^{<} = G^{R} - G^{A}$ that:

$$I_L = \frac{e}{\hbar}\int \frac{dE}{2\pi} \sum_{e,d,d'}^{\substack{e:0\to-\infty \\ d:1\to N}} t_{ed} t_{ed'}^{*} \left[ g_{ee}^{<} \left( G_{dd'}^{R} - G_{dd'}^{A} \right) + \left( g_{ee}^{<} - g_{ee}^{>} \right) G_{dd'}^{<} \right].$$

### *Solution*

We begin with the expression of Eq. 3.131. Since $d$ and $d'$ are dummy indices, we can switch them in the second term:

$$I_L = \frac{e}{\hbar}\int \frac{dE}{2\pi} \sum_{e,d,d'}^{\substack{e:0\to-\infty \\ d:1\to N}} \left[ t_{ed'}^{*} \left[ g_{ee}^{<} t_{ed} G_{dd'}^{t} - g_{ee}^{\bar{t}} t_{ed} G_{dd'}^{<} \right] - t_{ed'}^{*} \left[ g_{ee}^{t} t_{ed} G_{dd'}^{<} - g_{ee}^{<} t_{ed} G_{dd'}^{\bar{t}} \right] \right]$$

$$= \frac{e}{\hbar}\int \frac{dE}{2\pi} \sum_{e,d,d'}^{\substack{e:0\to-\infty \\ d:1\to N}} t_{ed} t_{ed'}^{*} \left[ g_{ee}^{<} \left( G_{dd'}^{t} + G_{dd'}^{\bar{t}} \right) - \left( g_{ee}^{\bar{t}} + g_{ee}^{t} \right) G_{dd'}^{<} \right]$$

$$= \frac{e}{\hbar}\int \frac{dE}{2\pi} \sum_{e,d,d'}^{\substack{e:0\to-\infty \\ d:1\to N}} t_{ed} t_{ed'}^{*} \left[ g_{ee}^{<} \left( G_{dd'}^{>} + G_{dd'}^{<} \right) - \left( g_{ee}^{>} + g_{ee}^{<} \right) G_{dd'}^{<} \right].$$

Rearranging the terms further and using the identity $G^{>} - G^{<} = G^{R} - G^{A}$ yields:

$$I_L = \frac{e}{\hbar}\int \frac{dE}{2\pi} \sum_{e,d,d'}^{\substack{e:0\to-\infty \\ d:1\to N}} t_{ed} t_{ed'}^{*} \left[ g_{ee}^{<} \left( G_{dd'}^{>} - G_{dd'}^{<} \right) + \left( g_{ee}^{<} - g_{ee}^{>} \right) G_{dd'}^{<} \right]$$

$$= \frac{e}{\hbar}\int \frac{dE}{2\pi} \sum_{e,d,d'}^{\substack{e:0\to-\infty \\ d:1\to N}} t_{ed} t_{ed'}^{*} \left[ g_{ee}^{<} \left( G_{dd'}^{R} - G_{dd'}^{A} \right) + \left( g_{ee}^{<} - g_{ee}^{>} \right) G_{dd'}^{<} \right].$$

---

Note that straightforward direct analogy with real space discretization ends here, when the following identities are used:

$$g_{ee}^{<}(E)=2\pi i f(E_e)\delta(E-E_e);$$
$$g_{ee}^{>}(E)=-2\pi i\left[1-f(E_e)\right]\delta(E-E_e); \qquad [3.132]$$
$$g_{ee}^{<}(E)-g_{ee}^{>}(E)=2\pi i\delta(E-E_e).$$

In the real space, summation over emitter sites (indexed by $e$, $e'$ in $t_{de}g_{ee'}^{R}\,t_{e'd'}$ or $t_{de}g_{ee'}^{<}\,t_{e'd'}$) are absorbed into the self-energy terms. However, summation over emitter sites is restricted to site 0 as one needs only to consider coupling of site 0 in the emitter and site 1 in the device. Finally, integration with respect to energy is carried out; effectively this corresponds to considering transmission at all energy levels for an emitter represented by one site only that is site 0. In the momentum representation, emitter discrete states are momentum states but not sites, so one can consider the emitter to be already effectively represented by one site.

We now proceed to determine that integration over energy is in fact restricted to the energy level of one emitter state only as below:

$$I_L=\frac{ie}{\hbar}\int dE\sum_{e,d,d'}^{\substack{e:0\to-\infty\\ d:1\to N}} t_{ed}t_{ed'}^{*}\left[f_L(E)\left(G_{dd'}^{R}-G_{dd'}^{A}\right)+G_{dd'}^{<}\right]\delta(E-E_e)$$
$$=\frac{ie}{\hbar}\sum_{e,d,d'}^{\substack{e:0\to-\infty\\ d:1\to N}} t_{ed}t_{ed'}^{*}\left[f_L(E_e)\left(G_{dd'}^{R}(E_e)-G_{dd'}^{A}(E_e)\right)+G_{dd'}^{<}(E_e)\right]. \qquad [3.133]$$

Note that the term $\delta\,(E-E_e)$ appears because $e$ represents the momentum states in the lead here. As the term $\delta\,(E-E_e)$ results in the suppression of the energy integral, in momentum representation, current is summed over the states of the emitter. But, since the emitter consists of a continuum of momentum states, one is allowed to convert the summation over $e$ to an integral following the standard procedure of $\sum_k f_k \to \int \frac{dk}{2\pi/L} f_k$, thus recovering the energy integral commonly present in the NEGF current.

The above analysis allows one to understand that, in momentum representation, current is integrated over the energy continuum of the emitter. The next step is to carry out the summation over all the energy levels on the emitter, which means summing over all momentum states in the emitter. (Note the contrast to the real space analog, where summation over all emitter sites is restricted to one spatial site.)

$$I_L=\int dE_e\frac{ie}{\hbar}\sum_{\sigma,d,d'}^{d:1\to N}\rho_{E_e}^{\sigma L}\,t_{dE_e}^{\sigma L}\left(t_{d'E_e}^{\sigma L}\right)^{*}\left[f_L(E_e)\left(G_{dd'}^{R}(E_e)-G_{dd'}^{A}(E_e)\right)\right.$$
$$\left.+G_{dd'}^{<}(E_e)\right]. \qquad [3.134]$$

Note that summation over $e$ runs over the momentum and spin states of the emitter, i.e. $k_e$ and $\sigma$; and summation over $k_e$ is now represented by $\int dE_e\,\rho^{\sigma L}$. Under the summation symbol, $e$ is removed; we have also inserted $\sigma$ and $\rho^{\sigma L}$ is the density of states on the emitter lead. Further simplification leads to:

$$I_L = \int dE_e \frac{ie}{\hbar} \sum_{\sigma,d,d'}^{d:1\to N} \Gamma_{d'd}^{\sigma L}(E_e)\Big[f_L(E_e)\big(G_{dd'}^{R}(E_e) - G_{dd'}^{A}(E_e)\big)$$

$$+ G_{dd'}^{<}(E_e)\Big]. \qquad [3.135]$$

Now the two representations (momentum and real space) have converged in form as both treat the emitter as one single site and involve summing over all energy levels due to the emitter site:

$$I_L = \int dE_e \frac{e}{\hbar} \sum_{\sigma} Trace\ i\Gamma^{\sigma L}\left[f_L(E_e)(\boldsymbol{G}^R - \boldsymbol{G}^A) + \boldsymbol{G}^<\right] \qquad [3.136]$$

where $\Gamma_{d'd}^{\sigma L}(E_e) = (t_{de}^{\sigma L})\rho^{\sigma L}(t_{ed'}^{\sigma L})^*$. Disregarding spin dynamic effects (e.g. spin flipping), Eq. 3.136 is reduced to the simple form of $I_L = \frac{2e}{\hbar}\int dE\ Trace\ i\Gamma^L$ $\left[f_L(\boldsymbol{G}^R - \boldsymbol{G}^A) + \boldsymbol{G}^<\right]$.

We would, however, like to stress here that the real space and momentum space current are mere analogies as the parameters of both representations have different physical meanings. For example, the coupling constant in the spatial representation relates to spatial wavefunction overlap; while in momentum representation, the coupling constant is related to momentum state transition. The Green's functions of both representations also have different interpretations. The spatial Green's function is suitable for large systems with many discrete sites; for small systems with many momentum states, the discrete momentum representation is suitable because of the distinct number of momentum states. It is hard to use both methods for the same system because of the difficulty in searching out two sets (local and momentum) of coupling constants to produce identical results. However, one can always tune the coupling constants of both representations to obtain the current-voltage (I–V) results that best coincide.

## 3.8 References


[1] Fetter A L and Walecka J D (2003), *Quantum Theory of Many Particle Systems*, Dover Publications, Mineola, NY.

[2] Datta S (1997), *Electronic Transport in Mesoscopic Systems*, Cambridge University Press, Cambridge; Datta S (2005), *Quantum Transport (Atom to Transistor)*, Cambridge University Press, Cambridge; Economou E N (2006), *Green's Functions in Quantum Physics*, Springer Verlag, Berlin; Langreth D C (1976), *Linear and Nonlinear Electron Transport in Solids (NATO Advanced Study Institutes Series: Series B, Physics)*, eds Devreese J T and van Doren V E, Plenum, New York.

[3] Lake R and Datta S (1992), 'Nonequilibrium Green's function method applied to double-barrier resonant-tunneling diodes', *Phys. Rev. B*, 45, 6670.

[4] Meir Y and Wingreen N S (1992), 'Landauer formula for the current through an interacting electron region', *Phys. Rev. Lett.*, 68, 2512; Meir Y, Wingreen N S and Lee P A (1993), 'Low-temperature transport through a quantum dot: The Anderson model out of equilibrium', *Phys. Rev. Lett.*, 70, 2601.
[5] Landauer R (1957), 'Spatial variation of currents and fields due to localized scatterers in metallic conduction', *IBM J. Res. Dev.*, 1, 233; M. Buttiker (1986), 'Four-terminal phase-coherent conductance', *Phys. Rev. Lett.*, 57, 1761.
[6] Jauho A-P, Wingreen N S and Meir Y (1994), 'Time-dependent transport in interacting and noninteracting resonant-tunneling systems', *Phys. Rev. B*, 50, 5528.
[7] Lake R, Klimeck G, Bowen R C and Jovanovic D (1997), 'Single and multiband modeling of quantum electron transport through layered semiconductor devices', *J. Appl. Phys.*, 81, 7845.
[8] Niu C, Lin D L and Lin T H (1999), 'Equation of motion for nonequilibrium Green functions', *J. Phys.: Condens. Matter*, 11, 1511.
[9] Datta S (2000), 'Nanoscale device modeling: the Green's function method', *Superlattices Microstruct.*, 28, 253.
[10] Cresti A (2006), 'Microscopic current imaging in quantum point contact devices', *J. Appl. Phys.*, 100, 053711.
[11] Nonoyama S and Oguri A (1998), 'Direct calculation of the nonequilibrium current by a recursive method', *Phys. Rev. B*, 57, 8797.
[12] Chang Y C and Kuo D M-T (2008), 'Theory of charge transport in a quantum dot tunnel junction with multiple energy levels', *Phys. Rev. B*, 77, 245412.
[13] Chen W, Jalil M B A and Tan S G (2008), 'Resonant spin transport through lateral ferromagnet-quantum dot-ferromagnet sandwich device', *J. Appl. Phys.*, 103, 07B732.
[14] Sergueev N, Sun F, Guo H, Wang B G and Wang J (2002), 'Spin-polarized transport through a quantum dot: Anderson model with on-site Coulomb repulsion', *Phys. Rev. B*, 65, 165303.
[15] Zhang P, Xie Q-K and Xie X C (2003), 'Spin current through a quantum dot in the presence of an oscillating magnetic field', *Phys. Rev. Lett.*, 91, 196602.
[16] Tan S G, Jalil M B A, Kumar S B and Liang G-C (2008), 'Spin tunneling in multilayer spintronic devices', *Phys. Rev. B*, 77, 085424.
[17] Mathon J and Umerski A (2001), 'Theory of tunneling magnetoresistance of an epitaxial Fe/MgO/Fe(001) junction', *Phys. Rev. B*, 63, 220403; Mathon J (1997), 'Tight-binding theory of tunneling giant magnetoresistance', *Phys. Rev. B*, 56, 11810.
[18] Zhu Z G, Su G, Zheng R and Jin B (2003), 'Time-dependent spintronic transport and current-induced spin transfer torque in magnetic tunnel junctions', *Phys. Rev. B*, 68, 224413.
[19] Wakabayahsi K, Fujita M, Ajiki H and Sigrist M (1999), 'Electronic and magnetic properties of nanographite ribbons', *Phys. Rev. B*, 59, 8271.
[20] Caroli C, Combescot R, Nozieres P and Saint-James D (1971), 'Direct calculation of the tunneling current', *J. Phys. C: Solid State Phys.*, 4, 916.

# 4
# Spin dynamics in nanoelectronic devices

**Abstract:** This chapter introduces the concept of transporting the spin degrees of freedom across the device, beginning with a simple two-current system. An introduction is also given to the effect of magnetic fields on electron dynamics, leading to classical and quantum Hall effects. Readers are then introduced to the use of non-equilibrium Green's functions (NEGFs) to study electron transport under magnetic fields and spin transport in magnetic systems. The second-quantized form of spin orbit coupling is discussed and a complete spin current formalism taking into account spin flip and spin orbit coupling is presented.

**Key words:** quantum Hall effects, spin transport, magnetic fields, spin orbit coupling, spin flip, non-collinear, spin current, NEGF.

## 4.1 Introduction: spin current and spin transport

In the previous chapter, we discussed the construction of the lattice Green's function for discrete nanoelectronic devices. We have turned off global many-body effects, but turned on inter-site perturbation. The Green's function formalism for one and more geometrical dimensions has also been discussed. The formalism for current was introduced using general discrete indices and adapted to discrete spatial and momentum states. The final current expressions in momentum and spatial representation share the same form. In the spatial adaptation, vertical and longitudinal perturbation effects were considered separately to provide a pedagogical picture of their origin, which leads to various useful expressions suitable for current imaging in nanostructure and consistent with electrical circuit law.

In modern nanoscale electronic systems (e.g. spintronics, graphene, quantum dot, orbital and superconducting electronics, and topological insulating electronics), current has an additional degree of freedom (DOF) which can be adequately described by the spinor algebra. These currents are sometimes known as color currents, which may not conserve due to their relation to the non-Abelian charges rather than the familiar scalar electronic charges in conventional metal–oxide–semiconductor field-effect transistor (MOSFET) electronics. For example in spintronics, the electron's spin plays the role of the color; in graphene, it would be the pseudospin as well as the valley degree of freedom; in orbital electronics, it is the orbital band.

Following the development of increasingly complex forms of nanoscale electronics, it is timely to develop a consistent and coherent form of transport formalism for the spinor or the color form of electronic current. The formalism here is general and provides for the rotation of spinor as well as its coupling with the electron's momentum. For simplicity and illustration of the importance of color electronic in applied physics and engineering, we will take the example of spintronics. The spin flip and spin orbital effects affect the transport of both charge and spin fluxes in spintronic devices. Understanding these effects goes a long way towards resolving many practical problems in technology such as spin transfer switching, spin oscillations, spin injection and spin valving. Similarly one can conjecture the pseudospin of graphene undergoing analogous effects of pseudospin transfer, oscillations and injection for possible future device usage. In fact it is not entirely far-fetched to conceive that pseudo-spintronic might become useful in the future. The same analogy applies to the valley degree of freedom, which triggers novel visions of valley based electronics, as well as the orbital degree of freedom which suggests orbital electronics.

## 4.2 Simple two-current system

We begin our studies of spin current with the Hamiltonian, which describes a system containing the flow of two branches of current. A two-current system here could refer to any quantum number, e.g. the $s$ and $d$ orbital bands of magnetic materials, the spin up/down bands of spintronic systems, and the pseudo up/down and valleys of graphene systems. Let us start with a one-dimensional nearest neighbor system at spatial point $i$ for a two-band system arbitrarily labeled as $s$ and $d$, which when written in second-quantized form is:

$$H^{v} = \sum_{i,\sigma \in\{s,d\}} 2t\, c^{\dagger}_{i\sigma} c_{i\sigma} - \sum_{i,\sigma \in\{s,d\}} \left( t c^{\dagger}_{i+1,\sigma} c_{i,\sigma} + t c^{\dagger}_{i,\sigma} c_{i+1,\sigma} \right)$$
$$- \sum_{i,\sigma \in\{s,d\}} \left( V^{\sigma\bar{\sigma}} c^{\dagger}_{i\sigma} c_{i\bar{\sigma}} \right). \qquad [4.1]$$

Alternatively, one can represent the above with the bra-ket operators. For a more physical visualization, we write the Hamiltonian to reflect the site energy operator of the two bands at specific site (e.g. $i = 3$) as well as the tunneling kinetic operator for both bands at the nearest neighboring sites of $x_1$ and $x_2$ under the tight binding approximation:

$$H^{v} = 2t|s,x_3\rangle\langle s,x_3| - t|s,x_3\rangle\langle s,x_2| - t|s,x_3\rangle\langle s,x_4|$$
$$+2t|d,x_3\rangle\langle d,x_3| - t|d,x_3\rangle\langle d,x_2|$$
$$-t|d,x_3\rangle\langle d,x_4| - V^{ds}|d,x_3\rangle\langle s,x_3| - V^{sd}|s,x_3\rangle\langle d,x_3|. \qquad [4.2]$$

The terms $V^{sd}\,|d, x_3\rangle\langle s, x_3|$ and $V^{ds}\,|s, x_3\rangle\langle d, x_3\,|$ give the coupling between the two bands of $s$ and $d$ with strength $V$ at site $x_3$. Thus, one notes that $\langle s, x_3 \mid H^{v}\, |d, x_3\rangle = V^{sd}$ and $\langle d, x_3|\, H^{v} \mid s, x_3\rangle = V^{ds}$. The Green's function expressions can be derived by taking note that $H^{v} = H^{v}_0 + V^{v}$ and that:

$$\left\langle x_3,\sigma'\middle|(E-H^v)(E-H^v)^{-1}\middle|x_3,\sigma''\right\rangle$$
$$=\left(E-H_0^f\right)G_{33}^{\sigma'\sigma''}-\sum_{i,\sigma\in\{s,d\}}V^{\sigma\bar{\sigma}}\delta_{\sigma\sigma'}\,\delta_{i3}\left\langle i\,\bar{\sigma}\middle|\left(E-H^v\right)^{-1}\middle|x_3,\sigma''\right\rangle$$
$$=\left(E-H_0^f\right)G_{33}^{\sigma'\sigma''}-V^{\sigma'\bar{\sigma}''}\left\langle x_3,\bar{\sigma}'\middle|\left(E-H^v\right)^{-1}\middle|x_3,\sigma''\right\rangle \qquad [4.3]$$

which generates a pair of useful Green's function expressions (Eq. 4.4 and Eq. 4.5). These may apply to any types of two-current systems (e.g. in spintronics, graphene, topological insulating, orbital, quantum dot, optical electronics and superconducting systems):

$$(E-H^f)G_{33}^{ss}=(E-2t)G_{33}^{ss}+tG_{23}^{ss}+tG_{43}^{ss}+V^{sd}G_{33}^{ds}=I \qquad [4.4]$$

$$(E-H^f)G_{33}^{dd}=(E-2t)G_{33}^{dd}+tG_{23}^{dd}+tG_{43}^{dd}+V^{ds}G_{33}^{sd}=I. \qquad [4.5]$$

In general form, the above is $(E - H^f)G^R(xx', E) = \delta_{xx'}$. We stick to the labeling convention of $G^{ab} = \langle a|\,(E - H^v)^{-1}|b\rangle$.

We draw particular comparison to the spintronic systems where spin relaxation is an important research topic. It is not hard to notice that in fact $V^{ds}$ mimics the spin flip constant. In this context, $V^{ds}$ has the physical outcome of measuring the strength, or the likelihood of spin flipping. This has far-reaching implications for technological applications like spin oscillation and current-induced magnetization switching. The above can be represented in matrix form with the off-diagonal sub-matrices containing the effects of electron spin flip or the electron *s–d* band coupling.

---

**Exercise 4.1**

Show that $(E-2t)G_{33}^{ss}+tG_{23}^{ss}+tG_{43}^{ss}+V^{sd}G_{33}^{ds}=1$.

***Solution***

$(E\mathbf{I} - \mathbf{H})\mathbf{G} = \mathbf{I}$

Bold denotes matrix representation, one thus has:

$$\langle s, 3|\,(E - H^v)\,G\,|s, 3\rangle = 1$$

where $G = (E - H^v)^{-1}$

$$\Rightarrow \langle s, 3|\,(|s, 3\rangle(E - 2t)\langle s, 3| + |s, 3\rangle t\langle s, 4| + |s, 3\rangle t\langle s, 2|$$
$$- |s, 3\rangle\, V^{sd}\,\langle d, 3|)\;G\,|s, 3\rangle = 1$$

$$\Rightarrow (E - 2t)\,\langle s, 3|\,G\,|s, 3\rangle + t\langle s, 4|\,G\,|s, 3\rangle + t\langle s, 2|G|\,s, 3\rangle$$
$$+ V^{sd}\,\langle d, 3|G|\,s, 3\rangle = 1$$

$$\Rightarrow (E-2t)G_{33}^{ss}+tG_{23}^{ss}+tG_{43}^{ss}+V^{sd}G_{33}^{ds}=1.$$

---

Below is an explicit matrix representation considering the band effect:

$$\begin{bmatrix} E-2t & t & 0 & V^{sd} & 0 & 0 \\ t & E-2t & t & 0 & V^{sd} & 0 \\ 0 & t & \cdots & 0 & 0 & \cdots \\ V^{sd} & 0 & 0 & \cdots & \dagger & 0 \\ 0 & V^{sd} & 0 & t & E-2t & t \\ 0 & 0 & \cdots & 0 & t & \cdots \end{bmatrix} \begin{bmatrix} G^{ss}_{22} & G^{ss}_{23} & G^{ss}_{24} & G^{sd}_{22} & G^{sd}_{23} & G^{sd}_{24} \\ G^{ss}_{32} & G^{ss}_{33} & G^{ss}_{34} & G^{sd}_{32} & G^{sd}_{33} & G^{sd}_{34} \\ G^{ss}_{42} & G^{ss}_{43} & \cdots & G^{sd}_{42} & G^{sd}_{43} & \cdots \\ G^{ds}_{22} & G^{ds}_{23} & G^{ds}_{24} & \cdots & G^{dd}_{23} & G^{dd}_{24} \\ G^{ds}_{32} & G^{ds}_{33} & G^{ds}_{34} & G^{dd}_{32} & G^{dd}_{33} & G^{dd}_{34} \\ G^{ds}_{42} & G^{ds}_{43} & \cdots & G^{dd}_{42} & G^{dd}_{43} & \cdots \end{bmatrix}$$

$$= \begin{bmatrix} I & 0 & 0 & 0 & 0 & 0 \\ 0 & I & 0 & 0 & 0 & 0 \\ 0 & 0 & I & 0 & 0 & 0 \\ 0 & 0 & 0 & I & 0 & 0 \\ 0 & 0 & 0 & 0 & I & 0 \\ -0 & 0 & 0 & 0 & 0 & I \end{bmatrix} \qquad [4.6]$$

Note that each sub-matrix belonging to a particular band is infinite in size. The number of sub-matrices is, however, finite and related directly to the number of bands. The matrix in Eq. 4.6 consists of four sub-matrices, namely the ss, ds, sd and dd sections. The off-diagonal sections account for coupling between the two bands or the two colors. The diagonal sections account for the individual bands. If one is interested in only a single-band system, one needs to consider only the top-left section, i.e. the ss section. The present matrix applies only to a one-dimensional system. For a two-dimensional system, one can further develop each matrix element into a sub-matrix block.

## 4.3 Spin and magnetic system

In nanoscale electronics, the application of magnetic fields is common. In metal and semiconductor spintronics, single and quantum dot, and graphene electronics, magnetic fields have been applied for various reasons including inducing Zeeman splitting, quantum Hall and band-induced magnetoresistance.[1] Below we give a brief introduction to electron dynamics under the effect of magnetic fields to help readers appreciate the physical pictures when applying non-equilibrium Green's function (NEGF) methods later.

### 4.3.1 Introduction to magnetic field and Hall effects

Electrons experience the Lorentz force in the presence of magnetic fields. The classical Hall effect was discovered more than a century ago by E. H. Hall. A magnetic field is applied vertical to the direction of current flow, and a voltage known as the Hall voltage can be detected across the transverse length. Equilibrium is achieved when the force due to the magnetic field is balanced by the countering one due to charge accumulation across the transverse length. The simple but illustrative mathematical description of this is that two opposing forces balance when:

$$E_y = v_x B_z . \quad [4.7]$$

One defines a Hall coefficient as follows:

$$R_H = E_y / B_z j_x \quad [4.8]$$

with $j_x = nev_x$ and $R_H = 1/ne$. It is well-known that the Hall coefficient can be measured to determine the carrier density in the material as shown in Fig. 4.1.

But the physics here is rather different from that of the integer quantum Hall effect (IQHE). In the classical case, the magnetic field is not strong enough with respect to the sample dimension to create bound cyclotron states. Hall resistance here depends on the free carrier density. In IQHE, Hall resistance depends on current-carrying edge states. The edge states are a multiple of the number of Landau levels below the Fermi energy. Increasing the magnetic field reduces the number of Landau levels below the Fermi energy in a discrete manner, resulting in quantized IQHE Hall resistance.

The application of the B field is a prelude to the interesting integer quantum Hall effect. From $R_H = E_y/B_z j_x$ and $j_x = nev_x$, one works out $R_H = 1/ne$. It is obvious $R_H$ is a measure of carrier density in a sample. This is one of the most well known applications of the classical Hall effect. One can also deduce that, since the Hall voltage is $V_H = I_x R_{xy}$, it measures the strength of the applied field if the carrier density is known.

A picture of the effect of B fields on the density of states below the Fermi energy is give in Fig. 4.2. For illustration, we consider a semiconductor two-dimensional electron gas (2DEG) with electron density of $n_s = 5 \times 10^{11}$ cm$^{-2}$. The

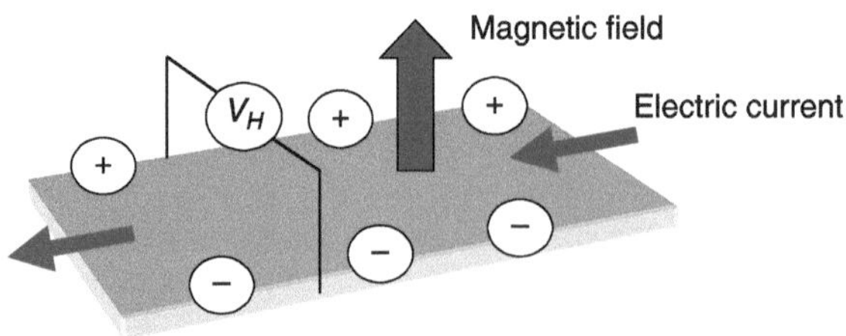


*4.1* Schematic illustration of the classical Hall effect. Electrons are deflected and accumulate on the two transverse sides of the sample, generating an electric field which balances the force of the magnetic field.

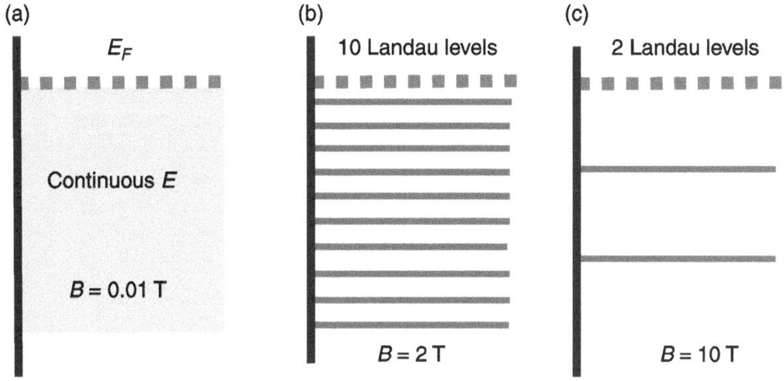


*4.2* Illustration of electron states under applied magnetic fields. (a) B field is weak and electrons are not bound. (b) B field is strong and Landau levels are created. (c) Under very strong B fields, there are fewer energy levels under the Fermi energy, implying higher degeneracy for each level.

energy states for B fields of 2 T and 10 T show a discrete number of Landau levels below the Fermi energy.

## 4.3.2 Introduction to quantum Hall effects

The quantum Hall effect is most representative of mesoscale systems or nanoscale devices because it exists only in two-dimensional (2D) systems, which do not normally exist in nature. Advancement in nanotechnologies made possible the fabrication of nanostructures only in recent times and the 2D structures turn out to host one of the most beautiful physics known to mankind.

Quantum Hall effects comprise the integer quantum Hall effect (IQHE) and the fractional quantum Hall effect (FQHE). Both have been Nobel-winning discoveries. In this book, only the underlying physics of the quantum Hall effects are discussed and they are introduced in simple, intuitive pictures. The fractional quantum Hall system is a collection of electrons which can no longer be treated as a non-interacting electron gas. The interaction is so strong that the fractional system can in some sense be regarded as a solid (Wigner crystal). In IQHE, semiclassical and conventional field theoretic methods could not fully explain the strange but interesting electronic behavior. Observables (e.g. energy levels and conductivity) exhibit topological properties. In the language of physics, these observables are not easily affected by the presence of impurities or scattering. This is unlike the electron gas systems for metals and semiconductors where conductivity depends on the scattering between electrons and phonons.

It is rather surprising that an elegant physical phenomenon like this exists only in systems that are man made. The quantum Hall effect like all others has its root in the classical Hall effect, which we have linked to the physical explanations based on the Lorentz force. IQHE was first observed by von Klitzing, Dorda and Pepper in 1980.[2] The experiment was carried out in a 2DEG formed by an inversion layer at a Si/$SiO_2$ interface. At a temperature of a few degrees Kelvin and a B field of a few Tesla, quantized conductivity was observed:

$$\sigma_{xy} = \frac{I_x}{V_y} = \frac{I_x}{V_L - V_R} = \frac{v}{R_q} \text{ where } R_q = \frac{h}{e^2} = 25.813\,\Omega. \qquad [4.9]$$

*Degeneracy and filling factor*

When the magnetic field is high, the trajectories of all free carriers curl up to form small circulating states in the sample. These circulating states are no longer free and technically speaking, the sample becomes an insulator in the bulk. These circulating states are equally distributed to each Landau level below $E_F$. On the edges, things are different because skipping orbit edge states are formed due to the boundary of the edges. In the presence of impurities in the system, these edge states are associated with the extended states of the Landau levels. We will not venture deep into the physics of localized and extended states. Instead we will restrict ourselves to understanding only the few important concepts in pure sample under the effect of *B* fields. Now we introduce the important quantity of:

$$N_D = \frac{eB}{h} \qquad [4.10]$$

known as the degeneracy strength or degeneracy factor of each Landau level per unit area of the 2D sample. For a particular Landau level, this is the number of available states per unit area. The larger the sample, the more degenerate states a Landau level could take. Figure 4.3 provides a crude schematic illustration of the circulating electron states in a 2DEG samples. We will refer to these states as cyclotron states for short.

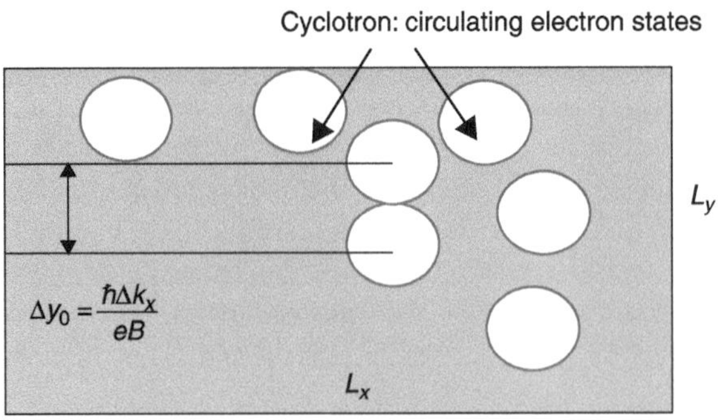


*4.3* Physical imagination of electron bound states (cyclotron states) and their distribution in a two-dimensional system based on the Landau gauge under a strong magnetic field.

Quantum mechanically, one could show that the shift of the cyclotron radius from a user-defined origin of the 2D sample is $y_0 = \frac{\hbar k_x}{eB}$, leading to the smallest separation between two cyclotrons of $\Delta y_0 = \frac{\hbar \Delta k_x}{eB}$. For a sample with geometric size of $A = L_x L_y$, one arrives at:

$$\Delta y_0 = \frac{2\pi\hbar}{eBL_x} = \frac{h}{eBL_x}. \qquad [4.11]$$

The degeneracy factor would thus be:

$$N_0 = \frac{L_y}{\Delta y_0} = \frac{eBA}{h}. \qquad [4.12]$$

The degeneracy factor per unit area is simply $N_D = \frac{N_0}{A} = \frac{eB}{h}$. When one writes:

$$N_0 = \frac{A}{h/eB} \qquad [4.13]$$

it is easy to see that *h/eB* would be the real space area occupied by one circulating electron state per Landau level. Thus *h/e* is also known as the flux quantum or the flux associated with one cyclotron electron (Fig. 4.4).

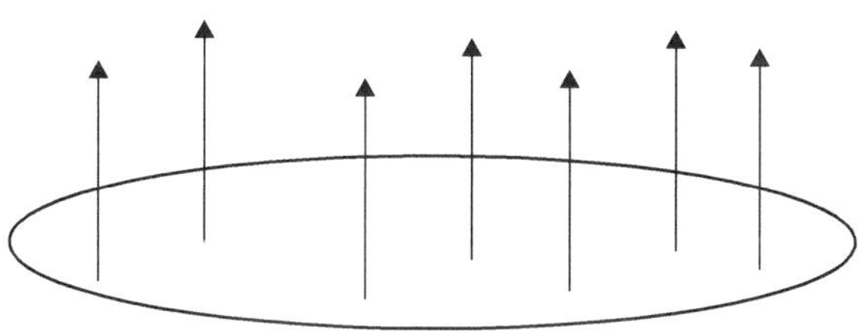

*4.4* An electron bound state circulates an area containing a flux quantum.

The flux quantum can also be written in terms of the area enclosed by the circulating electron:

$$\phi_0 = \frac{h}{e} = B\,a. \qquad [4.14]$$

An electron takes up a certain space depending on the strength of the *B* field in order to obey the rule of capturing with it a flux of $\frac{h}{e}$.

---

**Exercise 4.2**

Deduce in terms of the flux quanta and the tiny enclosures inscribed by a cyclotron what happens to the size of the cyclotron when B field increases. Does the size increase or decrease? This question tests the understanding of the qualitative picture given above.

---

A more rigorous representation of the real space distribution of the cyclotrons would require some quantum mechanics and the understanding of the wavefunctions. Here we only show the final results without derivations. The wavefunction in the spherical coordinates of a 2DEG system under a magnetic field is given by:

$$\psi_{nm}(r) = C_{nm} \exp\left(i(n-m)\theta - \frac{r^2}{4l^2}\right)\left(\frac{r}{l}\right)^{|m-n|} L^{nm}\left(\frac{r^2}{2l^2}\right) \quad [4.15]$$

where $C_{nm}$ is the normalization constant and $L^{nm}$ is the Laguerre polynomial. The quantum numbers are $n$ for energy and $m$ for angular momentum. We will look at the wavefunction of the lowest Landau level, i.e. $n = 0$:

$$\psi_{0m}(r) = \frac{1}{\sqrt{2\pi 2^m m!\, l}} z^m \exp\left(-\frac{|z|^2}{4}\right)$$

$$= \frac{1}{\sqrt{2\pi 2^m m!\, l}} \left(\frac{x-iy}{l}\right)^m \exp\left(-\frac{r^2}{4l^2}\right). \quad [4.16]$$

Here, $l = \sqrt{\hbar / e B}$ is the magnetic length; $(z = x - iy)$, $z^* = x + iy$. The magnetic length is also the radius of the cyclotron motion for $n = 0$. This wavefunction is obtained with the symmetric gauge of $\boldsymbol{A} = \left(-\frac{By}{2}, \frac{Bx}{2}, 0\right)$, and represents a circular distribution of electrons (Fig. 4.5).

Since all wavefunctions on the lowest Landau level have the same energy, the following wavefunction is also an eigenfunction of the lowest Landau level:

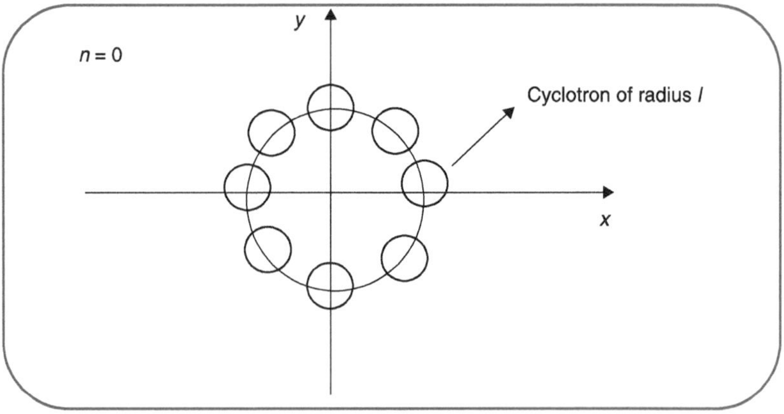


*4.5* Electron bound states and distribution in space can be imagined differently when the symmetric gauge is used. Observables are, however, invariant irrespective of the gauges used.

$$\psi(r) = f(z)e^{\frac{-|z|^2}{4}} \quad [4.17]$$

where $f(z)$ is an arbitrary function. If for a particular B field, the number of Landau levels below $E_F$ is $v$, there will be $v\frac{eB}{h}$ electron states below the Fermi energy.

---

### Exercise 4.3

Show that if $n_s$ is the electron density ($m^{-2}$) in the 2D sample:

$$v = \frac{n_s}{N_D} = \frac{electron\ density}{cyclotron\ state\ density\ per\ LL} = \frac{hn_s}{eB}$$

where $v$ is known as the filling factor.

---

If the B field strength is such that $v$ is an integer number, $v$ would then indicate the number of Landau levels below $E_F$. When $\upsilon$ is an integer, the Landau levels are completely filled. The implication is that the presence of a B field changes the density of states from a constant (for a usual 2D system) to a set of delta functions as shown in Fig. 4.6.

One sees that as B field increases, the number of electron states for each Landau level increases except for the highest Landau level, which is losing electrons to the lower Landau levels. Taking a digression, we note that this can be measured as an increase in the longitudinal resistance $\rho_{xx}$, as there are more empty states at the highest Landau level for scattering to take place. Such manifestation is also known as the Shubnikov–de IIaas (SdII oscillations), which wc will dcscribc only bricfly hcrc. As the highest Landau level loses all its electrons, it also moves above $E_F$; the next highest Landau level would now take the place of the outgoing one and become the highest Landau level below $E_F$ and which is fully occupied. The number of electron states below the Fermi energy remains a constant. Due to magnetic depopulation, the Fermi energy of the 2D system will thus follow a sawtooth shape. Scattering is low if the uppermost Landau level is either nearly full or nearly empty, resulting in low resistivity. Scattering is high when the uppermost level is half full, resulting in high resistivity. This variation with $B$ gives rise to SdH oscillations in resistivity.

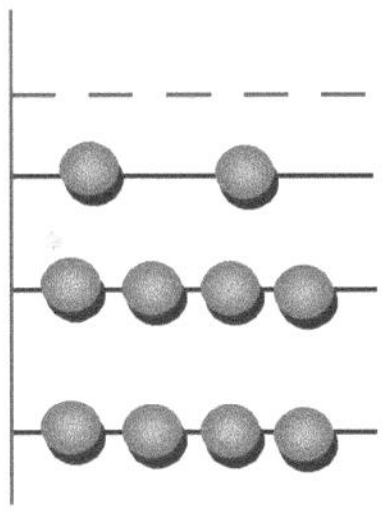

*4.6* Density of states in the IQHE system where each energy level accommodates a large number of electrons.

The important question to ask in IQHE is: How does quantized Hall resistance $\rho_{xy}$ arise? The answer would be simple if one can associate the Hall resistance to the number of Landau level below $E_F$. But how then does this relation arise? Here, one needs the concept of localized and extended states. Indeed, there is a fixed number (at least one) of these states for each Landau level. We will not explain the formation of extended and localized states here. Let us accept the fact that the number of extended states to each Landau level is fixed and depends on the temperature and the impurities in the sample. Keeping this idea of the extended states in mind (i.e. its association with the Landau level) and recalling that $\nu = \frac{hn_s}{eB}$ which gives the number of Landau level below the Fermi energy, it would be easy to deduce that the quantum Hall resistance simply increases with the number of Landau levels below the Fermi energy. The relation can be expressed as follows:

$$R_{xy}^{QH} = \frac{1}{\nu}\frac{h}{e^2} = \frac{h}{e^2}\frac{1}{\sum_i^N \theta\left(\frac{hn_s}{eB} - i\right)} \quad [4.18]$$

ensuring that $\nu$ takes on an integer number.

---

**Exercise 4.4**

Deduce that $\sum_i^N \theta\left(\frac{hn_s}{eB} - i\right)$ is always the largest integer number smaller than $\frac{hn_s}{eB}$.

Show that when $\nu = \frac{hn_s}{eB}$ takes on integer value, the quantum Hall resistance is equal to the classical Hall resistance, i.e.

$$R_{xy}^{QH} = R_{xy}^{CH} = \frac{B}{n_s e}.$$

***Solution***

When $\nu = \frac{hn_s}{eB}$ is an integer number, $\sum_i^N \theta\left(\frac{hn_s}{eB} - i\right) = \frac{hn_s}{eB}$. But when $\nu$ is not an integer number, $\sum_i^N \theta\left(\frac{hn_s}{eB} - i\right)$ will only be the largest integer number smaller than $\frac{hn_s}{eB}$.

---

By contrast the classical Hall resistance is:

$$R_{xy}^{CH} = \frac{B}{n_s e}. \quad [4.19]$$

Note that in two-dimensional systems $I_x = j_x l_y$ where $j_x$ ($\mathrm{Am^{-1}}$) is the current per unit of transverse length and $n_s$ is the carrier density with dimension of $\mathrm{m^{-2}}$.

### 4.3.3 Non-equilibrium electron transport under magnetic vector potential

The above is a semiclassical depiction of electron dynamics under the magnetic field and we have understood the physics which will give rise to IQHE. It is now interesting to image the distribution of these electrons under the non-equilibrium condition. We will use the simplest one-dimensional system in a 2DEG system having translational invariance along the transverse $y$ direction, with external magnetic field perpendicular to the 2DEG, and electric fields applied spatially along $x$ as shown in Fig. 4.7.

In the presence of the magnetic and electric fields, the Hamiltonian operator is:

$$H^f = \frac{(i\hbar\nabla + A)^2}{2m^*} + U + \frac{g\,\hbar\sigma B_z}{4m_0} \qquad [4.20]$$

where $m^*$ is the electron effective mass, $m_0$ is the electron mass in vacuum, $g$ is the effective Landé $g$-factor and $U$ is the total potential energy of the electron in the semiconductor. $B_z$ is the vertical magnetic field at $x$ due to the perpendicular magnetization of the ferromagnetic gates, $A$ is the corresponding vector potential chosen in the Landau gauge $(0, A_y(x), 0)$, and $\sigma = +/-1$ denotes spin up/down, respectively, and where the spin quantization axis is defined along the vertical magnetic fields. The electron conduction path in the $x$ direction is discretized into $n$ lattice points of equal distance $a$ apart, i.e. the discrete points are denoted by, $x_j = ja$, where $1 \le j \le n$ as shown in Fig. 4.7. We can now derive the matrix representation of the Hamiltonian by applying the finite difference approach in the limit of small $a$. If $\psi(x)$ is the eigenfunction of the system, which need not be solved analytically, then we have:

$$H^f\,\psi(x) = -\frac{\hbar^2}{2m}\psi''(x) + \frac{i\hbar A_x}{m}\psi'(x) + \left[\frac{A_x^2}{2m} + U + \frac{g\hbar\sigma B_z}{4m} + E_y\right]\psi(x). \qquad [4.21]$$

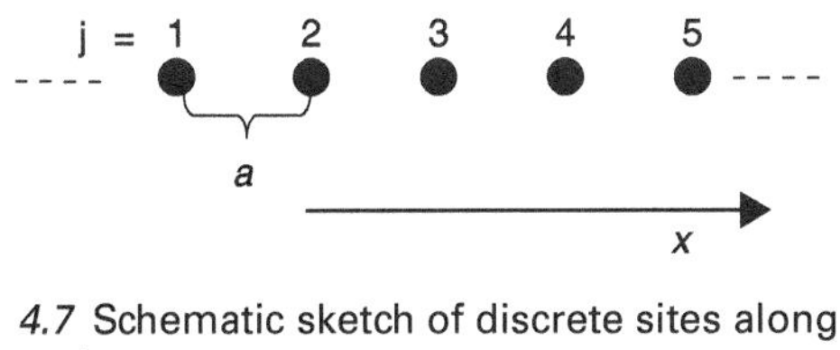


4.7 Schematic sketch of discrete sites along the electron conduction path.

For small $a$ (i.e. $a \to 0$), the first and second derivatives of $\psi(x)$ can be approximated as:

$$\psi'(x)\Big|_{ja} \approx \frac{\psi_{j+1/2} - \psi_{j-1/2}}{a}$$
$$\psi''(x)\Big|_{ja} \approx \frac{\psi_{j+1} - 2\psi_j + \psi_{j-1}}{a^2} \qquad [4.22]$$

where for brevity, $a$ is sometimes omitted from $j$ which always implies $ja$. With the above approximations, we rewrite the energy equation as:

$$(H^f\psi)_{ja} = \frac{-\hbar^2}{2m}\left(\frac{\psi_{j+1} - 2\psi_j + \psi_{j-1}}{a^2}\right) + \frac{i\hbar A_j}{m}\left(\frac{\psi_{j+1/2} - \psi_{j-1/2}}{a}\right)$$
$$+\left(\frac{A_j^2}{2m} + U_j + \frac{g\hbar\sigma B_j}{4m} + E_y\right)\psi_j \qquad [4.23]$$

or equivalently:

$$H^r\psi_j = (-t_j)\psi_{j-1} + (-iq_j)\psi_{j-1/2} + (2t_j + \omega_j + Z_j + U_j + E_y)\psi_j$$
$$+ (iq_j)\psi_{j+1/2} + (-t_j)\psi_{j+1} \qquad [4.24]$$

where $t = t_j = \dfrac{\hbar^2}{2ma^2}$, $q_j = \dfrac{\hbar A_j}{2ma} = \dfrac{A_j a}{\hbar} t$, $\omega_j = \dfrac{A_j^2}{2m} = \left(\dfrac{A_j a}{\hbar}\right)^2 t$ and $Z_j = \dfrac{g\hbar\sigma B_j}{4m}$.
By taking $q_j\psi_{j+\frac{1}{2}} \approx q_{j+\frac{1}{2}}\psi_{j+1}$ and $q_j\psi_{j-\frac{1}{2}} \approx q_{j-\frac{1}{2}}\psi_{j-1}$, one can show that:

$$H^r\ \psi_j = Q_j^\dagger\psi_{j-1} + P_j\psi_j + Q_j\psi_{j+1}. \qquad [4.25]$$

where $P_j = 2t_j + \omega_j + Z_j + U_j + E_y$ and $Q_j = -t_j + iq_{j+1/2}$. It is worth noting that that $q_{j+\frac{1}{2}}$ is to be obtained by evaluating $A(x)$ at halfway between $x_j$ and $x_{j+1}$. Thus in matrix form, the Hamiltonian is given by:

$$(H^f\psi(x))_{x=ja} = \sum_i C_i\psi_i$$
$$C_i = \begin{cases} Q_j^\dagger & i = j+1 \\ P_j & i = j \\ Q_j & i = j-1 \\ 0 & otherwise. \end{cases} \qquad [4.26]$$

Thus in matrix form, the Hamiltonian is given by:

$$H = \begin{bmatrix} P_1 & Q_1^* & 0 & \cdots & 0 \\ Q_1 & P_2 & \ddots & \ddots & \vdots \\ 0 & \ddots & \ddots & \ddots & 0 \\ \vdots & \ddots & \ddots & P_{n-1} & Q_{n-1}^* \\ 0 & \cdots & 0 & Q_{n-1} & P_n \end{bmatrix} \qquad [4.27]$$

where $n$ is the number of discrete points in the 2DEG. We model our device as a finite-sized conduction path in the 2DEG, which is connected on either side to semi-infinite leads. Thus, from the discretized Hamiltonian, we can then derive the retarded Green's function matrix for the 2DEG region, following standard methods.

$$G^R = \left[ EI - H - \left( \Sigma_L^R + \Sigma_R^R \right) \right]^{-1}$$

$$= \begin{bmatrix} E_x - P_1 - \Sigma_L^R(1,1) & -Q_1^* & 0 & \cdots & 0 \\ -Q_1 & E_x - P_2 & \ddots & \ddots & \vdots \\ 0 & \ddots & \ddots & \ddots & 0 \\ \vdots & & \ddots & E_x - P_{n-1} & -Q_{n-1}^* \\ 0 & & \cdots\ 0 & -Q_{n-1} & E_x - P_n - \sum_R^R(n,n) \end{bmatrix}, \qquad [4.28]$$

where $E$ represents the total electron energy, $E_x$ is the electron energy in the propagation ($x$)–direction, and $\Sigma_L^R$ and $\Sigma_R^R$ are the self-energies of the left and right lead, respectively. Note that $G^R$ is independent of the kinetic energy in the transverse direction $E_y = \hbar^2 k_y^2 / 2m^*$, due to the cancellation of the $E_y$ component in $E$ and $H$. Since electron transport is constrained to the $x - y$ plane of the 2DEG, the confinement in the $z$ direction will lead to the formation of sub-bands. We do not discuss the details concerning the sub-bands and assume that transport occurs only within the lowest sub-band. The component $G_{ij}^R$ describes the propagation of electron between the points $x_i$ and $x_j$ within the 2DEG channel, while the self-energy terms of $\Sigma^R = \Sigma_L^R + \Sigma_R^R$ represent the perturbative effect of the entire semi-infinite leads on the electron propagation within the channel. $\Sigma_{L,R}^R$ is related to the surface Green's function of the isolated leads, $g_{L,R}$, and can be expressed as follows:

$$\sum_L^R(j,k) = \begin{cases} t_L g_L t_L & j = k = 1 \\ 0 & elsewhere \end{cases}, \qquad \sum_R^R(j,k) = \begin{cases} t_R g_R t_R & j = k = n \\ 0 & elsewhere \end{cases} \qquad [4.29]$$

---

**Exercise 4.5**

In a magnetic field system, the Green's function needs to be modified due to the changes in the Hamiltonian. In the following:

(1) Show that in a system with magnetic field, the discretized Hamiltonian can be written as follows:

$$H^r\psi_j = (-t_j)\psi_{j-1} + (-iq_j)\psi_{j-1/2} + (2t_j + \omega_j + Z_j + U_j + E_y)\psi_j + (iq_j)\psi_{j+1/2} + (-t_j)\psi_{j+1}$$

where $t = t_j = \dfrac{\hbar^2}{2ma^2}$, $q_j = \dfrac{\hbar A_j}{2ma} = \dfrac{A_j a}{\hbar}t$, $\omega_j = \dfrac{A_j^2}{2m} = \left(\dfrac{A_j a}{\hbar}\right)^2 t$, and $Z_j = \dfrac{g\hbar\sigma B_j}{4m}$.

(2) By taking $q_j\psi_{j+\frac{1}{2}} \approx q_{j+\frac{1}{2}}\psi_{j+1}$ and $q_j\psi_{j-\frac{1}{2}} \approx q_{j-\frac{1}{2}}\psi_{j-1}$, show that:

$$H^r\psi_j = Q_j^\dagger\psi_{j-1} + P_j\psi_j + Q_j\psi_{j+1}$$

where $P_j = 2t_j + \omega_j + Z_j + U_j + E_y$ and $Q_j = -t_j + iq_{j+\frac{1}{2}}$. It is worth noting that $q_{j+\frac{1}{2}}$ is to be obtained by evaluating $A(x)$ at halfway between $x_j$ and $x_{j+1}$.

## 4.3.4 Non-collinear electron spin and magnetic field

We now inspect a system where the two currents might appear in a linear superposition of one another using the Green's function method.[3,4] The state of such system is a rotated spinor, which satisfies a rotated Hamiltonian and preserves the eigenenergies. This description corresponds to the physical case of a magnetic/spin moment oriented at a particular angle with respect to the laboratory frame (Fig. 4.8). When an electron spin is not aligned to the magnetic moment, it is not a good quantum number and the spin is not a conserved quantity. A classical spin moment might precess about the effective field of the moment, relax towards the field, shrink in magnitude, or even travel away from the field.

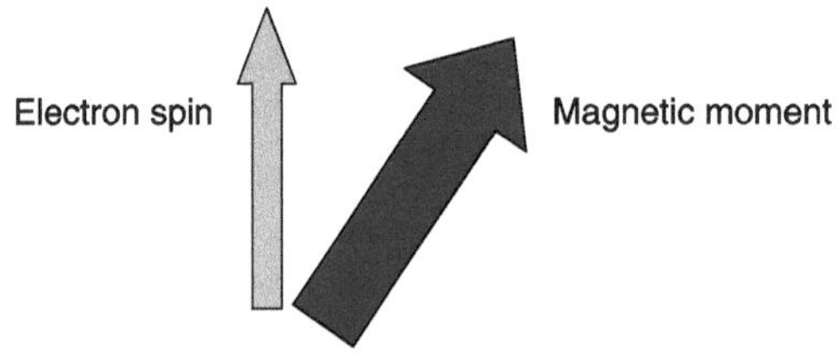


*4.8* Electron spin is not initially aligned to the magnetic moment. It is not a good quantum number.

### Exercise 4.6

The second-quantized Hamiltonian for a spinor system is:

$$\sigma_j^2 = \sum_{kk'\eta\eta'} \langle k\eta|\sigma_j^1|k'\eta'\rangle a_{k\eta}^\dagger a_{k'\eta'}$$

where superscripts indicate the type of quantization (e.g. '2' indicates second quantization) and the Pauli matrices in single-particle forms are:

$$\sigma_x^1 = \begin{pmatrix} 0 & 1 \\ 1 & 0 \end{pmatrix}, \quad \sigma_y^1 = \begin{pmatrix} 0 & -i \\ i & 0 \end{pmatrix}, \quad \sigma_Z^1 = \begin{pmatrix} 1 & 0 \\ 0 & -1 \end{pmatrix}.$$

Show that the Pauli in second-quantized forms are:

$$\sigma_y^2 = -i\sum_k a_{k\uparrow}^\dagger a_{k\downarrow} - a_{k\downarrow}^\dagger a_{k\uparrow} \Rightarrow -i\sum_k a_k^\dagger a_k \left(|\uparrow\rangle\langle\downarrow| - |\downarrow\rangle\langle\uparrow|\right)$$
$$\sigma_z^2 = 1\sum_k a_{k\uparrow}^\dagger a_{k\uparrow} - a_{k\downarrow}^\dagger a_{k\downarrow} \Rightarrow \sum_k a_k^\dagger a_k \left(|\uparrow\rangle\langle\uparrow| - |\downarrow\rangle\langle\downarrow|\right)$$
$$\sigma_x^2 = \sum_k a_{k\uparrow}^\dagger a_{k\downarrow} + a_{k\downarrow}^\dagger a_{k\uparrow} \Rightarrow \sum_k a_k^\dagger a_k \left(|\uparrow\rangle\langle\downarrow| + |\downarrow\rangle\langle\uparrow|\right).$$

---

In single-particle form, the spin Hamiltonian in the presence of a magnetic moment is:

$$H^f = J_{sd}\frac{\hbar}{2}\boldsymbol{\sigma}.\boldsymbol{M} + \frac{\hbar^2 k^2}{2m} = gM(\sigma_x \sin\theta + \sigma_z \cos\theta) + \frac{\hbar^2 k^2}{2m}. \qquad [4.30]$$

For simplicity, we have taken Ø = 0 and focused on the effect of one angular dimension only, i.e. $\theta$. In second-quantized form and momentum space, Eq. 4.30 is:

$$H^v = gM\left(\sum_k \left(a_{k\uparrow}^\dagger a_{k\downarrow} + a_{k\downarrow}^\dagger a_{k\uparrow}\right)\sin\theta + \left(a_{k\uparrow}^\dagger a_{k\uparrow} - a_{k\downarrow}^\dagger a_{k\downarrow}\right)\cos\theta\right)$$
$$+ \sum_{k\sigma} \varepsilon_{k\sigma} a_{k\sigma}^\dagger a_{k\sigma}. \qquad [4.31]$$

Noting the first term $H_1^v$ of the Hamiltonian in Eq. 4.31 is related to s-d coupling, the second term $H_2^v$ is kinetic energy and taking the Fourier transform, one has in real space:

$$H_1^v = \sum_m gM\left(c_{m\uparrow}^\dagger c_{m\downarrow}\sin\theta + c_{m\downarrow}^\dagger c_{m\uparrow}\sin\theta + c_{m\uparrow}^\dagger c_{m\uparrow}\cos\theta - c_{m\downarrow}^\dagger c_{m\downarrow}\cos\theta\right) \qquad [4.32]$$

or:

$$H_1^v = \sum_m gM\Big[\left(|x^m,\uparrow\rangle\langle x^m,\downarrow| + |x^m,\downarrow\rangle\langle x^m,\uparrow|\right)\sin\theta$$
$$+\left(|x^m,\uparrow\rangle\langle x^m,\uparrow| - |x^m,\downarrow\rangle\langle x^m,\downarrow|\right)\cos\theta\Big]. \qquad [4.33]$$

The second part of the Hamiltonian in real space is:

$$H_2^v = \sum_m \left(2c_m^\dagger c_m - c_m^\dagger c_{m-1} - c_m^\dagger c_{m+1}\right) \otimes \left(t^\uparrow |\uparrow\rangle\langle\uparrow| + t^\downarrow |\downarrow\rangle\langle\downarrow|\right)$$
$$+ \sum_m \left(2c_m^\dagger c_m\right) \otimes \left(U^\uparrow |\uparrow\rangle\langle\uparrow| + U^\downarrow |\downarrow\rangle\langle\downarrow|\right) \qquad [4.34]$$

where $t^\sigma$ is the strength of spin-dependent site coupling and $U^\sigma$ is the spin-dependent site potential. Physically $t^\sigma$ could mean spin-dependent tunneling between electrons in neighboring sites. On the other hand, $U^\sigma$ could be local magnetic impurities. It is obvious from Eq. 4.34 that the kinetic energy term has angular significance only when the site coupling or site potential are spin-dependent in strength.

---

**Exercise 4.7**

Show that $H = \sum_{k\sigma} \varepsilon_{k\sigma} a^\dagger_{k\sigma} a_{k\sigma}$ in momentum space can be written in real space as:

$$H = \sum_{m\sigma} \left( 2t^\sigma c^\dagger_{m\sigma} c_{m\sigma} - t^\sigma c^\dagger_{m\sigma} c_{m-1\sigma} - t^\sigma c^\dagger_{m\sigma} c_{m+1\sigma} \right).$$

As an aside:

$$\sum_k a^\dagger_k a_k \rightarrow \sum_m c^\dagger_m c_m$$

$$\sum_k k\, a^\dagger_k a_k \rightarrow \sum_m c^\dagger_m c_{m+1} - c^\dagger_m c_{m-1}$$

$$\sum_k k^2 a^\dagger_k a_k \rightarrow \sum_m 2c^\dagger_m c_m - c^\dagger_m c_{m+1} - c^\dagger_m c_{m-1}.$$

---

## 4.4 Second-quantized spin orbit coupling

In this section, we focus on a predominantly semiconductor effect known as the spin orbit coupling[5,6] based on the Green's function method. An analogous version of orbital coupling with pseudospin exists in graphene. For simplicity and illustration, we will look specifically at the linear conduction band type of spin orbit coupling due to Rashba and Dresselhaus. These effects have been studied extensively in semiconductors and more recently in metal spintronics. The first-quantized Hamiltonian of the Rashba and Dresselhaus type is given by:

$$H^f = \frac{p^2}{2m} + \alpha(k_x\sigma_y - k_y\sigma_x) + \beta(k_y\sigma_y - k_x\sigma_x). \qquad [4.35]$$

For simplicity, below is a discretized Hamiltonian with the Rashba spin orbit coupling (SOC), recalling that $H^r$ is an operator we defined earlier to act on one specific site only:

$$H^r\phi^m_v = 4t\phi^m_v - t\phi^{m+1}_v - t\phi^{m-1}_v - t\phi^m_{v+1} - \phi^m_{v-1} - i\alpha\sigma_y\left[\frac{\phi^{m+1}_v - \phi^{m-1}_v}{2a}\right]$$
$$+ i\alpha\sigma_x\left[\frac{\phi^m_{v+1} - \phi^m_{v-1}}{2a}\right] \qquad [4.36]$$

which by rearrangement would lead to:

$$H^r\phi_v^m = 4t\phi_v^m - \left(t + \frac{i\alpha\sigma_y}{2a}\right)\phi_v^{m+1} - \left(t - \frac{i\alpha\sigma_y}{2a}\right)\phi_v^{m-1} - \left(t - \frac{i\alpha\sigma_x}{2a}\right)\phi_{v+1}^m - \left(t + \frac{i\alpha\sigma_x}{2a}\right)\phi_{v-1}^m. \quad [4.37]$$

Incorporating the effect of spin, the Hamiltonian can be written in vector space:

$$H^v = \sum_{m,v,\sigma} 4t\left|x_v^m,\sigma\right\rangle\left\langle x_v^m,\sigma\right| - \sum_{m,v,\sigma} t\left|x_v^m,\sigma\right\rangle\left\langle x_v^{m+1},\sigma\right| - \sum_{m,v,\sigma} t\left|x_v^m,\sigma\right\rangle\left\langle x_v^{m-1},\sigma\right| - \sum_{m,v,\sigma} t\left|x_v^m,\sigma\right\rangle\left\langle x_{v+1}^m,\sigma\right| - \sum_{m,v,\sigma} t\left|x_v^m,\sigma\right\rangle\left\langle x_{v-1}^m,\sigma\right| + H_{SOC}^v \quad [4.38]$$

where:

$$H_{soc}^v = -\frac{i\alpha}{2a}\sum_{m,v}\left(\left|x_v^m\right\rangle\left\langle x_v^{m+1}\right|\hat{\sigma}_y - \left|x_v^m\right\rangle\left\langle x_v^{m-1}\right|\hat{\sigma}_y\right) + \frac{i\alpha}{2a}\sum_{m,v}\left(\left|x_v^m\right\rangle\left\langle x_{v+1}^m\right|\hat{\sigma}_x - \left|x_v^m\right\rangle\left\langle x_{v-1}^m\right|\hat{\sigma}_x\right). \quad [4.39]$$

The above is the obvious result of discretizing the linear $k_x$ and $k_y$ in the SOC expression. Note that $\sigma_\mu$ is the Pauli matrix, but $\hat{\sigma}_\mu$ is not the usual form of Pauli matrix. It is the bra-ket operator form, e.g. $\hat{\sigma}_y = -i(|\uparrow\rangle\langle\downarrow| - |\downarrow\rangle\langle\uparrow|)$:

$$H_{soc}^v = -\frac{i\alpha}{2a}\sum_{m,v} -i\left(\left|x_v^m,\uparrow\right\rangle\left\langle x_v^{m+1},\downarrow\right| - \left|x_v^m,\uparrow\right\rangle\left\langle x_v^{m-1},\downarrow\right|\right) + i\left(\left|x_v^m,\downarrow\right\rangle\left\langle x_v^{m+1},\uparrow\right| - \left|x_v^m,\downarrow\right\rangle\left\langle x_v^{m-1},\uparrow\right|\right) + \frac{i\alpha}{2a}\sum_{m,v}\left(\left|x_v^m,\uparrow\right\rangle\left\langle x_{v+1}^m,\downarrow\right| - \left|x_v^m,\uparrow\right\rangle\left\langle x_{v-1}^m,\downarrow\right|\right) + \left(\left|x_v^m,\downarrow\right\rangle\left\langle x_{v+1}^m,\uparrow\right| - \left|x_v^m,\downarrow\right\rangle\left\langle x_{v-1}^m,\uparrow\right|\right). \quad [4.40]$$

In second quantization:

$$H_{SOC}^v = \sum_{k,k'r,s} \alpha\left\langle kr\left|k_x\sigma_y - k_y\sigma_x\right|k's\right\rangle c_{kr}^\dagger c_{k's} = -\frac{i\alpha}{2a}\sum_{m,r,s}\left\langle r\left|\sigma_y\right|s\right\rangle\left(c_{v,r}^{m\dagger}c_{v,s}^{m+1} - c_{v,r}^{m\dagger}c_{v,s}^{m-1}\right) + \frac{i\alpha}{2a}\sum_{m,r,s}\left\langle r\left|\sigma_x\right|s\right\rangle\left(c_{v,r}^{m\dagger}c_{v+1,s}^{m} - c_{v,r}^{m\dagger}c_{v-1,s}^{m}\right). \quad [4.41]$$

In Eq. 4.41, a similar discretization of the linear $k_x$ and $k_y$ has been performed. Equation 4.41 can also be written as:

$$H^{v}_{SOC} = -\frac{i\alpha}{2a}\left(\sum_{m,v,r,s} c^{m\dagger}_{v,r}c^{m+1}_{v,s} - c^{m\dagger}_{v,r}c^{m-1}_{v,s}\right)\langle r|\sigma_y|s\rangle$$

$$+\frac{i\alpha}{2a}\left(\sum_{m,v,r,s} c^{m\dagger}_{v,r}c^{m}_{v+1,s} - c^{m\dagger}_{v,r}c^{m}_{v-1,s}\right)\langle r|\sigma_x|s\rangle. \qquad [4.42]$$

Noting that $\langle\uparrow|\sigma_y|\uparrow\rangle = \langle\downarrow|\sigma_y|\downarrow\rangle = 0$, while $\langle\uparrow|\sigma_y|\downarrow\rangle = -i$ and $\langle\downarrow|\sigma_y|\uparrow\rangle = i$; $\langle\uparrow|\sigma_x|\uparrow\rangle = \langle\downarrow|\sigma_x|\downarrow\rangle = 0$ and $\langle\uparrow|\sigma_x|\downarrow\rangle = \langle\downarrow|\sigma_x|\uparrow\rangle = 1$, one could find that:

$$H^{v}_{SOC} = -\frac{i\alpha}{2a}\sum_{mv}\left(-i\left[c^{m\dagger}_{v,\uparrow}c^{m+1}_{v,\downarrow} - c^{m\dagger}_{v,\uparrow}c^{m-1}_{v,\downarrow}\right] + i\left[c^{m\dagger}_{v,\downarrow}c^{m+1}_{v,\uparrow} - c^{m\dagger}_{v,\downarrow}c^{m-1}_{v,\uparrow}\right]\right)$$

$$+\frac{i\alpha}{2a}\sum_{mv}\left(\left[c^{m\dagger}_{v,\uparrow}c^{m}_{v+1,\downarrow} - c^{m\dagger}_{v,\uparrow}c^{m}_{v-1,\downarrow}\right] + \left[c^{m\dagger}_{v,\downarrow}c^{m}_{v+1,\uparrow} - c^{m\dagger}_{v,\downarrow}c^{m}_{v-1,\uparrow}\right]\right). \qquad [4.43]$$

One can now compare Eq. 4.43 to Eq. 4.40. As a matter of taste, one can reshuffle Eq. 4.43 by applying the rule of $\sum_j c^{\dagger}_j c_{j+1} + c^{\dagger}_j c_{j-1} \equiv \sum_j c^{\dagger}_j c_{j+1} + c^{\dagger}_{j+1} c_j$. We thus note that in general:

$$\sum_{m,v,r,s} c^{m\dagger}_{v,r}c^{m+1}_{v,s} + c^{m\dagger}_{v,r}c^{m-1}_{v,s} \equiv \sum_{m,v,r,s} c^{m\dagger}_{v,r}c^{m+1}_{v,s} + c^{m+1\dagger}_{v,r}c^{m}_{v,s}$$

$$\equiv \sum_{m,v,r,s} c^{m\dagger}_{v,r}c^{m+1}_{v,s} + c^{m+1\dagger}_{v,s}c^{m}_{v,r} \equiv \sum_{m,v,r,s} c^{m\dagger}_{v,r}c^{m+1}_{v,s} + h.c. \qquad [4.44]$$

The Green's function is, by definition, $G^{mm'}_{vv',\sigma\sigma'} = \left\langle x^{m}_{v},\sigma\middle|(E - H^{v}_{SOC})^{-1}\middle|x^{m'}_{v'},\sigma'\right\rangle$. Using the SOC Hamiltonian above, but replacing the running indices of $m$ and $v$ with $a$ and $b$, respectively, one can derive the matrix equation for the Green's function. The reason for replacing the running indices is because we wish to reserve $m$ and $v$ as a particular fixed lattice point for the Green's function below. The general Green's function expression is:

$$(E - H^{f}_{SOC})G^{mm'}_{vv'\sigma\sigma'}$$

$$= G^{mm'}_{vv'\sigma\sigma'}E + \frac{\alpha}{2a}\left[\delta_{\sigma\uparrow}G^{m+1,m'}_{vv'\downarrow\sigma'} - \delta_{\sigma\uparrow}G^{m-1,m'}_{vv'\downarrow\sigma'} - \delta_{\sigma\downarrow}G^{m+1,m'}_{vv'\uparrow\sigma'} + \delta_{\sigma\downarrow}G^{m-1,m'}_{vv'\uparrow\sigma'}\right]$$

$$+\frac{i\alpha}{2a}\left[-\delta_{\sigma\uparrow}G^{m,m'}_{v+1,v'\downarrow\sigma'} + \delta_{\sigma\uparrow}G^{m,m'}_{v-1,v'\downarrow\sigma'} - \delta_{\sigma\downarrow}G^{m,m'}_{v+1,v'\uparrow\sigma'} + \delta_{\sigma\downarrow}G^{m,m'}_{v-1,v'\uparrow\sigma'}\right]. \qquad [4.45]$$

We present two specific cases of the above:

$$(E - H^{f}_{SOC})G^{mm}_{vv\uparrow\uparrow} = G^{mm}_{vv\uparrow\uparrow}E + \frac{\alpha}{2a}\left[G^{m+1,m}_{vv\downarrow\uparrow} - G^{m-1,m}_{vv\downarrow\uparrow}\right] + \frac{i\alpha}{2a}\left[G^{m,m}_{v-1,v\downarrow\uparrow} - G^{m,m}_{v+1,v\downarrow\uparrow}\right] \qquad [4.45a]$$

$$(E - H^{f}_{SOC})G^{mm}_{vv\downarrow\downarrow} = G^{mm}_{vv\downarrow\downarrow}E + \frac{\alpha}{2a}\left[G^{m-1,m}_{vv\uparrow\downarrow} - G^{m+1,m}_{vv\uparrow\downarrow}\right] + \frac{i\alpha}{2a}\left[G^{m,m}_{v+1,v\uparrow\downarrow} - G^{m,m}_{v-1,v\uparrow\downarrow}\right]. \qquad [4.45b]$$

To give an even clearer illustration, we focus on a one-dimensional system, with just two horizontal points (i.e. $m = 1,2$), where index $v$ is irrelevant, in the central region of the device. It can then be worked out that the matrix equation is:

$$\begin{pmatrix} E & 0 & 0 & \frac{\alpha}{2a} \\ 0 & E & \frac{-\alpha}{2a} & 0 \\ 0 & \frac{-\alpha}{2a} & E & 0 \\ \frac{\alpha}{2a} & 0 & 0 & E \end{pmatrix} \begin{pmatrix} G^{11}_{\uparrow\uparrow} & G^{12}_{\uparrow\uparrow} & G^{11}_{\uparrow\downarrow} & G^{12}_{\uparrow\downarrow} \\ G^{21}_{\uparrow\uparrow} & G^{22}_{\uparrow\uparrow} & G^{21}_{\uparrow\downarrow} & G^{22}_{\uparrow\downarrow} \\ G^{11}_{\downarrow\uparrow} & G^{12}_{\downarrow\uparrow} & G^{11}_{\downarrow\downarrow} & G^{12}_{\downarrow\downarrow} \\ G^{21}_{\downarrow\uparrow} & G^{22}_{\downarrow\uparrow} & G^{21}_{\downarrow\downarrow} & G^{22}_{\downarrow\downarrow} \end{pmatrix} = I. \qquad [4.46]$$

It is not hard to see that, in general, all indices in the Green's function matrix are related to the central region of the device only. This Green's function matrix can be found by applying the inversion technique of the Hamiltonian as described previously. Since all spin (color) related effects are present in the Hamiltonian, the spin Green's function below contains all information needed to describe spin propagation or spin current:

$$\left(\begin{array}{cc|cc:cc|cc}
G^{mm}_{vv,\uparrow\uparrow} & G^{mm}_{vv+1,\uparrow\uparrow} & G^{m,m+1}_{vv,\uparrow\uparrow} & G^{m,m+1}_{vv+1,\uparrow\uparrow} & G^{mm}_{vv,\uparrow\downarrow} & G^{mm}_{vv+1,\uparrow\downarrow} & G^{m,m+1}_{vv,\uparrow\downarrow} & G^{m,m+1}_{vv+1,\uparrow\downarrow} \\
G^{mm}_{v+1v,\uparrow\uparrow} & G^{mm}_{v+1,v+1,\uparrow\uparrow} & G^{m,m+1}_{v+1v,\uparrow\uparrow} & G^{m,m+1}_{v+1,v+1,\uparrow\uparrow} & G^{mm}_{v+1v,\uparrow\downarrow} & G^{mm}_{v+1,v+1,\uparrow\downarrow} & G^{m,m+1}_{v+1v,\uparrow\downarrow} & G^{m,m+1}_{v+1,v+1,\uparrow\downarrow} \\
\hline
G^{m+1,m}_{vv,\uparrow\uparrow} & G^{m+1,m}_{vv+1,\uparrow\uparrow} & G^{m+1,m+1}_{vv,\uparrow\uparrow} & G^{m+1,m+1}_{vv+1,\uparrow\uparrow} & G^{m+1,m}_{vv,\uparrow\downarrow} & G^{m+1,m}_{vv+1,\uparrow\downarrow} & G^{m+1,m+1}_{vv,\uparrow\downarrow} & G^{m+1,m+1}_{vv+1,\uparrow\downarrow} \\
G^{m+1,m}_{v+1v,\uparrow\uparrow} & G^{m+1,m}_{v+1,v+1,\uparrow\uparrow} & G^{m+1,m+1}_{v+1v,\uparrow\uparrow} & G^{m+1,m+1}_{v+1,v+1,\uparrow\uparrow} & G^{m+1,m}_{v+1v,\uparrow\downarrow} & G^{m+1,m}_{v+1,v+1,\uparrow\downarrow} & G^{m+1,m+1}_{v+1v,\uparrow\downarrow} & G^{m+1,m+1}_{v+1,v+1,\uparrow\downarrow} \\
\hdashline
G^{mm}_{vv,\downarrow\uparrow} & G^{mm}_{vv+1,\downarrow\uparrow} & G^{m,m+1}_{vv,\downarrow\uparrow} & G^{m,m+1}_{vv+1,\downarrow\uparrow} & G^{mm}_{vv,\downarrow\downarrow} & G^{mm}_{vv+1,\downarrow\downarrow} & G^{m,m+1}_{vv,\downarrow\downarrow} & G^{m,m+1}_{vv+1,\downarrow\downarrow} \\
G^{mm}_{v+1v,\downarrow\uparrow} & G^{mm}_{v+1,v+1,\downarrow\uparrow} & G^{m,m+1}_{v+1v,\downarrow\uparrow} & G^{m,m+1}_{v+1,v+1,\downarrow\uparrow} & G^{mm}_{v+1v,\downarrow\downarrow} & G^{mm}_{v+1,v+1,\downarrow\downarrow} & G^{m,m+1}_{v+1v,\downarrow\downarrow} & G^{m,m+1}_{v+1,v+1,\downarrow\downarrow} \\
\hline
G^{m+1,m}_{vv,\downarrow\uparrow} & G^{m+1,m}_{vv+1,\downarrow\uparrow} & G^{m+1,m+1}_{vv,\downarrow\uparrow} & G^{m+1,m+1}_{vv+1,\downarrow\uparrow} & G^{m+1,m}_{vv,\downarrow\downarrow} & G^{m+1,m}_{vv+1,\downarrow\downarrow} & G^{m+1,m+1}_{vv,\downarrow\downarrow} & G^{m+1,m+1}_{vv+1,\downarrow\downarrow} \\
G^{m+1,m}_{v+1v,\downarrow\uparrow} & G^{m+1,m}_{v+1,v+1,\downarrow\uparrow} & G^{m+1,m+1}_{v+1v,\downarrow\uparrow} & G^{m+1,m+1}_{v+1,v+1,\downarrow\uparrow} & G^{m+1,m}_{v+1v,\downarrow\downarrow} & G^{m+1,m}_{v+1,v+1,\downarrow\downarrow} & G^{m+1,m+1}_{v+1v,\downarrow\downarrow} & G^{m+1,m+1}_{v+1,v+1,\downarrow\downarrow}
\end{array}\right). \qquad [4.47]$$

The dashed lines divide the matrix Green's function in Eq. 4.47 into spin sub-spaces. The solid lines divide the spin sub-matrix into the sub-space of longitudinal dimension.

## 4.5 Non-equilibrium spin current

The general current expression describing the propagation of electron from the left part of the device to the right part is given by Caroli *et al.*[7] Electron current can be described in terms of the discrete parts of the device like the emitter ($e$), central region ($d$) and collector ($c$). The expression for a neutral current crossing over from the left lead (emitter) to the central region is:

$$I_{\mathcal{L}} = \frac{2e}{\hbar}\int\frac{dE}{2\pi}\sum_{e,d}^{\substack{e\in(emitter)\\ d\in(central)}}\left[t_{ed}G^{<}_{ed} - t_{de}G^{<}_{ed}\right] = \frac{2e}{\hbar A}\int\frac{dE}{2\pi}\sum_{e,d}^{\substack{e\in(emitter)\\ d\in(central)}}\left[-t_{de}(G^{<}_{ed})^{\dagger} + t_{de}G^{<}_{ed}\right]. \qquad [4.48]$$

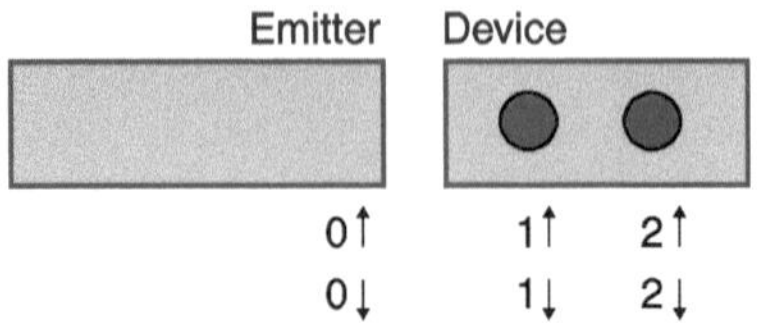


*4.9* One-dimensional schematic illustrating a nanostructure with discrete spatial site and spin states.

We show below that, in the presence of SOC or spin flipping effect, the current can still be described with Eq. 4.48 with modifications required for the parameters. Figure 4.9 shows a spatial distribution of the electrons.

Noting that $\sum^R_{dd'} = t_{de}\, g^R_{ee'}\, t_{e'd'}$ *or* $\Gamma^{ee'}_{dd'} = -i(t_{de}\, g^R_{ee'}\, t_{e'd'} - t_{de}\, g^A_{ee'}\, t_{e'd'})$, the $\Gamma$ matrix normally written for discrete spatial points only can now be written to reflect the spin, i.e. $(e, e', d, d') \rightarrow (e\sigma, e'\sigma', ds, d's')$ is followed in Eq. 4.49:

$$\Gamma^{ee'}_{dd'} = \begin{bmatrix} \Gamma^{ee'}_{11} & \Gamma^{ee'}_{12} \\ \Gamma^{ee'}_{21} & \Gamma^{ee'}_{22} \end{bmatrix} \Rightarrow \begin{bmatrix} \Gamma^{ee'}_{1\uparrow,1\uparrow} & \Gamma^{ee'}_{1\uparrow,2\uparrow} & \Gamma^{ee'}_{1\uparrow,1\downarrow} & \Gamma^{ee'}_{1\uparrow,2\downarrow} \\ \Gamma^{ee'}_{2\uparrow,1\uparrow} & \Gamma^{ee'}_{2\uparrow,2\uparrow} & \Gamma^{ee'}_{2\uparrow,1\downarrow} & \Gamma^{ee'}_{2\uparrow,2\downarrow} \\ \Gamma^{ee'}_{1\downarrow,1\uparrow} & \Gamma^{ee'}_{1\downarrow,2\uparrow} & \Gamma^{ee'}_{1\downarrow,1\downarrow} & \Gamma^{ee'}_{1\downarrow,2\downarrow} \\ \Gamma^{ee'}_{2\downarrow,1\uparrow} & \Gamma^{ee'}_{2\downarrow,2\uparrow} & \Gamma^{ee'}_{2\downarrow,1\downarrow} & \Gamma^{ee'}_{2\downarrow,2\downarrow} \end{bmatrix}. \qquad [4.49]$$

Restricting coupling constant to between the nearest neighbors only, all terms of $\Gamma^{ee'}_{dd'}$ with sites $d = 2$ to 4 would be zero, because site 2 of the central device has negligible coupling with any site on the emitter, or any point of $e$. The general expression for $\Gamma$ is:

$$\Gamma^{ee'}_{dd'} = \begin{bmatrix} \Gamma^{0\sigma,0\sigma'}_{1\uparrow,1\uparrow} & 0 & \Gamma^{0\sigma,0\sigma'}_{1\uparrow,1\downarrow} & 0 \\ 0 & 0 & 0 & 0 \\ \Gamma^{0\sigma,0\sigma'}_{1\downarrow,1\uparrow} & 0 & \Gamma^{0\sigma,0\sigma'}_{1\downarrow,1\downarrow} & 0 \\ 0 & 0 & 0 & 0 \end{bmatrix}. \qquad [4.50]$$

The general spin current expression is:

$$J_{\mathcal{L}} = \frac{2e}{\hbar A}\sum_k \int \frac{dE}{2\pi} Trace\left((\Sigma^R - \Sigma^A)\, \mathbf{G}^< - if^{\mathcal{L}}(\mathbf{G}^R - \mathbf{G}^A)\right)$$
$$= \frac{2e}{\hbar A}\sum_{k,e\sigma e'\sigma' ds\, d's'} \int -\frac{dE}{2\pi} Trace\, i\left(\Gamma^{e\sigma e'\sigma'}_{ds\, d's'}\right)\left(\mathbf{G}^< + f^{\mathcal{L}}(\mathbf{G}^R - \mathbf{G}^A)\right). \qquad [4.51]$$

The model above is suitable for computing spin current for systems with tunneling spin flip and spin flip within the dot. In these systems, the coupling constant $t$ will appear as various coefficients in the tunneling part of the Hamiltonian. While $t$ can be determined in a phenomenological or empirical manner, the lead Green's functions that need to be found are: $g^{0\uparrow,0\uparrow}, g^{0\uparrow,0\downarrow}, g^{0\downarrow,0\uparrow}, g^{0\downarrow,0\downarrow}$. Due to tight binding approximation, the device indices *d, d′* can only take on 1, which is the lattice site closest to the emitter and hence $e = 0$, $d = 1$.

One now considers the situation where there will not be spin flip on the electrode, i.e. $g^{0\uparrow0\downarrow} = g^{0\downarrow0\uparrow} = 0$. Thus, when lead interaction is ignored, $\sigma = \sigma'$, and the following is obtained for spin current:

$$J_{\mathcal{L}} = \frac{2e}{\hbar A}\sum_k \int -\frac{dE}{2\pi} Trace\ i\sum_\sigma \begin{bmatrix} \Gamma^{0\sigma,0\sigma}_{1\uparrow,1\uparrow} & 0 & \Gamma^{0\sigma,0\sigma}_{1\uparrow,1\downarrow} & 0 \\ 0 & 0 & 0 & 0 \\ \Gamma^{0\sigma,0\sigma}_{1\downarrow,1\uparrow} & 0 & \Gamma^{0\sigma,0\sigma}_{1\downarrow,1\downarrow} & 0 \\ 0 & 0 & 0 & 0 \end{bmatrix} \left(G^< + f^{\mathcal{L}}(G^R - G^A)\right). \quad [4.52]$$

In the event of no spin flip within the central device, $G^{\uparrow\downarrow} = G^{\downarrow\uparrow} = 0$, and $s = s'$ while $G^{\uparrow\uparrow}$ and $G^{\downarrow\downarrow}$ are finite. In the event of no tunneling spin flip, $\sigma = s$. The simplest spin current expression is thus:

$$J_{\mathcal{L}} = \frac{2e}{\hbar A}\sum_k \int -\frac{dE}{2\pi} Trace\ i\sum_\sigma \begin{bmatrix} \Gamma^{0\sigma,0\sigma}_{1\uparrow,1\uparrow} & 0 & 0 & 0 \\ 0 & 0 & 0 & 0 \\ 0 & 0 & \Gamma^{0\sigma,0\sigma}_{1\downarrow,1\downarrow} & 0 \\ 0 & 0 & 0 & 0 \end{bmatrix} \left(G^< + f^{\mathcal{L}}(G^R - G^A)\right). \quad [4.53]$$

Table 4.1 summarizes the necessary modifications one needs to apply to the spin current expressions under the different circumstances of interaction in the leads, across the leads, or within the device channel.

Table 4.1 Summary of spin current expressions modified by spin interaction in electrodes or central device

| Spin interaction | Spin current expression |
|---|---|
| Current in general form (considering the spin only) | $J = \sum_{\substack{\sigma\sigma'\sigma''\sigma''' \\ ss's''s'''}} \Gamma^{\sigma\sigma'}_{ss'} G^R_{s's''} \Gamma^{\sigma''\sigma'''}_{s''s'''} G^A_{s'''s}$ |
| In the absence of spin interaction in electrode: $\sigma = \sigma'$, $\sigma'' = \sigma'''$ | $J = \sum_{\substack{\sigma \\ ss's''s'''}} \Gamma^{\sigma\sigma}_{ss'} G^R_{s's''} \Gamma^{\sigma\sigma}_{s''s'''} G^A_{s'''s}$ |
| Assuming symmetrical electrode coupling to central device region: $\sigma = \sigma' = \sigma'' = \sigma'''$ | |
| In the absence of spin interaction in the central device: $G^R_{d'd''} = G^R_{d'}\delta_{d'd''}$ | $J = \sum_{\substack{\sigma \\ ss'}} \Gamma^{\sigma\sigma}_{ss'} G^R_{s's'} \Gamma^{\sigma\sigma}_{s's} G^A_{ss}$ |
| In the absence of tunneling spin flip from electrode to central device: $\Gamma^{\sigma\sigma}_{ss'} = \Gamma^{\sigma\sigma}_{ss}\delta_{\sigma s}$ | Simple scalar function: $J = \sum_{\sigma s} \Gamma^{\sigma\sigma}_{ss}\delta_{\sigma s} G^R_{ss} \Gamma^{\sigma\sigma}_{ss}\delta_{\sigma s} G^A_{ss}$ |

Note that $\Gamma^{ee'}_{dd'} = -i\left(t_{de}g^R_{ee'}t_{e'd'} - t_{de}g^A_{ee'}t_{e'd'}\right)$.

## 4.6 References

[1] Kumar S B, Jalil M B A, Tan S G and Liang G-C (2010), 'The effect of magnetic field and disorders on the electronic transport in graphene nanoribbons', *J. Phys.: Condens. Matter*, 22, 375303; Kumar S B, Jalil M B A, Tan S G and Liang G-C (2010), 'Magnetoresistive effect in graphene nanoribbon due to magnetic field induced band gap modulation', *J. Appl. Phys.*, 108, 033709.

[2] Klitzing K V, Dorda G and Pepper M (1980), 'New method for high-accuracy determination of the fine-structure constant based on quantized-Hall resistance', *Phys. Rev. Lett.* 45, 494.

[3] Zhu Z G, Su G, Jin B and Zheng Q R (2003), 'Spin-flip scattering effect on current-induced spin torque in ferromagnet-insulator-ferromagnet tunnel junctions', *Phys. Lett. A*, 306, 249; Chen X, Zheng Q R and Su G (2008), 'Spin transfer and critical current for magnetization reversal in ferromagnet-ferromagnet-ferromagnet double-barrier tunnel junctions', *Phys. Rev. B*, 78, 104410.

[4] Kalitsov A, Chshiev M, Theodonis I, Kioussis N and Butler W H (2009), 'Spin-transfer torque in magnetic tunnel junctions', *Phys. Rev. B*, 79, 174416.

[5] Pareek T P and Bruno P (2002), 'Spin coherence in a two-dimensional electron gas with Rashba spin-orbit interaction', *Phys. Rev. B*, 65, 241305(R).

[6] Xing M J, Jalil M B A, Tan S G and Jiang Y (2011), 'Spin torque switching in triple-quantum dot device with ferromagnetic contacts for memory applications', *J. Appl. Phys.*, 109, 07C702.

[7] Caroli C, Combescot R, Nozieres P and Saint-James D (1971), 'Direct calculation of the tunneling current', *J. Phys. C: Solid State Phys.*, 4, 916.

# 5
# Spintronics and spin Hall effects in nanoelectronics

**Abstract:** This chapter introduces spintronics as realized in metal and semiconductor materials. Particularly in semiconductors, the physics of spin orbit coupling and magnetic fields on the motion of electron and spin are discussed. The concept of gauge is introduced in the form of the spin orbit gauge and its relation to spin dynamics is studied heuristically. Spin Hall effects in the Rashba system are then discussed in connection with the gauge concept. Finally, the spin valve device is introduced alongside spin drift diffusion theory, which has been widely used to model spin transport in metal-based devices.

**Key words:** spin orbit coupling, gauge, Rashba, spin Hall effects, spin drift diffusion.

## 5.1 Introduction to spintronics

Spintronics[1–3] refers to a new class of electronics that exploits the spin together with the charge property of electrons for technological applications. Indeed the term 'spintronics' was first used in the 1990s to describe the spate of research activities focusing on manipulating or utilizing the electron's spin degree of freedom (DOF) to realize electronic functions, e.g. transistor, logic and memory. The materials that could be used for realizing spintronic functions include metal, carbon, semiconductors or rare earth. Spintronics is closely related to magnetism, which studies the collective spin behavior of the bound electrons. But in spintronics, emphasis is placed on the itinerant or moving electron whose spin coherence or collective transportation is controlled to generate a measurable conductance or current modulation. Thus, in general, what is minimally required of a spintronic device would be an apparatus to generate a collection of spin-polarized electrons, an effective channel to transport these 'spin' electrons to a separate apparatus which then measures the 'spin' in the form of resistance change. In nanoelectronics, this apparatus would have to be integrated into a single device, which is transistor like and nanoscale in size. The bulk of spintronic research, particularly metal-based research, has focused on spin asymmetric electron transport in magnetoresistive devices. However, semiconductor-based spintronics relies on spin orbit induced spin current in a paramagnetic semiconductor. Similar studies on magnetoresistance and spin orbit coupling have also been carried out in carbon nanotube and graphene based devices. In rare earth material where the Rashba effect (normally found in semiconductor two-dimensional electron gas) is strong, rare earth based spintronics may become a promising new field.

Besides spin current, transportation and measurement, of equal significance are studies of spin relaxation and semiconductor magnetism. Spin relaxation studies are an important subject in semiconductor spintronics, as a long spin coherent length is crucial for the successful operation of a spintronic transistor. In magnetic-based spintronics, studies have focused on a class of materials known as diluted magnetic semiconductors (DMS) where emphasis is placed upon inducing real magnetism by means of doping semiconductor with magnetic materials.

The spin Hall effect (SHE) refers to a special spin orbit coupling means which generates spin current in the transverse direction to the applied electric field. The SHE conductivity can be topological in nature and bears similarities to the quantum Hall effects. In fact in materials known as the topological insulators, spin orbit coupling is very strong, and quantum SHE has been shown to exist.

Spin torque refers to the transfer of spin angular momentum to the magnetic moment causing magnetization to be switched by the spin current instead of the magnetic field. This is an important subject in metal-based spintronics where the focus is on designing a class of magnetic memory known as the spin transfer switching (also known as spin transfer torque, STT) magnetic random access memory (MRAM). In the new MRAM, the application of spin transfer switching instead of a magnetic field allows memory cells to be packed closer in a magnetic field free system. This is an engineering feat that allows future non-volatile computer memory density to be scaled to the level which can meet consumer demand.

We have introduced a host of physical phenomena particularly relevant to the technology of spintronics. In this chapter, we focus on metal spintronics, semiconductor spintronics and SHE. We do not discuss DMS and spin relaxation. In Chapter 7, spin Hall and spin torque are discussed again using the more formal physics of quantum gauge theory.

## 5.1.1 Metal-based spintronics

It is worth noting that in the early years, research in spintronics was focused on metallic[4–6] spin-based devices. In fact in the late 1980s and early 1990s, when semiconductor spintronics was still largely at the conceptual stage, research into ferromagnetic (FM) metal devices based on the asymmetric scattering of spin current had already met with great experimental success. This spin-dependent transport and the devices (based on magnetic metal multilayers) are now generally known as metal spintronics, although their experimental success predates even the coining of 'spintronics'. Metal spintronics comprise mainly giant magnetoresistance (GMR), spin transfer torque and spin valve devices, much of which we introduced in Chapter 2. In this chapter we focus on the propagation of electron and spin flux in these devices. Many theoretical and simulation works have been carried out to investigate the GMR and spin transfer effects in both the current perpendicular-to-plane (CPP) and current-in-plane (CIP) (see Chapter 2) configurations. We introduce the phenomenological spin drift diffusion (SDD) equations, which have been widely used to model spin transport in spin valve and spin torque devices. An SDD-based model describes experimental observations very accurately and is the language understood by both theorists and experimentalists. Besides the Boltzmann phenomenological methods in spin electronics, the non-equilibrium Green's function (NEGF) method is also becoming popular and is increasingly being used to compute spin transport in tunneling magnetoresistance (TMR) or magnetic tunnel junction (MTJ) devices. Recent developments of spin Hall in platinum and Rashba effects in metal and rare earth materials may have great application potential, but these topics are not discussed in this book as much research is still being carried out at the time of writing.

## 5.1.2 Semiconductor spintronics

In semiconductor spintronics, the main idea is to induce magnetic-like behavior in non-magnetic semiconductors. It is an established fact that the versatility of semiconductors, especially in terms of their transport properties, has been a key factor in the astounding success of the metal-oxide-field-effect-transistor (MOSFET) technology since its invention. It is thus conceivable that semiconductor spintronics can lead to a new generation of spin devices based on the well-established MOSFET technologies. The most natural setting for implementing semiconductor-based spintronics is a transistor heterostructure consisting of a two-dimensional electron gas (2DEG) channel. A typical semiconductor 2DEG has a long mean free path (for ballistic transport) and large spin orbit coupling (SOC) effect, both of which are crucial requisites for the coherent transport and manipulation of spin current. Semiconductors are also much more versatile materials compared to FM metal elements and alloys. The physical properties of a semiconductor can be varied rather conveniently by changing its carrier concentration via either electrical or optical means. For instance, the magnetic properties (such as coercive field and Curie temperature) of a DMS[7,8] can be controlled either electrically or optically. Indeed the promise of semiconductor spintronics depends greatly on the important fact that spin conductivity can be varied over a wide range via numerous techniques, e.g. by doping, application of an external electric field or photo-excitation.

From the device perspective, it is essential not only to generate or inject spin-polarized current into the semiconductors, but also to manipulate the transport of electron spin electrically, such that some kind of 'spin transistor' function can be realized. This is certainly helped by the fact that the electron spin is relatively robust with respect to coherence in semiconductors. Kikkawa and Awschalom[3] have demonstrated that the spin relaxation time can be as long as 100 ns in bulk nGaAs with a doping density of $10^{16}$ $cm^{-3}$, translating to a macroscopic spin coherence length of the order of 100 μm. These remarkable observations lend further credence to the belief in the possibility of future spin transistor devices. The spin transistor or spin field-effect transistor (spinFET) will be a key element upon which other more complicated devices (e.g. logic gates, memories, etc.) can be built; the collection of these devices may thus constitute a whole new class of spin-based MOSFET electronic devices. Hence, it is not an understatement to say that semiconductor spintronics constitutes the next stage in the development of spintronic technology. Its main advantage over ferromagnetic metal-based spintronics is its integrability with the extensive semiconductor fabrication technology. That it is feasible to couple the spin and charge conductance property to light also opens up the possibility of the future convergence of electronics, photonics and spintronics in semiconductors.

## 5.2 Semiconductor spin transport

Semiconductor spintronic devices are generally based upon the concepts of generation, transportation and detection of spin current. This is the motivation behind the extensive review in this chapter on the various physical phenomena we can exploit to achieve the above. Table 5.1 summarizes the three important aspects of semiconductor spintronic devices, namely:

1. Generation of spin current.
2. Transportation and manipulation of spin current.
3. Detection and measurement of spin current.

The table contains details of the numerous well-established techniques to realize these three functions which are required for the proper working of spintronic devices. More details or variants of these techniques, as well as lesser known specialized methods, can be found in literature published in abundance in various journals.

Exchange coupling[9,10] arises due to the interaction of the s-electron spin with the collective presence of the d-electron moment in ferromagnetic materials. The exchange effect is a result of many-body physics; in spintronics, it suffices to understand that the exchange effect is a form of effective magnetic field felt by a single-particle spin moment. When the electron spin is aligned with this effective field, its energy is lower and vice versa is true. Thus in this context, electron transport through a ferromagnetic material can be treated like an electron passing through a magnetic field. On the other hand, the exchange energy is also related to the asymmetric density of states (DOS) at the Fermi energy. Thus, in the regime of drift-diffusive transport, spin current can be resulted from the more dominant effect of spin asymmetric scattering due to asymmetric DOS. In fact, most metal-based spintronic devices (e.g. spin valves, magnetic tunnel junctions) rely on the asymmetric scattering effect to generate spin current which is normally detected via resistance change. Electron spin interacts with actual magnetic fields producing Zeeman energy. It is thus natural to conceive that applying external magnetic field to a transistor device could generate spin current. In fact it has been shown in principle that external magnetic and electric fields induce spin polarization in various types of devices via wavevector filtering.[11–13] There has been a large body of work investigating the various possibilities of spin filtering using the external

*Table 5.1* Various techniques for the electrical generation of spin current, its transportation, detection and measurement

| Generation of spin current | Transportation and manipulation of spin current | Detection of spin current |
|---|---|---|
| **Exchange effects**<br>(1) Spin-dependent scattering in FM bulk/film<br>(2) Effective Zeeman effect of exchange field<br>**External magnetic and electric fields**<br>(1) Zeeman effect<br>(2) Landau effect<br>**Spin orbit coupling**<br>(1) Spin-dependent tunneling<br>(2) Spin Hall current<br>(3) Persistent spin helix<br>**Ferromagnetic resonance**<br>Precessing moment → spin current<br>**Spin seebeck**<br>Temperature gradient → spin current<br>**Optical methods**<br>Light polarization → spin polarization | **Spin injection**<br>(1) Spin dependent tunneling<br>(2) Spin drift diffusion<br>**Electrical gate pumping**<br>(1) Multiple gate channel<br>(2) Single-electronic quantum dots<br>**Non-local diffusion**<br>Induced flow of spin current in a non-local manner | **Spin valve**<br>Spin current → resistance change<br>**Optical detection**<br>Spin current → optical polarization<br>**Inverse spin Hall**<br>Spin current → charge current<br>**Spin torque oscillations**<br>Spin current → precessing moment |

Note: This table should be read vertically, but not across the columns. The symbol → means 'translates to' or 'results in'.

fields.[14–17] While these are good indications of the possible emergence of future spintronic devices, difficulties remain with many practical issues.[18–20] The need for a spatially concentrated magnetic field to approximate the delta function distribution fields in the channel of these devices is itself a formidable practical task.

On the other hand, there have been a slew of device propositions utilizing spin orbit coupling, foremost of which is the Datta–Das[21] transistor that utilizes Rashba[22–27] SOC effects to induce spin precession across the 2DEG conduction channel. In fact SOC is the only technique that generates spin current without the need for any form of magnetism, e.g. magnetic fields or magnetic moment. It is a purely electrical means of generating spin current. At a more fundamental level, it is a fascinating physical effect which has its root in relativistic quantum mechanics. Recently, it has even been perceived as related to a profound form of magnetic potential also known as the non-Abelian gauge potential, which explains SHE as well as longitudinal spin current from the perspectives of symmetry and topology.

More discussions of SOC are found in the later sections of this chapter. Other indirect ways to generate spin current includes the ferromagnetic resonance and the spin Seebeck, which generate spin current via a precessing moment and temperature gradient, respectively. Optical generation of spin current is also well-established. These techniques are mainly employed to generate an imbalance spin population for the studies of spin dynamics and spin relaxation.

Transporting spin current through a device channel and detecting its presence are crucial for the complete functioning of spintronic devices. Spin injection[28–31] refers to the coherent transfer of spin-polarized current from a normally ferromagnetic contact to the device channel. It is an engineering process that can be achieved via the physics of spin-dependent tunneling and spin drift diffusion. This process is, however, fraught with difficulty, ranging from resistance loading due to the high resistance semiconductor, to interfacial spin flip due to magnetic impurities or SOC. Besides spin injection, various techniques of electrical gate pumping, which rely on time-coordinated lowering and raising of electrical barriers across the electron conduction path to effect a net flow of electron or spin flux, have been studied.[32] Further understanding of these subjects requires the knowledge of single-electronics[33] and quantum dot spintronics. Non-local diffusion[34] is an interesting method with which to generate spin accumulation without needing to apply electrostatic potential along the spin conduction path. The electrical path for spin and charge flow is first established via applying an electrical potential along it. Pure spin current can be tapped from a pure spin flow branching off the electrical path.

The detection and measurement of spin current has drawn particularly high experimental interest due to its generic applicability to magnetism, optics, metal and semiconductor spintronics. This large body of research, which spans various disciplines, is not covered in this article beyond that presented in Table 5.1. But a general tenet remains true, i.e. spin detection can be viewed as the inverse effect of spin generation. For example, ferromagnetic resonance[35] has been employed to generate spin current via precessing magnetization. Thus when the opposite is performed, this technique can also be used to detect spin current.

As stated earlier, of particular interest in this chapter is the effect of SOC which is the underlying physics for many aspects of spintronics with device prospects, e.g. spin current generation, SHE, spin torque, spin oscillations and so forth. While the Zeeman and exchange effects require external fields and local moments, respectively, SOC enables purely electrical generation and manipulation of spin current. The following sections discuss in some detail the origin and the physics of SOC with respect to generating longitudinal as well as transverse spin current.

This chapter is intended to provide a brief review of the various aspects of spintronics with special emphasis on the underlying physics of SOC. Explanation of these effects is substantiated with basic quantum mechanics accessible to both experimentalists and theorists. The equations used here are meant to enhance the clarity of concepts discussed, not to amplify their abstractness. Elaborate explanations and useful mathematical identities are compiled separately in tables so that readers can focus on the physical pictures, reading the text with minimal distractions. The different sections are written in a self-contained manner such that they can be read and understood independently. For example, readers who are not familiar with relativistic quantum mechanics, which forms the basis of the vacuum SOC effect, can proceed to the later sections without much problem. The elucidation of the monopole-like magnetic field may, however, require a few other abstract concepts. But this should pose little hindrance to understanding its relevance to SHE, so long as readers are prepared to accept the existence of these effective monopole fields.

## 5.3 Spin orbit coupling (SOC) and Zeeman effects

Since SOC constitutes a major subject of this review, it makes sense to provide a more elaborate description of its physics and origin. We present a brief description of vacuum SOC in the context of relativistic quantum mechanics. Simply said, it is an effect in which an electric field appears to be an effective magnetic field when viewed in the rest frame of a moving electron. Dirac linearization of Einstein's energy equation to a minimally coupled momentum equation provides a natural means of deriving the SOC as well as the Zeeman terms. We start from the linear Dirac equation of:

$$\left(i\frac{\hbar}{c}\gamma^0\partial_0 - i\hbar\gamma^\mu\partial_\mu - mc\right)\psi = 0. \qquad [5.1]$$

For notational convenience, the $\hbar$ and $c$ accompanying the momenta are set to 1 and so the single-particle description of massive particle in the presence of electromagnetic field, $A_\mu$, is written as follows:

$$\left(i\gamma^\mu D_\mu - mc\right)\psi = 0 \qquad [5.2]$$

where $D_\mu = (\partial_\mu + ieA_\mu)$ and $\{\gamma^\mu, \gamma^\nu\} = 2g^{\mu\nu}$ obeys the commutative rule of Dirac algebra, $m$ is the rest mass, $\gamma^0$ and $\alpha^i$ are given in standard representation of $\gamma^0 = \begin{pmatrix} 1 & 0 \\ 0 & -1 \end{pmatrix}$, and $\alpha^i = \begin{pmatrix} 0 & \sigma^i \\ \sigma^i & 0 \end{pmatrix}$ and $\sigma^i$ are the 2×2 Pauli matrices. Pre-multiplying Eq. 5.2 with $(i\gamma^\nu D_\nu + mc)$, one obtains the second-order equation:

$$\left(\gamma^\nu\gamma^\mu D_\nu D_\mu + m^2c^2\right)\psi = 0. \qquad [5.3]$$

Since:

$$\gamma^\nu\gamma^\mu D_\nu D_\mu = \frac{1}{2}\gamma^\nu\gamma^\mu D_\nu D_\mu + \frac{1}{2}\gamma^\mu\gamma^\nu D_\mu D_\nu - \frac{1}{2}\gamma^\mu\gamma^\nu D_\nu D_\mu$$
$$+\frac{1}{2}\gamma^\mu\gamma^\nu D_\nu D_\mu = g^{\nu\mu} D_\nu D_\mu + \frac{1}{2}\gamma^\mu\gamma^\nu [D_\mu, D_\nu], \qquad [5.4]$$

one thus has $\left(g^{\nu\mu}D_\nu D_\mu + \frac{1}{2}\gamma^\mu\gamma^\nu[D_\mu, D_\nu] + m^2c^2\right)\psi = 0$. Note that $[D_\mu, D_\nu] = ieF_{\mu\nu}$, where $F_{\mu\nu} = -F_{\nu\mu}$, $\left(\frac{i\gamma^\mu\gamma^\nu}{2}\right)F_{\mu\nu} = i(\gamma^\mu\gamma^\nu - \gamma^\nu\gamma^\mu)F_{\mu\nu}$ and $F_{\mu\nu} = \partial_\mu A_\nu - \partial_\nu A_\mu$ is the electromagnetic field tensor. Equation 5.3 can now be expressed as:

$$\left(D^\mu D_\mu + \frac{e}{2}\Sigma^{\mu\nu}F_{\mu\nu} + m^2c^2\right)\psi = 0 \qquad [5.5]$$

where $\Sigma^{\mu\nu} = \frac{i}{2}[\gamma^\mu, \gamma^\nu]$. The explicit tensor form of $F_{\mu\nu}$ is given by:

$$F_{\mu\nu} = \begin{pmatrix} 0 & \frac{+E^1}{c} & \frac{+E^2}{c} & \frac{+E^3}{c} \\ \frac{-E^1}{c} & 0 & -B^3 & +B^2 \\ \frac{-E^2}{c} & +B^3 & 0 & -B^1 \\ \frac{-E^3}{c} & -B^2 & +B^1 & 0 \end{pmatrix}; \; \Sigma^{\mu\nu} = \begin{pmatrix} 0 & i\alpha^1 & i\alpha^2 & i\alpha^3 \\ -i\alpha^1 & 0 & \sigma^3 & -\sigma^2 \\ -i\alpha^2 & -\sigma^3 & 0 & \sigma^1 \\ -i\alpha^3 & \sigma^2 & -\sigma^1 & 0 \end{pmatrix}. \qquad [5.6]$$

Readers could verify that the matrix components can be summarized in Table 5.2 using compact notations for ease of referencing.

Table 5.2 is just the compact expression of $\frac{1}{2}\sigma^{\mu\nu}F_{\mu\nu} = -\boldsymbol{\sigma}.\boldsymbol{B} - i\boldsymbol{\alpha}\cdot\boldsymbol{E}$, which leads to the second-order equation of:

$$\left([\varepsilon - e\phi]^2 - [p_\mu - eA_\mu]^2 + e\boldsymbol{\sigma}\cdot\boldsymbol{B} + ie\boldsymbol{\alpha}\cdot\boldsymbol{E} - m^2\right)\psi = 0. \qquad [5.7]$$

Recall that we had earlier set $\hbar$, $c \to 1$. For correct dimensions, $D_0$ has to be accompanied by a $\hbar$, $D_\mu$ by a $c\hbar$, and $mc$ implies $mc^2$. One can then obtain:

$$\begin{pmatrix} [\varepsilon - e\phi]^2 - \pi_\mu^2 + e\boldsymbol{\sigma}\cdot\boldsymbol{B} - m^2 & ie\boldsymbol{\alpha}\cdot\boldsymbol{E} \\ ie\boldsymbol{\alpha}\cdot\boldsymbol{E} & [\varepsilon - e\phi]^2 - \pi_\mu^2 + e\boldsymbol{\sigma}\cdot\boldsymbol{B} - m^2 \end{pmatrix}\cdot\begin{pmatrix}\chi \\ \theta\end{pmatrix} = 0 \qquad [5.8]$$

where $\pi_\mu = p_\mu - eA_\mu$. Letting $(\psi_1\,\psi_2\,\psi_3\,\psi_4)^T = \begin{pmatrix}\chi \\ \theta\end{pmatrix}$, and using Eq. 5.1, one obtains:

$$\begin{pmatrix} (\varepsilon - e\phi) - m & -\boldsymbol{\sigma}\cdot\boldsymbol{\pi} \\ \boldsymbol{\sigma}\cdot\boldsymbol{\pi} & -(\varepsilon - e\phi) - m \end{pmatrix}\cdot\begin{pmatrix}\chi \\ \theta\end{pmatrix} = 0. \qquad [5.9]$$

From the above, one has:

$$\theta = \left[\frac{\boldsymbol{\sigma}\cdot\boldsymbol{\pi}}{(\varepsilon - e\phi + m)}\right]\chi. \qquad [5.10]$$

*Table 5.2* Matrix components of $F_{\mu\nu}$ and $\Sigma^{\mu\nu}$ in compact notations

| | Matrix components |
|---|---|
| 1. | $F_{k0} = E^k$ |
| 2. | $F_{0k} = -E^k$ |
| 3. | $F_{ij} = -\varepsilon_{ijk}B^k$ |
| 4. | $\sigma^{0k} = i\alpha^k$ |
| 5. | $\sigma^{k0} = -i\alpha^k$ |
| 6. | $\sigma^{ij} = \varepsilon_{ijk}\sigma^k$ |

In the non-relativistic limit, the norm of $\theta$ approaches zero, $\theta^{\dagger}\theta = \left(\frac{\pi^2}{(mc)^2}\right)\chi \approx \frac{v^2}{c^2}\chi \approx 0$ assuming small values of the vector potential. Careful substitution, however, leads to:

$$\left[[\varepsilon - e\phi]^2 - \pi_{\mu}^2 + e\boldsymbol{\sigma}\cdot\boldsymbol{B} - m^2 + \left(\frac{(ie\boldsymbol{\sigma}\cdot\boldsymbol{E})(\boldsymbol{\sigma}\cdot\boldsymbol{\pi})}{\varepsilon - e\phi + m}\right)\right]\chi = 0. \qquad [5.11]$$

Taking the realistic approximation of $\varepsilon = \varepsilon + mc^2$, in the non-relativistic limit where $\varepsilon$ and $e\phi$ are very small, and using the following identities based on the interchangeable mix of vector and tensor products as shown in Table 5.3, one now has:

$$\left[\frac{\left(p - \frac{eA}{c}\right)^2}{2m} + e\phi - \frac{e\hbar}{2mc}\boldsymbol{\sigma}\cdot\boldsymbol{B} + \frac{e\boldsymbol{\sigma}\cdot(\boldsymbol{E}\times\boldsymbol{\pi})}{4m^2c} - \frac{ie\boldsymbol{E}\cdot\boldsymbol{\pi}}{4m^2c}\right]\chi = \varepsilon\chi. \qquad [5.12]$$

In Eq. 5.12, the first term represents the minimal-coupled kinetic momentum, the third term represents the Zeeman effects, the fourth represents the SOC and the fifth represents the non-hermitian term that causes negative energy. It is worth noting that keeping the exact non-zero term of $\theta$ in the eigenspinor, the resulting $\chi$ manifest both the Zeeman effect and SOC in the low energy limit. The above analysis from Eq. 5.1 to Eq. 5.12 is about the vacuum SOC, which has a small coupling constant. In order for this effect to be strong enough without requiring a huge electric field, a large effective coupling constant is required.

*Table 5.3* Useful vector and tensor identities that simplify the derivation of SOC effects

| | Simple vector and tensor identities and relations |
|---|---|
| 1. | $\sigma_i\sigma_j = i\varepsilon_{ijk}\sigma_k + \delta_{ij}I$ |
| 2. | $(\boldsymbol{\sigma}\cdot\boldsymbol{E})(\boldsymbol{\sigma}\cdot\boldsymbol{\pi}) = \sigma_i\sigma_j E_i\pi_j$ |
| 3. | $(\boldsymbol{\sigma}\cdot\mathrm{E})(\boldsymbol{\sigma}\cdot\boldsymbol{\pi}) = i\sigma_k(\varepsilon_{ijk}E_i\pi_j) + E_i\pi_j\delta_{ij} = i\boldsymbol{\sigma}\cdot(\boldsymbol{E}\times\boldsymbol{\pi}) + \boldsymbol{E}\cdot\boldsymbol{\pi}$ |
| 4. | $\sigma_i$ and $E_j$ commute |

In the case of the Rashba SOC, such amplification of the coupling constant is found in a 2DEG semiconductor system with inversion asymmetry,[22–27] which gives rise to bandstructure that can be associated with a large coupling constant. In the case of the Dresselhaus SOC,[36] the effect is enhanced in semiconductors due to bulk inversion asymmetry (BIA). In a semiconductor 2DEG, due to the so-called structural inversion asymmetry (SIA) and the electric field penetrating the 2DEG plane vertically, the Rashba SOC can be described by:

$$H_R = \alpha\left(\sigma_y k_x - \sigma_x k_y\right). \quad [5.13]$$

The eigenstates can be found by explicitly expanding Eq. 5.13. We give in Eq. 5.14 the eigenstate and the eigenenergy corresponding to the ± Rashba SOC bands:

$$\left|\pm, k\right\rangle = \frac{1}{\sqrt{2}}\begin{pmatrix} \mp i e^{-i\phi} \\ 1 \end{pmatrix}; \; E_\pm = \frac{\hbar^2 k^2}{2m} \pm \alpha k, \quad [5.14]$$

where $k = \sqrt{k_x^2 + k_y^2}$. Using the relationship in Eq. 5.14, one can obtain the wavevector difference $\Delta k$ between the two Rashba bands at energy $E$:

$$\frac{\hbar^2 k_+^2}{2m} - \frac{\hbar^2 k_-^2}{2m} = -\alpha(k_+ + k_-) \;\Rightarrow \Delta k = (k_+ - k_-) = \frac{-2m\alpha}{\hbar^2}. \quad [5.15]$$

It is worth noting that, in spintronics, wavevector differences at a fixed energy have been commonly exploited to study the differential transmission between two spin eigenstates of a spintronic device. Figure 5.1 shows the band diagram of the Rashba SOC system. This diagram is important for device design and engineering, e.g. one could determine which portions of the double cone are to contribute to device current or conductance.

Besides Rashba SOC, there are a few other types of SOC affecting the electrons in the conduction band which can be utilized in spintronics, e.g. the Dresselhaus

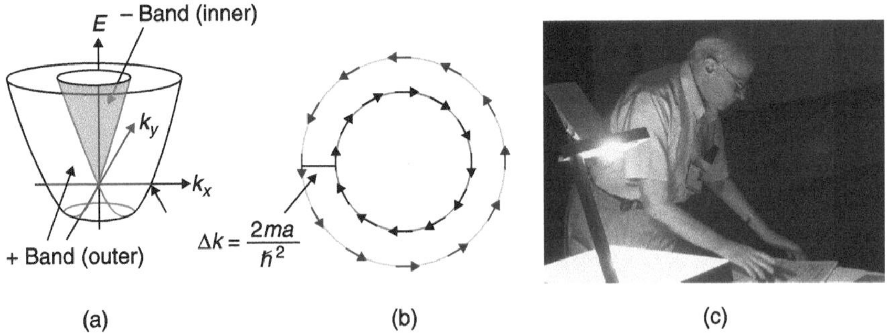


*5.1* (a) Dispersion relation for the two Rashba bands. (b) Cross-section of the Rashba bands consists of two concentric circles. The arrows depict the spin orientations of the spin eigenstates of the two Rashba bands, which point tangentially along the band surfaces. (c) Prof. E. I. Rashba, 2004.

*Table 5.4* Dresselhaus SOC constants for some common semiconductor materials

| | *GaSb* | *InAs* | *GaAs* | *InP* | *InSb* |
|---|---|---|---|---|---|
| ***β^C* (eV A^3)** | 187 | 130 | 24 | 8 | 220 |
| ***m^r = m*/m^0*** | 0.041 | 0.023 | 0.067 | 0.081 | 0.013 |
| ***β^C × m^r*** | 7.667 | 2.99 | 1.608 | 0.648 | 2.86 |
| g factor | 8 | | 0.44 | | −50 |

type as mentioned earlier. Table 5.4 provides a summary of some semiconductor materials with known, measured or predicted values of the Dresselhaus SOC constant.

We now consider the general case of a combined Rashba and Dresselhaus SOC. For simplicity, we consider the linear case of Dresselhaus SOC, so the Hamiltonian has the simple form of:

$$H = H_0 + H_R + H_D = \frac{-\hbar^2}{2m}\nabla^2 + \alpha(\sigma_y k_x - \sigma_x k_y) + \beta^L(\sigma_x k_x - \sigma_y k_y). \qquad [5.16]$$

The corresponding eigenvalue equation is:

$$\begin{pmatrix} E_0 & (\beta^L k_x - \alpha k_y) - i(\alpha k_x - \beta^L k_y) \\ (\beta^L k_x - \alpha k_y) + i(\alpha k_x - \beta^L k_y) & E_0 \end{pmatrix}\begin{pmatrix} u \\ v \end{pmatrix} = E\begin{pmatrix} u \\ v \end{pmatrix}, \qquad [5.17]$$

where $E_0 = \frac{\hbar^2}{2m}\left(k_x^2 + k_y^2\right) \equiv \frac{\hbar^2 k^2}{2m}$. Taking the determinant of $\begin{vmatrix} E_0 - E & \rho e^{-i\theta} \\ \rho e^{i\theta} & E_0 - E \end{vmatrix} = 0$

where $\rho = \sqrt{\left(\alpha^2 + (\beta^L)^2\right)k^2 - 4\alpha\beta^L k_x k_y}$ and $\tan\theta = \frac{\alpha k_x - \beta^L k_y}{\beta^L k_x - \alpha k_y}$, the eigenenergy is $E = E_0 \pm \rho$. With simplification, one arrives at:

$$E = E_0 \pm k\sqrt{\left(\alpha^2 + (\beta^L)^2\right) - 2\alpha\beta^L \sin 2\phi}, \qquad |\pm, k\rangle = \frac{1}{\sqrt{2}}\begin{pmatrix} e^{-i\phi} \\ \pm i \end{pmatrix} \qquad [5.18]$$

where $\tan\phi = (k_y/k_x)$.

In Table 5.5 we summarize a few common SOC systems in spintronics. It should now be clear by inspecting the relativistic equations of Eq. 5.12 that a SOC system can be viewed as some kind of effective momentum-dependent magnetic fields that the electron spin could 'feel'. But it is important to note that this field is only 'felt' by the spin but not the charge of the electron, which means that electron spin precesses about this effective field, but one should not reason that the electron experiences the corresponding Lorentz force. However, the Lorentz force does arise but in a more complex manner in SOC systems, e.g. via adiabatic spin relaxation to the momentum-dependent fields. Some aspects of this theory are presented in Section 5.6 on spin Hall effects. The eigenfunction of the Dresselhaus cubic system is:

$$\psi = \frac{1}{\sqrt{2}}\begin{pmatrix} i\dfrac{k_z\left(k_x^2 - k_y^2\right) \pm \sqrt{\left(k_y^2 + k_z^2\right)\left(k_x^4 + k_y^2 k_z^2\right) + k_x^2\left(k_y^4 - 6k_y^2 k_z^2 + k_z^4\right)}}{k_y\left(k_x^2 - k_z^2\right) + ik_x\left(k_y^2 - k_z^2\right)} \end{pmatrix}. \quad [5.19a]$$

*Table 5.5* SOC systems commonly studied in spintronics

| Spin orbit coupling types (material system) | Hamiltonian and effective ***b*** fields | Eigenstates |
|---|---|---|
| Linear Dresselhaus[37] | $H = \beta^L(\sigma_x k_x - \sigma_y k_y)$ <br> $\mathbf{b} = \frac{\beta^L}{\mu_B}\begin{pmatrix} k_x \\ -k_y \\ 0 \end{pmatrix}$ | $\frac{1}{\sqrt{2}}\begin{pmatrix} \pm\frac{k_x + ik_y}{k} \\ 1 \end{pmatrix}$ |
| Linear Rashba[22] | $H_R = \alpha(\sigma_x k_y - \sigma_y k_x)$ <br> $\mathbf{b} = \frac{\alpha}{\mu_B}\begin{pmatrix} k_y \\ -k_x \\ 0 \end{pmatrix}$ | $\frac{1}{\sqrt{2}}\begin{pmatrix} \pm\frac{k_y + ik_x}{k} \\ 1 \end{pmatrix}$ |
| Dresselhaus cubic (Bulk III-V)[36] | $H_D = \beta^C\left[k_x\left(k_y^2 - k_z^2\right)\sigma_x + k_y\left(k_z^2 - k_x^2\right)\sigma_y + k_z\left(k_x^2 - k_y^2\right)\sigma_z\right]$ <br> $\mathbf{b} = \frac{\beta^C}{\mu_B}\begin{pmatrix} k_x\left[k_y^2 - k_z^2\right] \\ k_y\left[k_z^2 - k_x^2\right] \\ k_z\left[k_x^2 - k_y^2\right] \end{pmatrix}$ | See Eq. 5.19a |
| Dresselhaus–Perel (collimated) ($k_z >> k_{\parallel}$), non-tunneling regime)[38] | $H_D = \beta^{Pc}\left(-\sigma_x k_x k_z^2 + \sigma_y k_y k_z^2 + \sigma_z k_z\left[k_x^2 - k_y^2\right]\right)$ <br> $\mathbf{b} = \frac{\beta^{pc}}{\mu_B}\begin{pmatrix} -k_x k_z^2 \\ k_y k_z^2 \\ k_z\left[k_x^2 - k_y^2\right] \end{pmatrix}$ | See Eq. 5.19b |
| Dresselhaus–Perel (tunneling electrons along ***z*** direction)[38] | $H_D = \beta^P k_z^2(\sigma_y k_y - \sigma_x k_x)$ <br> $\mathbf{b} = \frac{\beta^P}{\mu_B}\begin{pmatrix} -k_x k_z^2 \\ k_y k_z^2 \\ 0 \end{pmatrix}$ | $\frac{1}{\sqrt{2}}\begin{pmatrix} e^{i\frac{\varphi}{2}} \\ \mp e^{-i\frac{\varphi}{2}} \end{pmatrix}$ <br> where $\varphi$ is contained within the plane perpendicular to the tunneling direction along z |

Note: The Hamiltonians and eigenstates for the respective systems are presented categorically.

The eigenfunction of the Dresselhaus–Perel system is:

$$\psi = \frac{1}{\sqrt{2}}\left( i\frac{k_z\left(k_x^2 - k_y^2\right) \pm \sqrt{k_z^2\left(\left(k_x^2 - k_y^2\right)^2 + k_z^2\left(k_x^2 + k_y^2\right)\right)}}{k_z^2\left(k_x - iky\right)} \right). \qquad [5.19b]$$

The other important aspect of SOC in spintronics is the interesting fact that a finite geometric phase can be constructed in the momentum space. Details of the geometric phase, also known as the Berry's phase, are discussed in Chapter 7. It suffices to understand that the assumption of adiabatic evolution of spin eigenstates in the SOC system is necessary. Electrons completing a closed trajectory under the adiabatic condition in the momentum space will acquire such a geometric phase.

In the following we present a simple example of how, with the Hamiltonian and the eigenstate, spin-dependent transmission can be estimated. What is provided below can only be taken to be indicative of possible spin transport and is not intended to be conclusive. A conclusive approach would be one that evaluates its conductivity with the ballistic Kubo method. To consider the effect of impurity scattering, a more elaborate two-particle Green's function method would be required. For a simple, illustrative understanding of SOC effects on spin current in devices, we use the Dresselhaus–Perel system as an example and derive the electron momentum for a particular energy in terms of the chiral eigenstates. Spin current is studied in the context of mesoscopic boundary conditions, which increasingly define today's nanoscale devices. The Hamiltonian in the region of SOC and electric barrier $V_0$ is:

$$\left[\frac{p^2}{2m} + V_0 + \gamma(\sigma_y k_y - \sigma_x k_x)k_z^2\right]\phi(z)\chi^+ = E\phi(z)\chi^+. \qquad [5.20]$$

Figure 5.2 shows a sketch of the concept device, which consists of three parts like most ordinary MOSFET devices. Current is injected from source (Region 1) to drain (Region 3), passing through the SOC central part (Region 2). As a result of spin-dependent tunneling via the SOC region with electrical barrier, spin current can in principle be generated. Figure 5.3 shows the distribution of electron spin in momentum space of the Dresselhaus–Perel SOC region over an equi-energy circumference for a particular $k_z$. The cross section taken below is perpendicular to the z-axis.

Under a normally square potential barrier in the device channel where SOC is present, there are four eigenstates for a particular energy $E$. Let us choose one eigenstate $\phi_{k1}(z)\chi^+ = e^{k_1 z}\frac{1}{\sqrt{2}}\left(e^{i\frac{\varphi}{z}} \; -e^{i\frac{\varphi}{z}}\right)^T$ and let the energy operators act on the eigenstate as follows:

$$\left[\frac{p^2}{2m}+V_0-\gamma k\begin{pmatrix}0 & e^{i\varphi}\\ -e^{i\varphi} & 0\end{pmatrix}\frac{d^2}{dZ^2}\right]\phi_{kz}(z)\chi^+$$

$$=\left[\frac{\hbar^2k^2}{2m}+\frac{\hbar^2k_z^2}{2m}+V_0-\gamma kk_z^2\right]\phi_{kz}(z)\chi^+=E\phi_{kz}(z)\chi^+ \qquad [5.21]$$

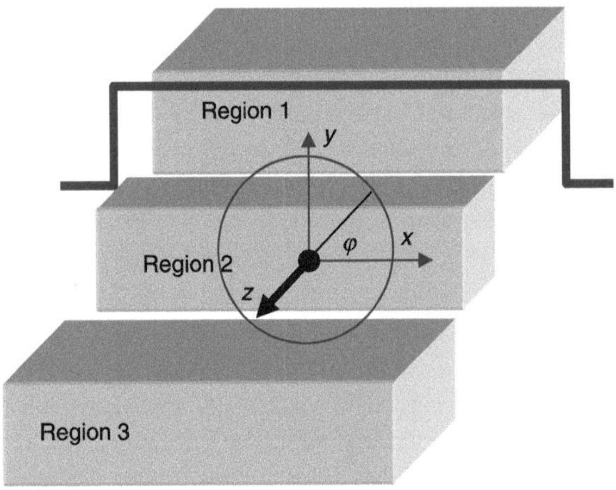


*5.2* Schematic diagram of a device nanostructure (Region 2) where the channel contains the square potential electric barrier and the SOC effect. The Fermi cross-section sliced at a particular value of $k_z$ is superimposed on the device structure as a contrast against the real space.

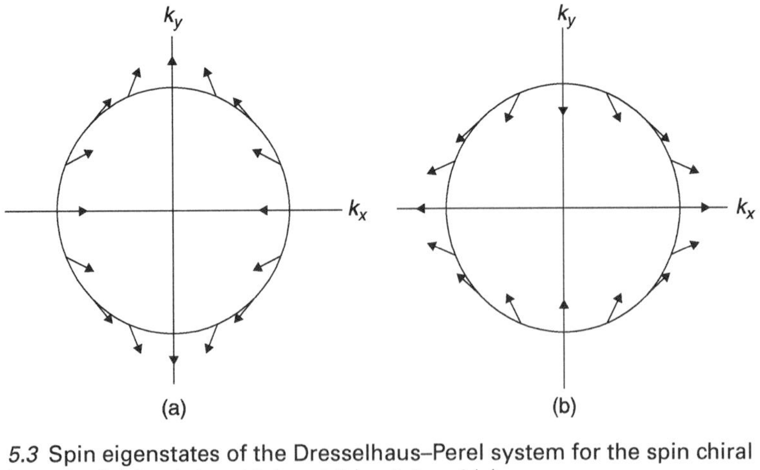


*5.3* Spin eigenstates of the Dresselhaus–Perel system for the spin chiral branch of: (a) sub-band (+) and (b) sub-band (–).

where $k = \sqrt{k_x^2 + k_y^2}$. By inspection, $E = \frac{\hbar^2 k^2}{2m} + \frac{\hbar^2 k_z^2}{2m} + V_0 - \gamma k k_z^2$. The solutions are $k_z = \pm k_z^\pm$ where:

$$k_z^\pm = \sqrt{\frac{\frac{2m}{\hbar^2}(E - V_0) - k^2}{1 \mp \frac{2m}{\hbar^2}\gamma k}}. \quad [5.22]$$

It is now obvious that, for a particular energy level in the SOC region and a fixed $k$, there will be four degenerate solutions associated with the traveling wavevectors: $(k_z^+, \chi^+)$, $(-k_z^+, \chi^+)$, $(k^-_z, \chi^-)$, $(-k_z^-, \chi^-)$. Thus momentum for a particular energy is chiral dependent.

Another important quantum mechanics method in spintronics is the rotation of the laboratory axis. As seen earlier, the linear SOC eigenstate of Rashba or Dresselhaus is chiral with respect to the laboratory axis. It has a combination of up and down states along the laboratory axis. If one is to view this eigenstate in its own frame, the physical picture becomes clearer as now the eigenstate is up or down. This can be achieved by rotating the laboratory frame to the eigenstate. We give one example below to illustrate the applications of spin rotation in the Rashba or Dresselhaus spintronic system. We now show that the above when viewed and solved in a rotated frame yield the same measureable results. There is nothing strange here, in fact it is something to be expected. Regardless of how one views the system, or how a system should be described under different frame or coordinate system, the physics remain unchanged. Hence a measurable outcome of any system is invariant under coordinate transformation. The above merely states the principle of symmetry.

### Exercise 5.1

Show, using $k_{z1}$ and $k_{z2}$ to denote $k_z^+$ and $k_z^-$, respectively, that rotating the z-axis to the spin quantization direction is a change of basis which preserves the eigenvalues.

#### *Solution*

$$U\left[\frac{p^2}{2m} + V_0 - \gamma k \begin{pmatrix} 0 & e^{i\varphi} \\ -e^{i\varphi} & 0 \end{pmatrix} \frac{d^2}{dz^2}\right] U^+ \left(U\phi_{k_{z1}}(z)\chi^+\right)$$

$$= \left[\frac{p^2}{2m} + V_0 - \gamma k \begin{pmatrix} 1 & 0 \\ 0 & -1 \end{pmatrix} \frac{d^2}{dz^2}\right]\left(e^{k_{z1}z}\begin{pmatrix} 1 \\ 0 \end{pmatrix}\right)$$

$$= \left[\frac{\hbar^2 k^2}{2m} + \frac{\hbar^2 k_{z1}^2}{2m} + V_0 - \gamma k k_{z1}^2\right]\left(e^{k_{z1}z}\begin{pmatrix} 1 \\ 0 \end{pmatrix}\right) = E\left(e^{k_{z1}z}\begin{pmatrix} 1 \\ 0 \end{pmatrix}\right).$$

We will now match the flux of two regions, i.e. $v_1\psi_1 = v_2\psi_2$. Region 1 corresponds to the non-magnetic contact of the device (source) where the electron flux consists of a linear combination of a SOC-free incident flux and a SOC affected reflecting flux. Region 2 is the SOC region as shown in Fig. 5.2. The rotated wavefunction in Region 2 can be written as:

$$\psi_2' = U\psi_2 = \left(B_1 e^{k_{z1}z} + C_1 e^{-k_{z1}z}\right)\begin{pmatrix} 1 \\ 0 \end{pmatrix} + \left(B_2 e^{k_{z2}z} + C_2 e^{-k_{z2}z}\right)\begin{pmatrix} 0 \\ 1 \end{pmatrix} \quad [5.23]$$

which is a linear combination of the four degenerate eigenstates, whose amplitudes $(B_1, B_2, C_1, C_2)$ can be found by performing flux matching at the device boundaries between Regions 1, 2 and 3. It is important to note that the system has been rotated to the quantization axis of the chiral eigenstate, and that the eigenstate of $(1\ 0)^T$ is with respect to the rotated axis but not the laboratory axis. Region 3 is the contact (drain) where detection of spin current can be carried out.

In spintronic devices based on wavefunction treatment, flux continuity has to be observed across the device[39] as has been specifically shown above. Generally for a three-region device, the standard current operator reads $j(z) = \frac{\hbar}{2mi}\left(\psi^\dagger \frac{d\psi}{dz} - \psi \frac{d\psi^\dagger}{dz}\right)$. For Region 1, inserting $\psi_1$ into the operator yields $j_1(z) = \frac{p_z}{m}\left(1 - |R_1|^2 - |R_2|^2\right)$ whilst, for Region 3, the result is $j_3(z) = \frac{p_z}{m}\left(|T_1|^2 + |T_2|^2\right)$ where $R$ and $T$ are the reflection and transmission coefficients, respectively, related to the wavefunction amplitude ratio. The current is just the product of the probability amplitudes of the component wavefunctions and their respective velocities. The flux operator in Region 2 is thus $j_2(z) = |B_1|^2 \langle \psi_2' | v_2' | \psi_2' \rangle + |C_1|^2 \langle \psi_2' | v_2' | \psi_2' \rangle + |B_2|^2 \langle \psi_2' | v_2' | \psi_2' \rangle + |C_2|^2 \langle \psi_2' | v_2' | \psi_2' \rangle$, where $v'_2$ is the SOC-modified velocity operator. Upon obtaining the transport coefficients, one can verify that the flux is conserved across the trilayer structure, i.e. $j_1(z) = j_2(z) = j_3(z)$.

---

### Exercise 5.2

The expression below has the physical meaning of rotating electron spin about a particular axis. Prove this expression:

$$exp\left(-i\frac{\hbar\theta}{2}\boldsymbol{\sigma}.\boldsymbol{a}\right) = I\, cos\frac{\theta}{2} - i\, \boldsymbol{\sigma}.\boldsymbol{a}\, sin\frac{\theta}{2}$$

The above has the physical meaning of rotating about axis along $\boldsymbol{a}$. Show that when $\boldsymbol{a} = -sin\phi \boldsymbol{i} + cos\phi \boldsymbol{j}$, the above can then be written as:

$$U = \begin{pmatrix} cos\frac{\theta}{2} & -sin\frac{\theta}{2}e^{-i\phi} \\ sin\frac{\theta}{2}e^{i\phi} & cos\frac{\theta}{2} \end{pmatrix}.$$

What is the physical meaning of the $U$ matrix?

Prove the following

$\sigma_\mu \sigma_\nu = i\varepsilon_{\mu\nu\kappa}\sigma_\kappa + I\delta_{\mu\nu}$

**Hint**: Use the SU(z) algebra of Pauli matrices in tensor representation.

---

Spin polarization is normally defined as $P = (T^+ - T^-)/(T^+ + T^-)$, which can be found in terms of the amplitude of the incident and outgoing wave amplitudes which are themselves scalar coefficients. Figure 5.4 shows a typical numerical result of a Dresselhaus–Perel system, which allows one to determine spin-polarized current at the Fermi energy.

What is discussed above is spin transportation in Regions 1 and 2, which correspond to the two important spintronic functions of spin generation and spin transportation. The third crucial spintronic function of spin current detection is normally carried out in Region 3. We present briefly below some techniques useful for understanding the detection of spin current. Let us take a particular eigenstate of the Dresselhaus SOC $|\chi_\theta^{\sigma'}\rangle$, where $\sigma' = \pm$ is the index which denotes the chiral eigenstates of the Dresselhaus system. In the frame of the laboratory, the Dresselhaus eigenstate is a linear combination of the spin up and down eigenstates. The probability amplitude of finding an up or down state for $|\chi_\theta^{\sigma'}\rangle$, where spin up/down is denoted by the index $\sigma = \pm$, is:

$$\langle\sigma|\chi_\theta^{\sigma'}\rangle = \frac{1}{\sqrt{2}}(1-\sigma i) = \frac{1}{\sqrt{2}}\begin{pmatrix} 1 \\ -\sigma' e^{-i\theta} \end{pmatrix} = \frac{1}{2}(1+\sigma\sigma' i e^{-i\theta}). \qquad [5.24]$$

The probability is simply $\left|\langle\sigma|\chi_\theta^{\sigma'}\rangle\right|^2 = \frac{1}{2}(1+\sigma\sigma' sin\theta)$. In the SOC electron gas Fermi system, there is a spectrum of momentum with different chiral states, each making a different angle with the laboratory quantization axis. The above

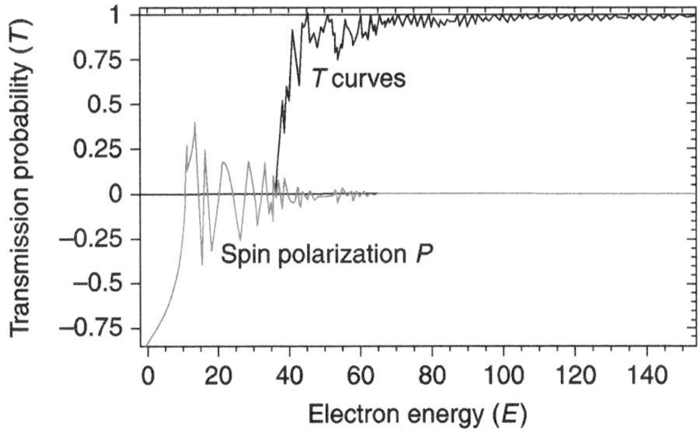


*5.4* Spin polarization of a Dresselhaus–Perel system along the quantization axis of a particular chiral eigenstate. This curve makes sense for one transport mode only. Proper spin polarization should be one summed over all transport modes.

determines the probability to be in a laboratory up/down state $|\uparrow/\downarrow\rangle$ denoted by quantum number $\sigma$ for a particular chiral state denoted by quantum number $\sigma'$. Thus, the transmission probability for a particular energy of a laboratory up/down state over all modes can be estimated as follows:

$$T^{\sigma}(k_z) = \sum_{\theta} \left| \left\langle \sigma \middle| \chi_{\theta}^{\sigma'} \right\rangle \right|^2 f(\theta, k_z) \qquad [5.25]$$

where $f(\theta, k_z)$ consists of flux amplitude in Regions 1 and 3, which can be found by flux matching. Note that $f(\theta, k_z)$ is a function of $\theta$ due to the effect of $k_x$ and $k_y$ on $k_z$. Readers are probably aware that spin-dependent tunneling discussed above is most often discussed in the context of a device system with local boundaries that restrict spin transport to a short 2DEG channel. This is because spin current vanishes globally in a 2DEG. Nevertheless, actual detection of spin current in semiconductor spintronics remains difficult. Impurities scattering has been shown to destroy spin current. Even in clean, ballistic limit, multi-mode transport could still greatly neutralize the spin polarization which arises due to individual mode. Other factors that complicate spin current detection include non-uniform SOC strength, interfacial scattering, and spin-injection mismatch between semiconductor channel and contact detectors. The promises brought upon by long spin relaxation length and electrical generation of spin polarization cannot be realized due to these reasons. A stable semiconductor spinFET working at room temperature has yet to be demonstrated.

The following section discusses the effects of magnetic fields and spin orbit coupling on spin current. Before moving on, we mention again that to make a spintronic device work, the coherence of spin transported over a long distance (at least 100 nm) is crucial. This brings in the important topic of spin relaxation. Nonetheless, we will not discuss this topic in detail apart from stating that the underlying physics of spin relaxation has various origins. Spin relaxation can be attributed to three mechanisms: Dyakonov–Perel (DP),[40] Bir–Aronov–Pikus (BAP)[41] and Elliott–Yafet (EY).[42] Readers interested in more details of these mechanisms can refer to the review given in reference 43, which discussed these effects in some detail.

## 5.4 Spin current under magnetic fields and spin orbit coupling

We now give a quantum mechanical[44,45] illustration of a spin polarized wavefunction distributing spatially in a 2DEG SOC. The following provides an early indication of the possible existence of spin current (longitudinal or Hall) in a SOC system. The single-electron Hamiltonian of a 2DEG in the presence of perpendicular magnetic fields and linear SOC effects is given by:

$$\frac{-\hbar^2}{2m}\phi''(z)\psi(x,y) + ezF_z\phi(z)\psi(x,y) + \left( \frac{(p_x + eA_x)^2}{2m} + \frac{p_y^2}{2m} + U + H_z + H_{SO} \right)\phi(z)\psi(x,y) = E\phi(z)\psi(x,y) \qquad [5.26]$$

where the gauge of $A = (-By, 0, 0)$ is chosen. $F_z$, $H_{so}$, $H_z$ *and* $U$ are terms corresponding to the 2DEG triangular confining potential, the linear spin orbit coupling, the Zeeman split and the applied electric potential, respectively. The solution to the bound wavefunction $\phi(z)$ is given by a linear combination of Airy functions, i.e.

$$\phi(z) = e^{-\frac{bz}{2a}} AiryAi\left[\frac{b^2 + 4ac - 4aE_n + 4adz}{4a^{4/3}d^{2/3}}\right]C_1 + e^{-\frac{bz}{2a}} AiryBi\left[\frac{b^2 + 4ac - 4aE_n + 4adz}{4a^{4/3}d^{2/3}}\right]C_2 \quad [5.27]$$

where $C_1$, $C_2$ are constants, $a = \dfrac{\hbar^2}{2m}, b = 0, c = \dfrac{p_x^2}{2m} + \dfrac{(p_y - eA_y)^2}{2m} + U + H_Z + H_{SO}$, and $d = eF_z$. For the above solution to be a well-behaved function as, $z \to \infty$, $C_2 = 0$. To determine the sub-band energies within the 2DEG potential well, one imposes the boundary condition of $\phi(z = 0) = 0$, where $z = 0$ is taken to be the bottom of the 2DEG in the vertical direction. Noting that $\phi(z) = e^{-\frac{bz}{2a}} AiryAi\left[\dfrac{b^2 + 4ac - 4aE_n + 4adz}{4a^{4/3}d^{2/3}}\right]$ has roots −2.338, −4.008 ........ and solving for the Hamiltonian at $z = 0$ results in the lowest energy eigenvalue ($z$) of $E_1 = c + 2.338a^{1/3}d^{2/3}$, i.e. the energy of the system is:

$$E = \frac{(p_x + eA_x)^2}{2m} + \frac{p_y^2}{2m} + U + H_Z + H_{SO} + 2.338\left(\frac{[[eF_z\hbar]]^2}{2m}\right). \quad [5.28]$$

In the energy levels above the sub-band of $E_S = 2.338\left(\dfrac{[eF_z\hbar]^2}{2m}\right)$, electrons define a Fermi circle ($x$, $y$) in the momentum space and the radius of the circle depends on the sheet electron density. In the conduction band, an electron in a two-dimensional system is subject to the linear spin orbit coupling effects of Rashba and Dresselhaus, respectively. In explicit spinor form, one has:

$$\begin{bmatrix} \dfrac{(p_x + eA_x)^2}{2m} + \dfrac{p_y^2}{2m} + U + H_Z + E_S & -(\alpha + i\beta)k_y - (\beta + i\alpha)\left(k_x + \dfrac{e}{\hbar}A_x\right) \\ -(\alpha - i\beta)k_y - (\beta - i\alpha)\left(k_x + \dfrac{e}{\hbar}A_x\right) & \dfrac{(p_x + eA_x)^2}{2m} + \dfrac{p_y^2}{2m} + U - H_Z + E_S \end{bmatrix} \times \psi(x, y) = E\psi(x, y) \quad [5.29]$$

The conduction band wave function of the Landau spin orbit coupling system that corresponds to energy eigenvalue $E$ is:

$$\psi = \sum_{n=1} \varphi_n (y - y_a) \begin{pmatrix} a_n \\ b_n \end{pmatrix} \phi(x) \qquad [5.30]$$

where the $y$ dimension wavefunction is taken to be a linear combination of eigensolutions to a 2DEG with only vertical magnetic field, i.e. without the SOC.

---

**Exercise 5.3**

Show that

$$\int \varphi_m(x)\,(\partial_x + x) \sum_{n=1}^{N} \varphi_n a_n \, dx = \sqrt{2(m+1)}\, a_{m+1}.$$

***Solution***

$$\int \varphi_m(x) \sum_{n=1}^{N} \frac{a_n}{\sqrt{c_n}} x H_n(x) \exp\left(\frac{-x^2}{2}\right) dx$$

$$= \frac{1}{2} \int \varphi_m(x) \sum_{n=1}^{N} \frac{a_n}{\sqrt{c_n}} \left[ H_{n+1}(x) + 2n H_{n-1}(x) \right] \exp\left(\frac{-x^2}{2}\right) dx$$

$$= \frac{1}{2} \left[ \frac{a_{m-1}}{\sqrt{c_{m-1}}} \sqrt{c_m} + 2(m+1) \frac{a_{m+1}}{\sqrt{c_{m+1}}} \sqrt{c_m} \right]$$

where $\sqrt{c_m} = \sqrt{\sqrt{\pi}\, 2^m m! r}$. Hence:

$$\int \varphi_m(x)(\partial_x + x) \sum_{n=1}^{N} \varphi_n a_n \, dx = 2(m+1) \frac{a_{m+1}}{\sqrt{c_{m+1}}} \sqrt{c_m} = \sqrt{2(m+1)}\, a_{m+1}.$$

---

**Exercise 5.4**

Show that:

$$\int \varphi_m(x)(\partial_x - x) \sum_{n=1}^{N} \varphi_n a_n \, dx = -\sqrt{2m}\, a_{m-1}.$$

***Solution***

$$\int \varphi_m(x)\sum_{n=1}^{N}\frac{a_n}{\sqrt{c_n}}\partial_x\left[H_n(x)\exp\left(\frac{-x^2}{2}\right)\right]dx$$

$$=\int \varphi_m(x)\sum_{n=1}^{N}\frac{a_n}{\sqrt{c_n}}\exp\left(\frac{-x^2}{2}\right)\left[nH_{n-1}(x)-\frac{1}{2}H_{n+1}(x)\right]dx$$

$$=\left[\frac{a_{m+1}}{\sqrt{c_{m+1}}}(m+1)\sqrt{c_m}-\frac{1}{2}\frac{a_{m-1}}{\sqrt{c_{m-1}}}\sqrt{c_m}\right]$$

where $\sqrt{c_m}=\sqrt{\sqrt{\pi}2^m m!r}$. Hence:

$$\int \varphi_m(x)(\partial_x - x)\sum_{n=1}^{N}\varphi_n a_n dx=\frac{-a_{m-1}}{\sqrt{c_{m-1}}}\sqrt{c_m}=-\sqrt{2m}\,a_{m-1}.$$

---

With the integral equations[46,47] above, refer to Appendix 5.A for details on the derivation of the wavefunction. The $x$ dimension wavefunction is a non-decaying oscillatory traveling function. One can deduce that the final wave function for the (+) and (−) branches are:

$$\psi^+(y-y_0)=\frac{e^{ik_x x}}{\sqrt{1+(D_S^+)^2}}\begin{pmatrix} D_s^+\varphi_{s-1}(y-y_0)\\ \varphi_s(y-y_0)\end{pmatrix}$$

$$\psi^-(y-y_0)=\frac{e^{ik_x x}}{\sqrt{1+(D_S^-)^2}}\begin{pmatrix} \varphi_{s-1}(y-y_0)\\ D_S^-\varphi_s(y-y_0)\end{pmatrix} \qquad [5.31]$$

where:

$$D_s^+=D_S^-=\frac{-i\alpha\sqrt{2s}/r}{\left(\xi_0+\sqrt{\xi_0^2+2s\left(\frac{\alpha}{r}\right)^2}\right)} \qquad [5.32]$$

and $\xi_0 = 1/2\ (\hbar\omega) = g\mu B$. One can now study the distribution of spin current in the bulk of the 2DEG. Note that, the choice of gauge might present slightly different results for $y_0$ whereas $y_a = y_0 - \frac{eEy}{m\omega}$ will be true for gauge potentials of $(A_x, 0,0)$ and $(0, A_y, 0)$ as well.

---

## Exercise 5.5

Show that the current expression for a Rashba system without external magnetic field is:

$$j_x = \left( \frac{\hbar k_x}{m} \mp \frac{2}{\hbar} \alpha \frac{k_x}{k} \right).$$

***Solution***

Current expression in the absence of external magnetic–electric fields is:

$$j_x = -\frac{i\hbar}{2m}\left( \psi^* \frac{\partial \psi}{\partial x} - \psi \frac{\partial \psi^*}{\partial x} \right) = \frac{1}{2}\left( \phi^*(a^*\, b^*) v_x \phi \begin{pmatrix} a \\ b \end{pmatrix} - \phi(a\, b) v_x\, \phi^* \begin{pmatrix} a^* \\ b^* \end{pmatrix} \right)$$

where $v_x$ is the electron velocity that can be found using the Hamilton equation that provides the first momentum derivative of the Hamiltonian matrix, i.e. $v_x = \frac{\partial H}{\partial p_x}$. Explicitly:

$$v_x = \begin{bmatrix} -\frac{i\hbar^2}{m}\frac{\partial}{\partial x} & i\alpha \\ -i\alpha & -\frac{i\hbar^2}{m}\frac{\partial}{\partial x} \end{bmatrix}.$$

Expanding the above leads to $j_x = \left( \frac{\hbar k_x}{m}(a^2 + b^2) - \frac{2}{\hbar}\alpha\, Im[a^* b] \right)$. Substituting with the Rashba eigenstates, the current expression is:

$$j_x = \left( \frac{\hbar k_x}{m} \mp \frac{2}{\hbar} \alpha \frac{k_x}{k} \right).$$

For $k_y = 0$, a common approach taken in the design of spintronic devices, the above is:

$$j_x = \left( \frac{\hbar k_x}{m} \mp \frac{2}{\hbar} \alpha \right).$$

---

Using the current expression:

$$j_x = \left[ \omega y_0 - \frac{eBy}{m} - \frac{2}{\hbar}\alpha\, Im(a^* b) \right] = -\left[ \omega(y - y_a) - \frac{E_y}{B} + \frac{2}{\hbar}\alpha\, Im(a^* b) \right] \qquad [5.33]$$

and substituting the eigenfunctions of the Landau SOC system, one obtains the current expression whose spin-dependence arises from coupling constant $\neq 0$:

$$J_{xs}^{\sigma} = -\left[ \omega(y - y_0) - \frac{Ey}{B} + \sigma \frac{2}{\hbar}\alpha\, Im\left( \frac{D_s^{\sigma}\, \varphi_{s-1}\varphi_s}{1 + \left(D_s^{\sigma}\right)^2} \right) \right]. \qquad [5.34]$$

It is worth noting that the application of $E_y$ provides a base value for the average current in the form of $\frac{E_y}{B}$. This ensures that there would be an average current even when all the SOC effects are absent. We have thus shown analytically, in the Rashba SOC 2DEG system with cross electric magnetic field, that current is spin polarized. One can now at least have a crude feeling of how SOC distributes spin current spatially. The above naturally leads to the idea that Hall effect involving spin may be generated from the SOC effect. Indeed, we discuss below the finding that the so-called spin Hall effect (SHE) is ubiquitous in SOC systems. Various theoretical and conceptual methods have been put forth to study the physics of SHE and its applications.

## 5.5 Spin dynamics under the spin orbit gauge

In nanoscale devices, common measure of electron flow in current or voltage can be related closely to single-particle electron dynamics. An electron's precession about the Dirac effective magnetic field can be viewed to give rise to a non-integrable special unitary group of degree 2 (SU(2)) phase factor and a spin orbit gauge, which leads to the presence of non-commutative electromagnetism. In the language of gauge theory, the spin orbit gauge is simply the gauge invariant manifestation of the non-integrable phase factor. In spintronics, one can investigate electron dynamics based on such non-commutative electromagnetism, which has also been shown to be one of the underlying mechanisms for spin Hall effects. For a clearer visualization of the strength of the SOC, we convert specific SOC systems into the vacuum form, i.e. all different SOC systems take the form of the vacuum coupling constant such that the SOC strength (energy) is reflected in the effective electric fields. Thus the effective electric field is simply the field that would generate the same SOC strength if these specific systems were to be vacuum. The single-particle form of the Hamiltonian $H = \int \psi^\dagger \left( \frac{1}{2} D_\mu^2 + V \right) \psi \, dx_\mu$ is given by:

$$H = \sum_{\mu}^{3} \frac{1}{2m} \left( p_\mu - \frac{e\hbar}{2gmc^2} (\boldsymbol{E} \times \boldsymbol{\sigma}) . \hat{\boldsymbol{n}}_\mu \right)^2 + e\, \boldsymbol{E} . \boldsymbol{r} \qquad [5.35]$$

where $A_r = \frac{\hbar}{2gmc^2} (\boldsymbol{E} \times \boldsymbol{\sigma}) = G(\boldsymbol{E} \times \boldsymbol{\sigma})$ is the spin orbit gauge and $G$ is a time coupling constant. From the non-trivial curvature of the spin orbit gauge, one can construct a magnetic field. The main feature about the spin orbit gauge here is its non-trivial curvature, which provides a magnetic field effect on particles. To obtain further insights, we will investigate the curvature and its implications to electron dynamics in the SOC system. Since the momentum conjugate to $\psi$ is $\frac{\partial L}{\partial (D_0 \psi)} = i\psi^\dagger$, one obtains the Hamiltonian density $H = \frac{1}{2} (D_\mu \psi^\dagger)(D_\mu \psi) + V\psi^\dagger \psi$ after applying the Legendre transformation.

### 5.5.1 Single-particle electron dynamics in real space

By considering the curvature $\boldsymbol{\Omega}_r = \nabla_r \times \boldsymbol{A}_r - \frac{ie}{\hbar}\boldsymbol{A}_r \times \boldsymbol{A}_r$, which can be interpreted as the non-commutative magnetic field, one can borrow the classical picture of Lorentz to derive the average transverse force that an electron particle will feel in this system. In classical physics, the general equation of motion (EOM) is $\langle f \rangle = m\frac{d(v)}{dt}$. The EOM for the Lorentz force would be $m\frac{d(v)}{dt} = \frac{e}{m}\langle \boldsymbol{p}\rangle \times \boldsymbol{B}$ where $\boldsymbol{B}$ is a physical magnetic field. We now replace $\boldsymbol{B}$ with $\boldsymbol{\Omega}_r$, an effective magnetic field derived from quantum mechanics. Thus below is an equation based on classical EOM but a quantum mechanics effective magnetic field:

$$\langle f \rangle = \frac{e}{m}\langle \boldsymbol{p}\rangle \times \boldsymbol{B} = \left\langle \frac{e\boldsymbol{p}}{m} \times \boldsymbol{\Omega}_r \right\rangle = \left\langle \frac{e\boldsymbol{p}}{m} \times \left( \nabla_r \times \boldsymbol{A}_r - \frac{ie}{\hbar}\boldsymbol{A}_r \times \boldsymbol{A}_r \right) \right\rangle. \quad [5.36]$$

The first term on the right-hand-side can be expressed as:

$$\langle f_1 \rangle = \frac{e}{m}\langle \nabla_r(\boldsymbol{p}.\boldsymbol{A}_r) - (\boldsymbol{p}.\nabla_r)\boldsymbol{A}_r \rangle \quad [5.37]$$

since the gauge is in real space, $\boldsymbol{A}_r$ is used to denote the spin orbit gauge. This term vanishes when the electric field is spatially uniform, which is assumed in this case to simplify our analysis.

In spintronics, a uniform electric field implies that the Rashba or Dresselhaus coupling parameter is a constant. In the context of device electronics (e.g. spintronics based on the Rashba or Dresselhaus effects), the term $\langle f_1 \rangle$ should capture the imagination of electronic engineers designing new devices. This term gives the heuristic indication that a deliberate engineering of the Rashba or Dresselhaus coupling to be inhomogeneous within the electron conduction path, such as at the metal–semiconductor interface (where SOC constants emerge abruptly in the semiconductors), might result in additional electron dynamics driven by forces related to SOC gradient. The second term is:

$$\langle f_2 \rangle = \frac{-ie^2G^2}{m\hbar}\langle \boldsymbol{p} \times ((\boldsymbol{E} \times \boldsymbol{\sigma}) \times (\boldsymbol{E} \times \boldsymbol{\sigma})) \rangle \quad [5.38]$$

which can be simplified by using the general identity of $\boldsymbol{A} \times (\boldsymbol{B} \times \boldsymbol{C}) = \boldsymbol{B}\,(\boldsymbol{A}.\boldsymbol{C}) - (\boldsymbol{A}.\boldsymbol{B})\,\boldsymbol{C}$ with the assignment rules of $\boldsymbol{E} \times \boldsymbol{\sigma} \rightarrow \boldsymbol{A}$, $\boldsymbol{E} \rightarrow \boldsymbol{B}$, and $\boldsymbol{\sigma} \rightarrow \boldsymbol{C}$, and the spin algebra of $\boldsymbol{\sigma} \times \boldsymbol{\sigma} = 2i\boldsymbol{\sigma}$. With the above, one can readily derive the expression $\langle f_2 \rangle = \frac{2e^2G^2}{m\hbar}(\boldsymbol{\sigma}.\boldsymbol{E})\,\boldsymbol{p} \times \boldsymbol{E}$. This expression can be interpreted as a type of spin-dependent transverse force.

To express $\langle f_2 \rangle$ in a more compact form, a different set of assignment rules $\boldsymbol{p} \rightarrow \boldsymbol{A}$, and $\boldsymbol{A}_r \rightarrow \boldsymbol{B}, \boldsymbol{C}$ is applied and the following is obtained:

$$\langle f_2 \rangle = \frac{-ie^2}{m\hbar}\langle \boldsymbol{A}_r\,(\boldsymbol{p}.\boldsymbol{A}_r) - (\boldsymbol{p}.\boldsymbol{A}_r)\boldsymbol{A}_r \rangle = \frac{-ie^2}{m\hbar}\langle [\boldsymbol{A}_r, \boldsymbol{p}.\boldsymbol{A}_r] \rangle. \quad [5.39]$$

Noting, $\boldsymbol{p}.\boldsymbol{A}_r = G\boldsymbol{\sigma}.(\boldsymbol{p} \times \boldsymbol{E})$, one is able to derive, albeit rather tediously a more explicit expression for $\langle f_2 \rangle$ that is not discussed here. The above expressions for $\langle f_1 \rangle$ and $\langle f_2 \rangle$, which describe the forces experienced by the electron due to the spin orbit gauge field, have been obtained from semiclassical analysis.

We now study the time dependence of the momentum or velocity using the operator method. One can compare that identical force expressions can be obtained using the operator algebra method. Once again borrowing $\langle f \rangle = m\frac{d(v)}{dt}$ and treating $v$ as an operator in Hilbert space, one could write down the 'force' felt by an electron treated like a particle with mass $m$ but zero volume as:

$$\langle \boldsymbol{f} \rangle = m\frac{d(\boldsymbol{v})}{dt} = \frac{d(\boldsymbol{p}-e\boldsymbol{A})}{dt}. \qquad [5.40]$$

Note when one starts with a general EOM, the velocity operator of $\boldsymbol{v} = \boldsymbol{p} - e\boldsymbol{A}$ is used. Previously, in the Lorentz EOM, the canonical $\boldsymbol{p}$ operator is used instead of $\boldsymbol{v}$. In fact the general EOM will lead to the Lorentz-related terms with the use of $= \boldsymbol{p} - e\boldsymbol{A}$. One can now let these operators obey the expectation formalism in quantum mechanics as follows:

$$i\hbar\frac{d\langle \boldsymbol{O} \rangle}{dt} = \langle\, [\boldsymbol{O}, H]\,\rangle + i\hbar\left\langle \frac{\partial \boldsymbol{O}}{\partial t} \right\rangle. \qquad [5.41]$$

The expression for 'force' would now be:

$$\langle \boldsymbol{f} \rangle = \left( \frac{1}{i\hbar}\langle\, [\boldsymbol{p}, H] \rangle + \left\langle \frac{\partial \boldsymbol{p}}{\partial t} \right\rangle \right) + \left( -\frac{e}{i\hbar}\langle [\boldsymbol{A}_r, H]\,\rangle - e\left\langle \frac{\partial \boldsymbol{A}_r}{\partial t} \right\rangle \right) \equiv \langle \boldsymbol{f}_1 \rangle + \langle \boldsymbol{f}_2 \rangle \qquad [5.42]$$

where $\left\langle \frac{\partial \boldsymbol{A}_r}{\partial t} \right\rangle = G\left\langle \boldsymbol{\sigma} \times \frac{\partial \boldsymbol{E}}{\partial t} + \frac{\partial \boldsymbol{\sigma}}{\partial t} \times \boldsymbol{E} \right\rangle$. Bold notations merely reflect their vectorial nature. In the above, only $\boldsymbol{p}$, $H$, $\boldsymbol{A}_r$, $\boldsymbol{\sigma}$ are operators. Referring to the Hamiltonian, one can thus deduce that:

$$\langle \boldsymbol{f}_1 \rangle = \frac{1}{i\hbar}\left\langle \left[ \boldsymbol{p}, \left( \frac{p^2}{2m} + \frac{e\,\boldsymbol{p}.\boldsymbol{A}_r}{m} + e\,\boldsymbol{E}.\boldsymbol{r} \right) \right] \right\rangle = \frac{e}{i\hbar}\left\langle \left[ \boldsymbol{p}, \frac{\boldsymbol{p}.\boldsymbol{A}_r}{m} \right] + \boldsymbol{E} \right\rangle \qquad [5.43]$$

where use has been made of the fact that $\left[ \boldsymbol{p}, \frac{p^2}{2m} \right] = 0$ and $[\boldsymbol{p}, e\,\boldsymbol{E}.\boldsymbol{r}] = eE_v[\boldsymbol{p}, r_v] = eE_v\delta_{iv}\hat{\boldsymbol{n}}_i$.

We will now ignore the electron acceleration due to the $E$-field since we want to focus on the electron dynamics arising due to the spin orbit gauge field but not the external electric field. Note that an external electric field in the context of a lateral device normally originates from applying source–drain bias across the current conduction path. Similarly, assuming no explicit time and spatial dependence of $E$-field, one would obtain:

$$\langle \boldsymbol{f}_2 \rangle = \frac{-e}{i\hbar}\left\langle \left[ \boldsymbol{A}_r, \left( \frac{p^2}{2m} + \frac{e\,\boldsymbol{p}.\boldsymbol{A}_r}{m} + e\,\boldsymbol{E}.\boldsymbol{r} \right) \right] \right\rangle = \frac{ie^2}{m\hbar}\langle [\boldsymbol{A}_r, \boldsymbol{p}.\boldsymbol{A}_r] \rangle. \qquad [5.44]$$

Since $\langle f_1 \rangle$ and $\langle f_2 \rangle$ are terms that affect the electron's equation of motion in the presence of SOC, we can relate them to the dynamics of spin Hall, spin torque, anomalous Hall, etc, which have been experimentally observed in various spintronic systems with SOC. For exact quantification of these effects and comparison with non-ideal practical systems, the semiclassical treatment given above may be extended to include the effects of scattering, either via the microscopic Kubo formalism or the semiclassical Boltzmann drift diffusion equation. The present approach discussed in this section is general and can be used to describe more complicated spin dynamics, e.g. in the presence of magnetic fields or spatially or temporally varying electric fields.

### 5.5.2 Single-particle electron dynamics in momentum space

We now derive the $k$-space equivalent of the above by applying Heisenberg algebra methods. The $k$-space gauge curvature is particularly useful in elucidating the quantum description of anomalous velocity under the influence of spin orbit coupling, which has been described as a $k$-space Lorentz force or Karplus velocity term in previous works. The anomalous velocity in the spin orbital system can be deduced from $v = \frac{1}{i\hbar}[r, H]$ or $v = \frac{\partial H}{\partial p}$, both yield the velocity operator of $\boldsymbol{v} = \frac{\boldsymbol{p}}{m} - \frac{e}{m}\boldsymbol{A}_r$. However, the physical interpretation for $\frac{e}{m}\boldsymbol{A}_r$ is unclear. We will now expand and rearrange the SOC Hamiltonian into the form:

$$H = H_0 + e\boldsymbol{E}.\left(\frac{G}{2m}(\boldsymbol{\sigma} \times \boldsymbol{p}) + \boldsymbol{r}\right) + \frac{e^2 G^2}{2m}(\boldsymbol{E} \times \boldsymbol{\sigma}).(\boldsymbol{E} \times \boldsymbol{\sigma}). \qquad [5.45]$$

We ignore the higher order $O(G^2)$ term and define $\boldsymbol{A}_k = \frac{G}{2m}(\boldsymbol{\sigma} \times \boldsymbol{p})$, which has the unit of length and can be interpreted as the $k$-space gauge field of the SOC system. In this way, the spatial coordinates acquire their non-commutativity from the $k$-space gauge field. The curvature of this $k$-space gauge field, which is given by $\boldsymbol{\Omega}_k = \boldsymbol{\nabla}_k \times \boldsymbol{A}_k - i\boldsymbol{A}_k \times \boldsymbol{A}_k$, is reminiscent of its real space counterpart described earlier. Expanding the first term of the curvature yields:

$$\boldsymbol{\nabla}_k \times \boldsymbol{A}_k = \frac{G}{2m}\boldsymbol{\nabla}_k \times (\boldsymbol{\sigma} \times \boldsymbol{p}) = \frac{G}{2m}\left(\boldsymbol{\sigma}(\boldsymbol{\nabla}_k \cdot \boldsymbol{p}) - (\boldsymbol{\sigma}.\boldsymbol{\nabla}_k)\boldsymbol{p}\right). \qquad [5.46]$$

A careful inspection shows that this leads to:

$$\boldsymbol{\nabla}_k \times \boldsymbol{A}_k = \frac{\hbar G}{2m}\left(\boldsymbol{\sigma}\left(\frac{\partial}{\partial k_\nu}k_\nu\right) - \sigma_\nu \frac{\partial}{\partial k_\nu}k_\mu \hat{\boldsymbol{n}}_\mu\right) = \frac{\hbar G}{m}\boldsymbol{\sigma}. \qquad [5.47]$$

Similarly, for the second term, we can show by using the same relation and assignment rules which were applied to $\langle f_2 \rangle$ that:

$$\boldsymbol{A}_k \times \boldsymbol{A}_k = \left(\frac{G}{2m}\right)^2 (\boldsymbol{\sigma} \times \boldsymbol{p}) \times (\boldsymbol{\sigma} \times \boldsymbol{p}) = -\left(\frac{G}{2m}\right)^2 (\boldsymbol{\sigma}.\boldsymbol{p})\boldsymbol{p}. \qquad [5.48]$$

Combining the above leads to the total curvature in $k$-space of:

$$\langle \Omega_k \rangle = \frac{\hbar G}{m}\sigma + i\left(\frac{G}{2m}\right)^2 (\sigma . p)p. \qquad [5.49]$$

In the presence of the gauge field, the velocity expression is given by:

$$\langle v \rangle = \left\langle \frac{p}{m} - \frac{1}{\hbar}\frac{dp}{dt} \times \Omega_k \right\rangle = \left\langle \frac{p}{m} - \frac{Ge}{m}(E \times \sigma) \right\rangle = \left\langle \frac{p}{m} - \frac{e}{m}A_r \right\rangle. \qquad [5.50]$$

The second velocity term can be formally interpreted as the anomalous velocity $v_a$ or the Karplus term. The anomalous velocity term is related to the SU(2) gauge field $A_r$ in coordinate space by the relation:

$$v_a = \frac{e}{m}A_r. \qquad [5.51]$$

To make a clear comparison between the two pictures, we note that both real and momentum analysis yields the identical relation of anomalous velocity effect in $mv = p - eA_r$.

## 5.6 Spin Hall effects (SHE)

One of the means to generate spin current in semiconductor spintronics is via the spin Hall effect (SHE) as stipulated in Table 5.1. The description of SHE has its origin in the early work of Russian theoretical physicists M. I. Dyakonov and V. I. Perel.[48] Their work was the first time that the notion of spin current was introduced. SHE in the form of transverse spin accumulation was predicted and equations describing such phenomenon were written. Experimental works[49] were later carried out which resulted in the measurement of what is known today as the inverse version of the SHE. In 1999, J. E. Hirsch[50] broke a long silence and once more raised discussion on the prospects of a SHE-based device. But experimentally,[51] SHE was observed in semiconductors only more than 30 years after the original prediction. Spin accumulation induces circular polarization of the emitted light, as well as the Faraday (Kerr) rotation of the transmitted (reflected) light polarization, which allows one to monitor SHE by optical means. There were soon other descriptions[52] relating SHE to classical Hall as well as the anomalous Hall effect (AHE).

This section is devoted to discussing in greater detail the important physics of spin Hall and its possible applications. The crucial elements for realizing SHE are the SOC effects and the application of electric field. The electric field is only an external means to generate a longitudinal current. The coupling between SHE current and longitudinal charge current is most probably due to the physics of spin orbit coupling. SHE also refers to the observed transverse separation of opposite spin across a nanoscale device system, which results in a net spin accumulation but zero Hall voltage across the two edges of the device. SHE is thus reminiscent of the classical Hall effect as well as the integer quantum Hall effect (IQHE), where charges accumulate on opposite lateral edges of a rectangular current carrying region in the presence of an external magnetic field applied perpendicular to the device. However, unlike classical Hall or IQHE, no external magnetic field is involved in SHE, which requires only SOC and the longitudinal charge current.

It is worth mentioning that most experimental observation of SHE is of extrinsic nature, or at least not conclusively indicative of intrinsic SHE. The processes of skew scattering and side jumping have been attributed to causing the extrinsic SHE. One can then think in terms of spin anisotropic scattering of electrons by impurities. Spin up electrons will scatter in a direction different from that of spin down electrons, hence leading to spin accumulation along the lateral edges. Recent works have spawned numerous versions which include SHE of photons,[53,54] phonons,[55] excitons[56] and in graphene.[57] In this chapter, discussion is restricted to the intrinsic SHE which can be observed in an SOC system with an applied electric field even in the absence of disorder or impurities scattering. In fact intrinsic SHE has been predicted to be several orders of magnitude larger than the extrinsic one. It will be intuitive to view the SOC and electric field system as the generator of spin-dependent electron motion.

In semiconductor spintronics, intrinsic SHE has been studied in different types of SOC systems as listed in Table 5.6. Specifically, we focus on the SHE of electron spin in the Rashba 2DEG. The Luttinger bulk system is also discussed briefly. One, however, needs to keep in mind that extrinsic SHE might be more promising in terms of device applications because of experimental success and its observation at room temperature.[58] In fact this recent spate of experimental successes involving the inverse SHE might be related to the extrinsic SHE.

Before we move on to the physics of spin Hall, we draw readers' attention to the interesting analogies and connections between the different types of Hall effects. Figure 5.5 classifies the various Hall systems based on the transverse separation of charge and spin current.

*Table 5.6* SHE in semiconductor spintronics and their respective universal conductivity

| | Semiconductor systems | Spin Hall conductivity |
|---|---|---|
| 1. | Rashba 2DEG[59] | $\sigma^z_{xy} = \frac{-e}{8\pi}$ |
| 2. | Luttinger (p-doped bulk)[60] (Note: 'not universal' because of its $k_F$ dependence; $H$ and $L$ stand for heavy and light holes, respectively) | $\sigma^z_{xy} = \frac{e\left(3k_F^H - k_F^L\right)}{12\pi^2}$ |
| 3. | Cubic Dresselhaus (n-doped bulk)[61] (Note: 'not universal' because of its $k_F$ dependence) | $\sigma^z_{xy} = \frac{-ek_F}{12\pi^2}$ |
| 4. | Rashba heavy holes[62] | $\sigma^z_{xy} = \frac{9e}{8\pi}$ |

Prof. V. I. Perel, 1976 Prof. M. I. Dyakonov, 1976

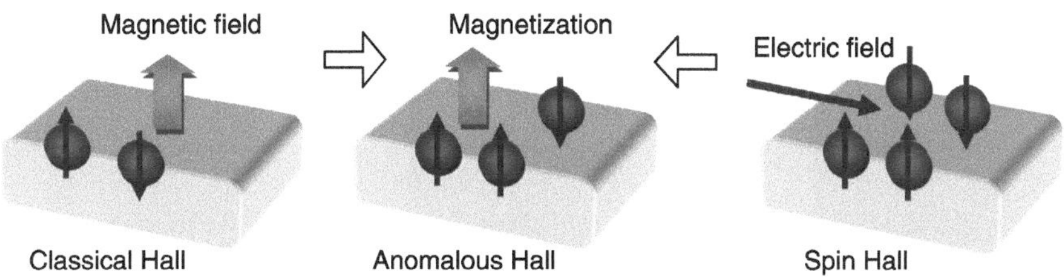


*5.5* Schematic illustration of three types of Hall effects involving the charge and spin degree of freedoms of electrons. The anomalous Hall effect (AHE) is both charge and spin Hall effect (SHE). Both AHE and SHE need spin orbit coupling.

## 5.7 SHE in the Rashba 2DEG system

We now take a look into a specific type of SHE, namely one with vertical spin (perpendicular to the two-dimensional system) separated oppositely to the two edges of the mesoscopic system. SHE in Rashba 2DEG was first discussed in reference 59, which shows the spin Hall conductivity with the Kubo method where substituting the eigenstates of the Rashba system into the Kubo conductivity equation shows spin dependent conductivity. But the Kubo method lacks transparency in terms of the underlying physical effects. We will thus discuss a few other more heuristic approaches[63,64] in sub-sections to add physical clarity to the results obtained with the Kubo formula. The numerous methods used here might partially describe physical effects which could possibly be related to SHE. But we would not pretend that these methods fully represent the SHE originally described in the work of Dyakonov and Perel in 1971.[48]

We first consider the velocity operators in the case of Rashba, i.e.

$$\hat{v}_x = \partial H / \partial p_x = \frac{\hat{p}_x}{m} - \frac{\alpha}{\hbar}\sigma_y,$$

$$\hat{v}_y = \partial H / \partial p_y = \frac{\hat{p}_y}{m} - \frac{\alpha}{\hbar}\sigma_x, \qquad [5.52]$$

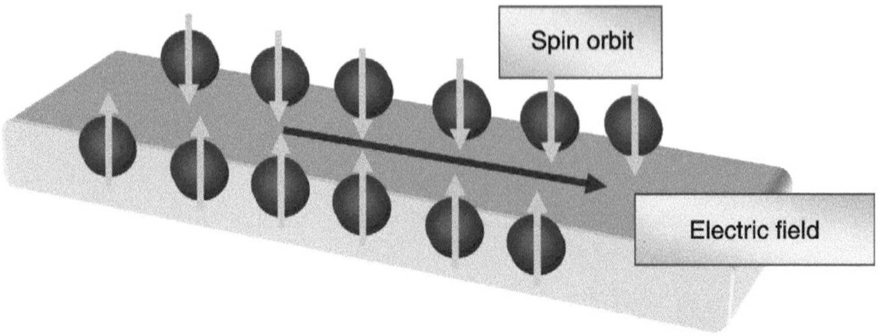


5.6 Schematic illustration of the vertical SHE in a semiconductor 2DEG system under the effect of applied electric field and Rashba SOC.

where the hat sign is used to emphasize its operator nature. Equation 5.52 leads to the useful spin current operator of:

$$\hat{J}_x^z = \frac{\hbar}{4}\{\sigma_z, \hat{v}_x\} = \frac{\hbar}{4}\left\{\sigma_z, \frac{p_x}{m} - \frac{\alpha}{\hbar}\sigma_y\right\} = \frac{\hbar}{2m}\sigma_z p_x, \quad [5.53]$$

Applying the Kubo–Greenwood relation, the direct current (DC) transverse spin Hall conductivity is given by:

$$\sigma_{xy}^{SH}(\omega \to 0) = \frac{e\hbar}{A}\sum_{k,n\neq n'}(f_{n'k} - f_{nk})\frac{\mathrm{Im}\left[\left\langle n'k\left|\hat{J}_x^z\right|nk\right\rangle\left\langle nk\left|\hat{v}_y\right|n'k\right\rangle\right]}{(E_{nk} - E_{n'k})^2}, \quad [5.54]$$

where $n$ refers to the sub-band index and $A$ is the area.

In the limit of $T \to 0$, the Fermi factor simplifies to $(f_{+k} - f_{-k}) = 1$, within the annular region in $k$-space of $k_- < k < k_+$. With the spin Hall conductivity shown to be $\sigma_{xy}^{SH}(0) = \frac{e\hbar^2\alpha}{mA}\Sigma_k(f_{+k} - f_{-k}) \times \frac{k_x^2/k}{(4k^2\alpha^2)}$, and replacing the discrete sum by an integral over the annular region, one has:

$$\sigma_{xy}^{SH}(0) = \frac{e\hbar^2}{4m\alpha A}\sum_k(f_{+k} - f_{-k})\frac{k_x^2}{k^3} = \frac{e\hbar^2}{4m\alpha A}\left(\frac{L^2}{4\pi^2}\right)\int_{k-}^{k+} d\boldsymbol{k}\left(\frac{k_x^2}{k^3}\right)$$

$$= \frac{e\hbar^2}{16m\pi^2\alpha}\int_{k-}^{k+} dk\int_0^{2\pi} d\varphi\cos^2\varphi. \quad [5.55]$$

But the difference in the wavevector of the two (+) and (−) Rashba sub-bands is given by $k = (k_+ - k_-) = \frac{2m\alpha}{\hbar^2}$. (see Fig. 5.1). One can thus determine that:

$$\sigma_{xy}^{SH} = \frac{e\hbar^2}{16m\pi^2\alpha}(k_+ - k_-)\pi = \frac{e\hbar^2}{16m\pi\alpha}\left(\frac{2m\alpha}{\hbar^2}\right) = \frac{e}{8\pi}. \quad [5.56]$$

This is a universal spin Hall conductivity value, obtained via linear response theory. However, SHE vanishes in the diffusive limit due to impurity scattering, causing a steady cancellation of the required acceleration. This effect is consistent with the prediction of Inoue *et al.*,[65] who showed that the spin Hall conductivity vanishes when vertex corrections are introduced to model the effects of impurity scattering. The suppression effect is, however, not generally true; for instance, in bulk p-type semiconductors and two-dimensional (2D) hole gases, SHE persists in the presence of impurities. What is remarkable here is that the SHE conductivity is a universal constant independent of material parameters. Nevertheless, the Kubo method lacks transparency in terms of physical elucidation. It is unclear how such independent conductivity arises, although SOC is accepted to be the underlying reason. There are numerous other theories which study the different aspects of SHE, e.g. the edge precession effects driving the spin component to be out-of-plane, resulting in observable SHE. Since it is hard to make a direct comparison of these methods with those described below, we do not discuss these methods in detail. However, for completeness, we refer readers to references 66–69 for more details. Of the many theories that have been put forward to explain SHE in Rashba 2DEG, we will discuss only three theories cast in the context of spin-dependent effects on transverse electron motion, where electron motion is described heuristically in terms of momentum, Lorentz force, position, or conductivity:

1. Electron acceleration and spin precession.[59]
2. Time–space gauge theoretic.[63]
3. Non-Abelian gauge and zitterbewegung.[64]

*Table 5.7* Theories that explain spin transverse separation of SHE in a 2DEG with Rashba SOC and applied electric fields

| | Physical elucidation of SHE in Rashba 2DEG | Spin polarization | Transverse effects | Quantification |
|---|---|---|---|---|
| 1. | Electron acceleration and spin precession | Internal Zeeman fields | Spin (Zeeman) dependent transverse momentum | Universal conductivity $\frac{e}{8\pi}$ |
| 2. | Time–space gauge theoretic | Internal Zeeman fields | Spin (Zeeman) dependent transverse momentum | Universal conductivity $\frac{e}{8\pi}$ |
| 3. | Non-Abelian gauge and Zitterbewegung | Cannot predict an internal field | Spin dependent transverse | Average Lorentz force |
| | | External means would be required to sustain a vertical spin polarization<br>(1) Spin injection<br>(2) Chiral magnetic field | Lorentz force | Average position |

Note: The first two methods yield the same universal conductivity value, while the third method gives a heuristic indication.

From Table 5.7, method 3 is not able to predict the existence of an internal Zeeman field which polarizes the spin in the vertical direction. Thus one may conjecture that external means of sustaining vertical spin polarization may be needed, e.g. via spin injection or external magnetic fields. But physical elucidations of types (1) and (2) in Table 5.7 could indeed predict a natural internal Zeeman field, which thus suggests that in SHE one may not need an external means to sustain spin polarization. In any case, all three methods predict spin-dependent momentum, which means all three methods point to the fact that transverse separation of spin can occur whenever vertical spin polarization is realized in the Rashba 2DEG system, whether it is via external means or the internal Zeeman field. However, transverse separation of spin should not be taken as the final proof of SHE. A conclusive indication of SHE should come from quantification in the form of conductivity. Methods 1 and 2 are able to predict a universal conductivity, while method 3 could only lead to average transverse velocity.

## 5.7.1 Electron acceleration and spin precession

This rather well-accepted method, which describes SHE as arising due to the spin precession and the acceleration of electrons by the electric field in the presence of Rashba SOC, was first proposed by Sinova *et al.*[59] It was explained that accelerated electrons in a Rashba 2DEG with a net left transverse velocity precess in the opposite direction to those traveling to the right, resulting in a transverse spin separation. Thus the system can be perceived to possess internal Zeeman fields as a result of electron acceleration under SOC. The 'special' Zeeman fields sustain polarization in the 2DEG. In the above, one can visualize electron spin precession as momentum dependent. By reciprocity, one can also say that the transverse momentum of electron is spin (Zeeman) dependent.

It is purely a matter of taste how one likes to visualize the underlying dynamics of SHE. Summing over all momenta over the annular Fermi surface yields a remarkable universal SHE conductivity identical to that obtained with the Kubo formula. It is imperative that electrons accelerate, i.e. that the momentum is time dependent. This imposes the strict requirement for ballistic transport. In the presence of impurities, the retardation force on electrons reduces the average acceleration to zero. Thus the SHE vanishes in the diffusive limit because the impurity scattering causing a steady cancellation of the required acceleration. This prediction is also consistent with the method of Inoue *et al.*,[65] which shows that SHE conductivity vanishes when vertex corrections are introduced to model the effects of impurity scattering in Rashba 2DEG. This suppression effect is, however, not generally true; for instance, in the bulk p-type semiconductors and the 2D-hole gas, SHE persists in the presence of impurities.

The electron acceleration and precession approach can essentially be viewed as a form of spin-dependent momentum. Below is a mathematical illustration of the spin-dependent momentum description. Electrons traveling in the transverse $y$-direction are spin-polarized in the vertical $z$-direction, i.e.

$$I_y^Z = \sum_{k\in\{annulus\}} \left(\frac{\hbar}{2}\right) n_z(\boldsymbol{k}) v_y = \sum_{k\in\{annulus\}} \left(\frac{\hbar}{2}\right)\left(\kappa\frac{p_y}{p^3}\right)\left(\frac{p_y}{m}\right), \qquad [5.57]$$

where $\kappa = -\frac{e\hbar^2 E_x}{2\alpha}$. The physics of spin-dependent momentum transport is captured in the vertical polarization term of $n_z(\boldsymbol{k})$ whose detailed derivation is not shown here, but can be found in references 59 and 70. Converting the above summation to an integral over the states within the annular region between the two Rashba sub-bands (Fig. 5.1), we have:

$$\begin{aligned} I_y^Z &= \int_{\boldsymbol{k}\in\{annulus\}} \frac{L^2}{(2\pi)^2} d^2\boldsymbol{k} \left(\frac{\hbar}{2}\right)\left(\kappa\frac{p_y}{p^3}\right)\left(\frac{p_y}{m}\right) \\ &= \frac{A}{4\pi^2}\int_{k\in\{annulus\}} (k\,dk\,d\varphi)\left(\frac{\kappa}{2m}\right)\left(\frac{k_y^2}{k^3}\right) \end{aligned} \qquad [5.58]$$

and thus, the current density is given by:

$$\begin{aligned} j_y^Z &= \frac{\kappa}{8\pi^2 m}\int_{k\in\{annulus\}} \left(\frac{k_y^2}{k^2}\right) dk\,d\varphi = \frac{\kappa}{8\pi^2 m}\int_{k_-}^{k_+} dk \int_0^{2\pi} d\varphi \sin^2\varphi \\ &= \frac{\kappa}{8\pi m}(k_+ - k_-). \end{aligned} \qquad [5.59]$$

Substituting the explicit expression for $\kappa$, we have:

$$j_y^Z = \frac{eE_x\hbar^2}{16\pi\alpha m}(k_+ - k_-). \qquad [5.60]$$

Referring to Fig. 5.1 once again for $k_+ - k_-$, the transverse spin current is thus:

$$j_y^Z = -\frac{eE_x}{8\pi}, \qquad [5.61]$$

resulting in the spin Hall conductivity of:

$$\sigma_{SH} = \frac{j_y^z}{E_x} = \frac{-e}{8\pi}. \qquad [5.62]$$

---

**Exercise 5.6**

Show that

(1) $\left\langle +k \right| \hat{v}_y \left| -k \right\rangle = \frac{i\alpha}{\hbar}\cos\varphi.$

(2) $\left\langle -k \right| \hat{J}_x^z \left| +k \right\rangle = -\frac{\hbar^2 k_x}{2m}.$

(3) Spin Hall conductivity is given by:

$$\sigma_{xy}^{SH}(0) = \frac{e\hbar^2\alpha}{mA}\sum_k (f_{+k} - f_{-k}) \times \frac{k_x^2 / k}{(4k^2\alpha^2)}.$$

***Solution***

(1) We evaluate $\langle nk|\hat{v}_y|n'k\rangle$ for $n = +$, and $n' = -$,

$$\left\langle +k\left|\hat{v}_y\right|-k\right\rangle = \left\langle +k\left|\frac{\hat{p}_y}{m} + \frac{\alpha}{\hbar}\sigma_x\right|-k\right\rangle = \frac{\alpha}{\hbar}\langle +k|\sigma_x|-k\rangle$$

$$= \frac{\alpha}{2\hbar}(ie^{i\varphi}, 1)\begin{pmatrix} 0 & 1 \\ 1 & 0 \end{pmatrix}\begin{pmatrix} ie^{i\varphi} \\ 1 \end{pmatrix} = \frac{i\alpha}{\hbar}\cos\varphi.$$

where $\varphi$ is the in-plane angle subtended by $k_x$ and $k_{//}$, where $k_{//} = \sqrt{k_x^2 + k_y^2}$.

(2) $$\left\langle -k\left|\hat{j}_x^Z\right|+k\right\rangle = \frac{\hbar}{2m}\langle -k|\sigma_z p_x|+k\rangle = \frac{\hbar^z k_x}{2m}\left[\frac{1}{2}(-ie^{i\varphi}, 1)\begin{pmatrix} 1 & 0 \\ 0 & -1 \end{pmatrix}\begin{pmatrix} -ie^{-i\varphi} \\ 1 \end{pmatrix}\right]$$

$$= -\frac{\hbar^z k_x}{2m}.$$

(3) Combining both, we obtain:

$$\mathrm{Im}\left[\left\langle -k\left|\hat{J}_x^z\right|+k\right\rangle\left\langle +k\left|\hat{v}_y\right|-k\right\rangle\right] = \mathrm{Im}\left[\left(\frac{i\alpha}{\hbar}\cos\varphi\right)\left(-\frac{\hbar^2 k_x}{2m}\right)\right] = -\frac{\hbar k_x\alpha}{2m}\cos\varphi.$$

Repeating the above calculation for the other combination, i.e. $n = -$, and $n' = +$, we find the correlation differs only in signs, i.e.

$$\mathrm{Im}\left[\left\langle +k\left|\hat{J}_x^Z\right|-k\right\rangle\left\langle -k\left|\hat{v}_y\right|+k\right\rangle\right] = \frac{\hbar k_x\alpha}{2m}\cos\varphi.$$

Contribution to the conductivity is, however, of the same sign, due to an additional (−) sign from the Fermi distribution factor $(f_{n'k} - f_{nk})$. Hence, the spin Hall conductivity is given by:

$$\sigma_{xy}^{SH}(0) = \frac{e\hbar}{A} \times 2\sum_k (f_{+k} - f_{-k}) \times \frac{\mathrm{Im}\left[\left\langle +k\left|\hat{J}_x^Z\right|-k\right\rangle\left\langle -k\left|\hat{v}_y\right|+k\right\rangle\right]}{(E_{-k} - E_{+k})^2}$$

$$= \frac{2e\hbar}{A}\sum_k (f_{+k} - f_{-k}) \times \frac{\left(\frac{\hbar k_x\alpha}{2m}\right)\cos\varphi}{(E_{-k} - E_{+k})^2}$$

$$= \frac{e\hbar^2\alpha}{mA}\sum_k (f_{+k} - f_{-k}) \times \frac{k_x^2 / k}{(4k^2\alpha^2)}.$$

In the last line of the above, we have made use of the fact that $(E_{-k} - E_{+k}) = 2k\alpha$.

## 5.7.2 Time-space gauge theoretic

While the acceleration and precession method is plausible, one wonders if a gauge theoretic approach will yield similar results. This is in view of the fact that the adiabatic spin alignment method has been employed rather convincingly to expound hole SHE in p-doped semiconductors. In fact taking a closer look at the 2DEG system reveals a remarkable connection[63] between the gauge theory and the precession approach. The clue lies again in the acceleration of electron and the necessary generation of a vertical effective magnetic field, which can be linked to those discussed by Aharonov and Stern.[70] Thus a time-dependent momentum is required similar to the above method, imposing the need for ballisticity. In fact it might have been implicit in the previous method[59] that electron spin eventually aligns with the additional vertical magnetic field.

In fact, casting the Hamiltonian of the system in the interaction picture and treating the SOC term as a perturbation, a local gauge transformation could impart an extra term to the Hamiltonian. It was found that this term is like a Zeeman magnetic field. For a right moving electron, this Zeeman field is vertical up generating spin up current; for a left moving electron, this Zeeman field is vertical down generating spin down current. The transverse momentum of electron is thus Zeeman (spin polarization) dependent, resulting in SHE, similar to the previous theory discussed above. Summing over all momenta over the annular Fermi surface yields a remarkable universal SHE conductivity identical to that obtained with the Kubo formula. In the following, we provide some details of this theoretical method and summarize important results in Table 5.8. This provides a summary of the Hamiltonian and its transformation in Schrödinger and interaction pictures. A local gauge transformation in the interaction picture is required to generate the effective magnetic field.

In the interaction picture where $V_1 = \gamma\boldsymbol{\sigma}.\boldsymbol{B(k)}$ is taken as a perturbation to the system, one then has:

$$V_I(t) = e^{iH_0t}V_Ie^{-iH_0t} = \gamma\boldsymbol{\sigma}.\left(\boldsymbol{B}(\boldsymbol{k}) - \frac{eE_vt}{\hbar}.\frac{\partial \boldsymbol{B}(\boldsymbol{k})}{\partial k_v}\right) = \gamma\boldsymbol{\sigma}.\boldsymbol{B}(t). \qquad [5.63]$$

*Table 5.8* Hamiltonian in the laboratory and transformed frame under the Schrödinger and interaction pictures reveal different additional energy terms

| | Schrödinger picture | Interaction picture |
|---|---|---|
| Lab frame | $H_S = \frac{\boldsymbol{p}^2}{2m} + e\boldsymbol{E}.\boldsymbol{r} + \gamma\boldsymbol{\sigma}.\boldsymbol{B}(\boldsymbol{k})$ | $H_I(t) = \frac{\boldsymbol{p}^2}{2m} + e\boldsymbol{E}.\boldsymbol{r} + V_I(t)$ |
| Transformed frame | $H'_s \equiv \left[\frac{\boldsymbol{p}^2}{2m} - e\boldsymbol{E}.(\boldsymbol{r} - iU^{\dagger}\partial_k U) + \gamma\sigma_z B\right]$ <br> $= i\hbar U^{\dagger}\partial_t U$ <br> $i\hbar U^{\dagger}\partial_t U = E$ | $H'_I = \left[\frac{\boldsymbol{p}^2}{2m} - e\boldsymbol{E}.(\boldsymbol{r} - iU^{\dagger}\partial_k U) + \gamma\sigma_z B(t)\right]$ <br> $= i\hbar U^{\dagger}\partial_t U$ <br> $i\hbar U^{\dagger}\partial_t U = E - \hbar\mathcal{A}_0(t)$ |

The term $i\hbar U^{\dagger}\partial_t U$ yields $E$ in the Schrödinger picture but $E - \hbar\mathcal{A}_0(t)$ in the interaction picture because, in the latter, momentum hence $U(\boldsymbol{p}(t))$ is time-dependent, resulting in non-vanishing $i\hbar(U^{\dagger}\partial_t U)$ as shown below:

$$i\hbar U^{\dagger}\partial_t U = i\hbar\partial_t + i\hbar(U^{\dagger}\partial_t U) = E - \hbar\mathcal{A}_0(t). \quad [5.64]$$

This need for a time-dependent momentum implies the need for ballistic transport. The process of local gauge transformation in the interaction picture has generated an extra gauge term of $\hbar\mathcal{A}_0(t)$, which is also expressible by $\hbar\mathcal{A}_0(t) = \hbar\boldsymbol{\sigma}.(\boldsymbol{m} \times \partial_t \boldsymbol{m})$, if $U = \boldsymbol{\sigma}.\boldsymbol{m}(t)$. It is not hard to see that as $U = \exp\left\{\frac{i}{2}\boldsymbol{\sigma}.\boldsymbol{\omega}t\right\}$, $\hbar\mathcal{A}_0(t) = -i\hbar(U^{\dagger}\partial_t U)$ yields by approximation:

$$\hbar\mathcal{A}_0(t) = \frac{\hbar}{2}U(\boldsymbol{\sigma}.\boldsymbol{\omega})U^{\dagger} \quad [5.65]$$

which, when viewed in the lab frame, is simply a Zeeman-like magnetic field. In other words, an effective magnetic field has been generated in this system. When the local gauge transformation involving $U = \exp\left\{\frac{i}{2}\boldsymbol{\sigma}.\boldsymbol{\omega}\,t\right\}$ is taken to transform the eigenvector of $\boldsymbol{n}(t)$ to $\boldsymbol{n}(t + dt)$, it also means rotating the $z$ axis from $\boldsymbol{n}(t)$ to $\boldsymbol{n}(t + dt)$ about $\boldsymbol{\omega}$. The explicit expression of $U$ is not unique; in Eq. 5.66 we give two possible unitary matrices that could perform the process described above:

$$U1 = \frac{1}{\sqrt{2}}\begin{bmatrix} 1 & e^{-i\varphi} \\ e^{i\varphi} & -1 \end{bmatrix};\ U2 = \frac{1}{\sqrt{2}}\begin{bmatrix} e^{i\varphi/2} & e^{-i\varphi/2} \\ -e^{i\varphi/2} & e^{-i\varphi/2} \end{bmatrix}. \quad [5.66]$$

The process of axis rotation can be illustrated by the schematic shown in Fig. 5.7.

For convenience, $\boldsymbol{n}_1 = \boldsymbol{n}(t)$, $\boldsymbol{n}_2 = \boldsymbol{n}(t + dt)$. Note that $\boldsymbol{n}$ is the unit vector of the spin orbit effective magnetic field, i.e. $\boldsymbol{B} = B\boldsymbol{n}$. From the above, one can work out $\partial_t\boldsymbol{n} = \boldsymbol{n} \times \boldsymbol{\omega}$, which leads naturally to:

$$\boldsymbol{\omega} = \boldsymbol{n} \times \partial_t\boldsymbol{n} + (\boldsymbol{\omega}.\boldsymbol{n})\boldsymbol{n} \quad [5.67]$$

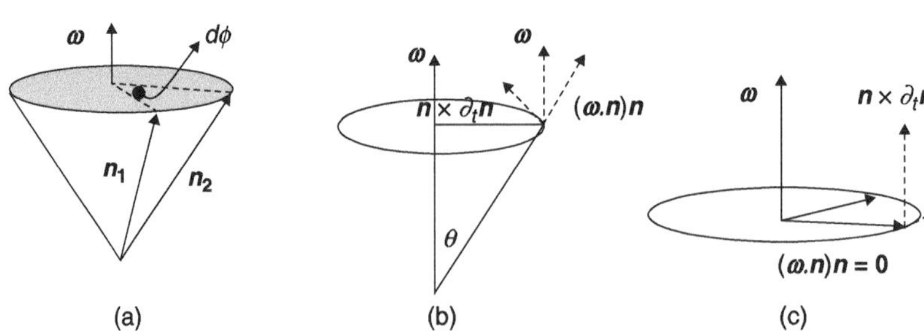


*5.7* (a) General schematic showing the rotation of an eigenvector about the rotation axis $\omega$. (b) For an arbitrary angle $\theta$, schematic illustrating the generation of an effective Zeeman field. (c) For $\theta = 90$, the Zeeman-like $\omega$ is now vertical to the plane containing $n$.

where when $\boldsymbol{n}_1$ and $\boldsymbol{n}_2$ are both contained in the 2DEG plane; $(\boldsymbol{\omega}.\boldsymbol{n})$ vanishes and one is left with $\boldsymbol{\omega} = \boldsymbol{n} \times \partial_t \boldsymbol{n}$ pointing vertical to the 2DEG plane. Thus the term $\boldsymbol{\sigma}.\boldsymbol{\omega}$ can be viewed as a form of internal Zeeman fields which sustain the vertical spin polarization in the Rashba 2DEG. This is not unlike the picture described in the precession–acceleration correlated mechanism discussed earlier.

Critical readers might argue that Eq. 5.67 does not pre-determine the orientation of $\boldsymbol{\omega}$. There are many possible solutions of $\boldsymbol{\omega}$ and $\boldsymbol{n}$ that can satisfy this equation, depending on how a rotation is performed. Simply put, $\boldsymbol{\omega}$ and $\boldsymbol{n}$ are not unique because there is a rotational freedom. Let us begin with an arbitrary axis and rotate this axis with $U = \exp\left\{\frac{i}{2}\boldsymbol{\sigma}.\boldsymbol{\omega}\, t\right\}$ to $\boldsymbol{n}_1$ about $\boldsymbol{\omega}$. Although $\boldsymbol{n}_1$ is a vector contained within the 2DEG plane, this fact does not predetermine $\boldsymbol{\omega}$. But now, the arbitrary axis is aligned along $\boldsymbol{n}_1$. The next instant is to rotate this axis to $\boldsymbol{n}_2$, which is equivalent to rotating $\boldsymbol{n}_1$ to $\boldsymbol{n}_2$. The fact that both $\boldsymbol{n}_1$ and $\boldsymbol{n}_2$ lie in the same 2DEG plane allows no other solution for $\boldsymbol{\omega}$ than to stand perpendicular to the 2DEG plane. This leaves $(\boldsymbol{\omega}.\boldsymbol{n})\boldsymbol{n}$ zero, and reduces $\boldsymbol{\omega} = \boldsymbol{n} \times \partial_t \boldsymbol{n} + (\boldsymbol{\omega}.\boldsymbol{n})\boldsymbol{n}$ to $\boldsymbol{\omega} = \boldsymbol{n} \times \partial_t \boldsymbol{n}$. This process is clearly illustrated in Fig. 5.7(c). Rotation back to the lab frame would leave the Hamiltonian with a Zeeman term of $\boldsymbol{\sigma}.\boldsymbol{w}$. Careful examination of the term $\boldsymbol{\omega} = \boldsymbol{n} \times \partial_t \boldsymbol{n}$ reveals that an electron's transverse momenta will take on a positive or a negative spin polarization depending on the Rashba band it is associated with. Careful selection of the band region, in this case the annular region of the Rashba band, is crucial for producing a SHE effect. The above shows that spin-dependent transverse separation is heuristically possible. In fact, with this modified Hamiltonian, SHE in Rashba 2DEG can be quantified with integration over the proper momenta region to yield a definitive spin transverse conductivity. The process is identical to the semiclassical approach used in the previous method. In fact the difference between the present and the former method lies in the vertical spin polarization $\boldsymbol{n}$. In this method, the vertical polarization arises due to symmetry involving the time coordinate. The universal value of $\sigma_{xy} = \frac{e}{8\pi}$ can be reproduced.

### 5.7.3 Non-Abelian gauge SHE

In the context of non-Abelian phase in spintronics,[71,72] a form of spin transverse separation can be predicted in which an up spin 'feels' a vertically up magnetic field, while a down spin 'feels' the down version. This method predicts two copies of effective magnetic fields. However, the effective magnetic fields here is of Lorentz nature and not Zeeman, which means it exerts force on electron but has no effect on precessing or locking the electron spin. This system can be viewed as generating spin-dependent Lorentz magnetic fields. By contrast, the previous two methods generate spin-dependent (Zeeman) momenta. One distinct difference is that, unlike the previous two methods, the non-Abelian gauge method could not predict a Zeeman field to sustain spin polarization. It is therefore still unclear at this point whether this mechanism concurs with the other two in terms of the universal SHE conductivity. Non-Abelian gauge effects in nanoscale devices were first discussed in the late 1980s.[73–75] Its relation to SHE was discussed only in earnest in recent years.[64,76–80] To study SHE under the non-Abelian gauge field, one may need a special 2DEG system with external magnetization or magnetic field vertical-to-plane. This is because the requisite vertically spin-polarized state is not an eigenstate of the system as the vertical Rashba electric field produces the spin orbit magnetic field in the in-plane direction. This leads to two effects, both detrimental to the spin current:

1. The large electric field hastens the relaxation of the initial vertical spins to the in-plane direction, thereby suppressing the SU(2) transverse force.
2. For channel lengths longer than the spin coherence length, the spin vector will precess about the in-plane relativistic magnetic field, causing a zitterbewegung-like motion and resulting in zero net transverse spin current.

To observe SHE in these systems, the problem of spin relaxation or precession has to be circumvented. As is well known, the local magnetization with net chirality could in the adiabatic limit generate a topological spin gauge field that separates conduction electrons of opposite spins in the transverse direction. Additionally, the SOC can be viewed to produce a SU(2) gauge field which reinforces or opposes the effect of the spin gauge. This system could thus provide a tunable spin separation effect, where an applied gate voltage modulates the transverse spin current.

To provide a theoretical description of such an SHE system, we write the Hamiltonian of the system as:

$$H = \frac{p_x^2}{2m} + \frac{(p_y + eA_y)^2}{2m} + \frac{eg\hbar}{4m}\boldsymbol{\sigma}.\boldsymbol{B} + \frac{\hbar e}{4m^2c^2}\boldsymbol{\sigma}.\left[(\boldsymbol{p} + e\boldsymbol{A}^{\boldsymbol{B}})\times \boldsymbol{E}\right], \qquad [5.68]$$

where $\boldsymbol{\sigma}$ is the vector of Pauli spin matrices, $\boldsymbol{A^B} = (0A_y^B,0)$ is the Landau gauge associated with the external $\boldsymbol{B}$ field and $g\hbar/4m$ is the Zeeman splitting strength. Equation 5.68 can be rewritten as:

$$H = \frac{1}{2m}\left(p_x + \left[\frac{\hbar e}{4mc^2}\sigma_i E_j \varepsilon_{ijx}\right]\right)^2 + \frac{1}{2m}\left(p_y + e\left[A_y^B + \frac{\hbar}{4mc^2}\sigma_i E_j \varepsilon_{ijy}\right]\right)^2$$
$$+ \frac{eg\hbar}{4m}\boldsymbol{\sigma}.\boldsymbol{B}, \qquad [5.69]$$

after ignoring the higher order terms. If one considers a 2DEG system which consists of both the local magnetization and SOC effects, SHE in this system is slightly more complex depending on how the system is treated. If one regards the SOC as weak and that spin aligns mainly with the local magnetization, then a unitary transformation can be performed to rotate the $z$-axis to the $\boldsymbol{B}$ field:

$$H' = UHU^{\dagger} = \sum_{k=x,y,z} \frac{1}{2m}\left( p_k + e\left[ A_k^B + \frac{\hbar}{4mc^2} U\sigma_i E_j \varepsilon_{ijx} U^{\dagger} + \frac{\hbar}{e} U\partial_k U^{\dagger} \right]\right)^2 + \frac{eg\hbar}{4m}\sigma_z B_z. \qquad [5.70]$$

Under the adiabatic condition, the 2×2 matrix $U\partial_k U^{\dagger}$ becomes diagonal, hereafter represented by matrix $A_k^M$ (whose elements are the monopole potential). Taking note that the electric field is vertical to the 2DEG plane in the Rashba system, the transformation $U(\sigma_i E_j \varepsilon_{ijk})U^{\dagger}$ is equivalent to rotating the laboratory $z$-axis to the $\boldsymbol{B}$ field. Thus, the gauge fields comprise the $A^{SU(2)}$ term from the SOC effects, and the topological term arising from the net chirality of the $\boldsymbol{B}$ field, $A_k^M = \begin{pmatrix} a_k & 0 \\ 0 & -a_k \end{pmatrix}$, where $a_k$ is the $U(1)$ monopole potential. The diagonal components of the monopole gauge matrix ($\pm a_k$) can be understood by the path-integral method, $\psi(x,t) = \int G(xt, x_0 t_0)\, \psi(x_0, t_0) dx_0$ for spin parallel/anti-parallel to the $\boldsymbol{B}$ field, where $G(xt, x_0 t_0)$ is the propagator between times $t_0$ and $t$, and $x$ and $t$ are $x_{N+1}$ and $t_{N+1}$, respectively. Explicitly, the evolution is described by:

$$\psi(x,t) = \int \left[ \left( \frac{m}{2\pi i\hbar\Delta t} \right)^{\frac{n+1}{2}} e^{\left[ \frac{m}{2}\left( \frac{x_{n+1} - x_n}{\Delta t} \right)^2 - V(x_n) \right] \frac{i\Delta t}{\hbar}} \cdots e^{\left[ \frac{m}{2}\left( \frac{x_1 - x_0}{\Delta t} \right)^2 - V(x_0) \right] \frac{i\Delta t}{\hbar}} dx_n \ldots dx_1 \right] \psi(x_0 t_0) dx_0, \qquad [5.71]$$

where $V(x_n)$ is the local potential. Equation 5.71 can be written in a simple form:

$$\psi(x,t) = \int \left[ \left( \frac{m}{2\pi i\hbar\Delta t} \right)^{\frac{n+1}{2}} e^{\frac{iS(t)}{\hbar}} dx_n \ldots dx_1 dx_0 \right] \psi(x_0 t_0), \qquad [5.72]$$

where $S(t) = \frac{1}{\hbar}\int_0^T \frac{m}{2}\left( \frac{dx}{dt} \right)^2 - V(x_n)\, dt$ is the action of the system. In the dynamic spinor system, neglecting the dynamic phase, $S$ would correspond to the action $S\left[ n(t) \right] = \frac{\hbar}{2}\int_0^t iz^{+}\partial_t z\, dt$, where $n(t)$ refers to the spinor vector at time $t$ and $z$ is the spinor. An expansion of the action leads to:

$$S = \frac{\hbar}{2}\int_0^t iz^{+}\partial_t z\, dt = \pm\frac{\hbar}{2}\int_0^t (1 - \cos\theta)\frac{\partial\phi}{\partial t}\, dt, \qquad [5.73]$$

for spin parallel (+) and anti-parallel (−) to field. Considering the evolution over a short time interval, the above integration leads to:

$$\int \boldsymbol{a}.\, d\boldsymbol{r} = \pm \int \frac{\hbar}{2e}(1-\cos\theta)\nabla_r \phi.\, d\boldsymbol{r}. \qquad [5.74]$$

It is unclear if the non-Abelian method will lead to the universal SHE conductivity, but at least we can show indication of SHE by two methods – the force operator and the Heisenberg equation of motion method.

*Spin-dependent Yang–Mills SU(2) force*

The effect of the gauge fields on the electron motion, for spin assuming out-of-plane eigenstates under the applied ***B*** field, can be reduced to examining the individual effect of $A^{SU(2)}$ and $A_k^M$. As $\boldsymbol{A^B}$ acts on both up and down spin species in the same transverse direction, assumption of a weak vertical polarization (i.e. that there are equal numbers of up and down spin) implies that $\boldsymbol{A^B}$ would not contribute to the transverse spin separation. We thus focus on the effective magnetic fields generally prescribed by the Yang–Mills field tensor of:

$$F_{\mu\nu} = \partial_\mu A_\nu - \partial_\nu A_\mu + \frac{ie}{\hbar}[A_\nu, A_\mu]. \qquad [5.75]$$

Focusing on the spin-dependent part of the curvature, and using the relation $[U\lambda_\nu U^\dagger, U\lambda_\mu U^\dagger] = U[\lambda_\nu, \lambda_\mu]U^\dagger$ where $\lambda_{\mu\nu}$ is an arbitrary vector component, an effective field for the SU(2) gauge field can be obtained. With $\left|\psi_{\uparrow,k}\right\rangle = \left|\psi_k\right\rangle \begin{pmatrix} 1 \\ 0 \end{pmatrix}$ and after some algebraic manipulation, one arrives at:

$$\left\langle \psi_{\uparrow,k} \middle| \widehat{F}^{SOC}_{y\,ym} n_y \middle| \psi_{\uparrow,k} \right\rangle = \frac{\hbar^2 e^2 E_z^2}{8m^3c^4}\cos\theta \left[ \frac{1}{\pi d} e^{-k_{ox}^2 d^2} \sum_{n=0}^{\infty} \frac{(2dk_{0x})^n}{n!} \int\int d\phi\, dq \cos^{n+1}\phi\, q^{n+2} e^{-q^2} \right] \qquad [5.76]$$

where the Gaussian wavepacket has been used. Further derivation leads to:

$$\begin{aligned} &\left\langle \psi_{\uparrow,k} \middle| \widehat{F}^{SOC}_{y\,ym} n_y \middle| \psi_{\uparrow,k} \right\rangle \left\langle \psi_{\uparrow,k} \middle| \widehat{F}^{SOC}_{y\,ym} n_y \middle| \psi_{\uparrow,k} \right\rangle \\ &= \frac{\hbar^2 e^2 E_z^2}{8m^3c^4}\cos\theta \left[ \frac{1}{\pi d} e^{-k_{0x}^2 d^2} \sum_{N=0}^{\infty} \frac{(2dk_{0x})^{2N+1}}{(2N+1)!} \frac{2\pi(2N+2)!}{\left(2^{N+1}(N+1)!\right)^2} \int_0^\infty dq q^{2N+3} e^{-q^2} \right] \\ &= \frac{\hbar^2 e^2 E_z^2}{8m^3c^4}\cos\theta \left[ \frac{2}{d} e^{-k_{0x}^2 d^2} \sum_{N=0}^{\infty} \frac{(dk_{0x})^{2N+1}}{N!(N+1)!} \int_0^\infty dq q^{2N+3} e^{-q^2} \right]. \end{aligned} \qquad [5.77]$$

The expression in the square brackets evaluates to a simple form, i.e. $k_0 x$. The non-Abelian nature of the gauge arises from the non-commutativity of the SU(2) spin algebra. The approximation $U(\sigma_i E_j \varepsilon_{ijk})U^\dagger \approx \sigma_i E_j \varepsilon_{ijk}$, which holds for small $\theta$, would lead to a force-like operator of:

$$\widehat{F}^{SOC}_y n_y = \frac{\hbar e^2}{8m^2c^4} \sigma_z \hat{v}_x E_z^2 n_y. \qquad [5.78]$$

*Heisenberg equation of motion*

In the following, we provide a more definitive indication of spin separation by deriving the position operator in the Heisenberg picture. In fact a force operator can be related to the time derivative of the velocity operator in the non-relativistic limit. This force operator is, however, different from the curvature force operator. It is remarkable that both force operators yield the same EOM by taking their expectations:

$$\hat{y}_H(t) = e^{i\hat{H}t/\hbar}\,\hat{y}(0)e^{-i\hat{H}t/\hbar} = \sum_{n=0}^{\infty}\left(\frac{\tau^n}{n!}\right)[\hat{H},[\hat{H}\,...[\hat{H},\hat{y}]...]], \qquad [5.79]$$

where $\tau = it/\hbar$. Using the Baker–Campbell–Hausdorff relation and summing the odd and even terms carefully, the position operator can be found. In fact it is interesting to note that the position operator arising due to the force operator, which is itself related to the curvature of the Yang–Mills gauge potential, of $\hat{y}_Y(t)$ is the same as the spin separation part of $\hat{y}_H(t)$ upon converting to the Rashba constant and removing the chirality strength of $\cos\theta$ based on the approximation above.

We now calculate the force expectation values for the case of a Gaussian wavepacket state with the spin aligned to the applied magnetic field. What is to be expected here is that the force expectation values arising from an injected Gaussian wavepacket state would have the same form as that would be obtained with the plane wavefunction, but with the generic $k_x$ of the latter being replaced by $k_{x0}$, which corresponds to the center of the Gaussian wavepacket along the $k_x$ axis. The application of the Heisenberg EOM formalism on the Gaussian wavepacket has an advantage from the physical point of view. It indicates that the force expectation values are dependent on the average net velocity of the Gaussian wavepacket, and can thus be related to the average drift velocity of electrons in the presence of a longitudinal electric field, $E_x$. The drift velocity in the 2DEG system under consideration, which consists of an InAs/InGaAs heterostructure, can typically range over a wide range of values depending on the mobility and the applied electric field.

To obtain a measurable spin separation, we represent the electron's probability amplitude with a Gaussian wavepacket of width $d$ in $k$-space, prepared in the spin up state, i.e. $\psi_{\uparrow_k}(r,t) = \dfrac{d}{2\pi\sqrt{\pi}}\int d^2k e^{-\frac{1}{2}d^2(k-k_0)^2} e^{ik.r}\begin{pmatrix}1\\0\end{pmatrix}$. Assuming that the expectation value of the transverse wavevector is zero, i.e. $\langle\psi_{\uparrow k}|k_y|\psi_{\uparrow k}\rangle = k_{y0} = 0$, the spin-dependent average transverse separation of electrons is:

$$\left\langle\psi_{\uparrow\downarrow}\middle|\hat{y}_H(t)\middle|\psi_{\uparrow\downarrow}\right\rangle = y(0) + \left(\frac{eB_z t^2}{m^2} \pm \frac{eb_z^M t^2}{m^2} \mp \frac{\alpha^2 t^2}{\hbar^3}\right)\hbar k_{x0}. \qquad [5.80]$$

If we look only into the SOC part, it is not hard to see that $\langle\psi_{\uparrow\downarrow}|\hat{y}_H^{SOC}(t)|\psi_{\uparrow\downarrow}\rangle = \langle\psi_{\uparrow\downarrow}|\hat{y}_{ym}^{SOC}(t)|\psi_{\uparrow\downarrow}\rangle$. Both Yang–Mills force and Heisenberg EOM methods give the same indication of possible SHE in Rashba 2DEG. Table 5.9 summarizes the position and force operators due to SOC as well as local magnetization.

*Table 5.9* Yang–Mills force due to SOC and the opposing force due to the chiral local magnetization

| | Spin orbit coupling | Local chiral magnetization |
|---|---|---|
| Effective magnetic fields and force operators | $\boldsymbol{b}_z^{soc} = \frac{\hbar e}{8\, m^2 c^4}(U\boldsymbol{\sigma}U^{\dagger}.\boldsymbol{E})E_z\boldsymbol{n}_z$ | $\boldsymbol{b}_z^{M} = \frac{\hbar}{2eR^2}\sigma_z\boldsymbol{n}_z$ |
| | $\widehat{F}_y^{SOC}\,\boldsymbol{n}_y = \frac{\hbar e^2}{8m^2c^4}(\sigma_z\cos\theta)\widehat{v}_x E_z^2\,\boldsymbol{n}_y$ | $\widehat{F}_y^{M}\,\boldsymbol{n}_y = -\frac{\hbar}{2R^2}\sigma_z\widehat{v}_x\boldsymbol{n}_y$ |
| Expectation forces | $\langle\psi_{\uparrow k}\lvert\widehat{F}_{y\ ym}^{SOC}\,\boldsymbol{n}_y\rvert\psi_{\uparrow k}\rangle = \frac{\hbar^2 e^2E_z^2\cos\theta}{8m^3c^4}k_{x0}\boldsymbol{n}_y$ | $\langle\psi_{\uparrow k}\lvert\widehat{F}_y^{M}\,\boldsymbol{n}_y\rvert\ \psi_{\uparrow k}\rangle = -\frac{\hbar}{2R^2}\langle\psi_{\uparrow k}\lvert\sigma_z\hat{v}_x\,\boldsymbol{n}_y\rvert\langle\psi_{\uparrow k}\rangle = -\frac{\hbar^2}{2mR^2}k_{x0}\boldsymbol{n}_y$ |
| Position operator derived from Heisenberg EOM | $\hat{y}_H(t) = \hat{y}(0) + \frac{\alpha\sigma_y}{2B_z} + \frac{\hat{p}_y t}{m} - \left(\frac{\alpha^2\hat{p}_x\sigma_z}{\hbar^3} - \frac{eB_z\hat{p}_x}{m^2} - \frac{eb_z^M\hat{p}_x\sigma_z}{m^2}\right)t^2$ | |
| | $-\frac{\alpha\sigma_y}{2B_z}\cos\left(\frac{2tB_z}{\hbar}\right) - \frac{\alpha\sigma_z}{2B_z}\sin\left(\frac{2tB_z}{\hbar}\right)$ | |
| | $\hat{y}_H^{SOC}(t) = -\frac{\alpha^2\hat{p}_x\sigma_z}{\hbar^3}t^2$ | $\hat{y}_H^{M}(t) = \frac{eb_z^M\hat{p}_x\sigma_z}{m^2}t^2$ |
| Position operator derived from Yang–Mills force | $\hat{y}_{ym}^{SOC}(t) = \frac{\hbar e^2}{16m^3c^4}(\sigma_z\cos\theta)\hat{v}_x E_z^2 t^2$ | |

## 5.8 Spin drift diffusion for collinear spin valve

Following the discovery of the GMR effect, the theoretical study of spin transport in CIP geometry was initiated by Camley and Barnas,[81] based on the Boltzmann equation. Subsequently, Valet and Fert[82] simplified the Boltzmann equation to the spin diffusion and the two-current models in order to calculate the GMR ratio in the CPP geometry. The SDD equations have been used[83,84] to study spin transport at interfaces between a ferromagnetic and a non-magnetic layer. It was pointed out that, at these interfaces, a spin-dependent split of the electrochemical potential $\Delta\mu = \mu\uparrow - \mu\downarrow$ can be observed or measured. The split can be ascribed to an effective interfacial resistance.

In metal spintronics, the most notable device is the spin valve, which consists of a superlattice of thin metallic layers. The ferromagnetic layer with fixed magnetization is known as the pinned layer and plays the role of a spin polarizer. The ferromagnetic layer with free magnetization is known as the free layer and plays the role of a spin detector. In the CPP configuration, one can imagine spin-polarized current travels from the pinned layer to the free layer and the magnetization orientation of the free layer determines the resistance of the device. Figure 5.8 shows a two-layer basic representation of an otherwise multilayer CPP spin valve.

The SDD equations describe the flow of spin flux in the device:

$$j_s = -\sigma_s \frac{\partial \mu_s}{\partial x} \qquad [5.81]$$

where $\mu$ is the electrochemical potential, $\sigma$ is the conductivity, $x$ is the spatial position and '$s$' refers to the up/down spin orientations. It is not hard to see that Eq. 5.81 is a more general description of Ohm's law for linear potential drop. The second order spin diffusion equation is:

$$\frac{\partial^2 \Delta\mu_s}{\partial x_2} = \frac{\Delta\mu}{\lambda^2} \qquad [5.82]$$

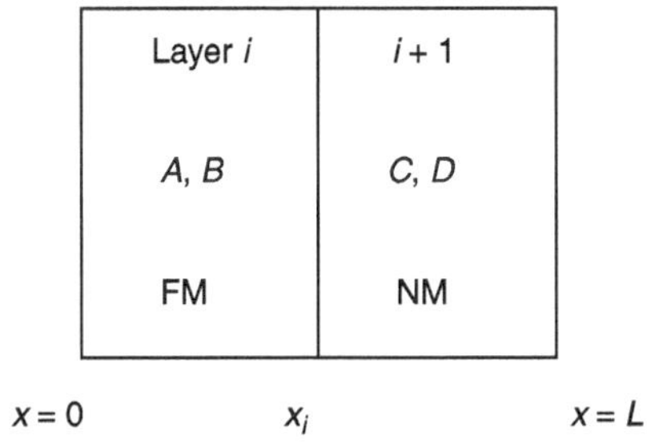


*5.8* Two-layer schematic illustration of a typical multilayer spin valve comprising a core ferromagnetic–non-magnetic–ferromagnetic part. Notations *A, B, C, D* are spin accumulation amplitude at the different layers.

where $\lambda$ is the spin diffusion length (SDL). These equations are to be solved subject to the boundary conditions:

- continuity of spin accumulation, $\Delta\mu_i = \Delta\mu_{i+1}$
- terminal boundary condition, $\Delta\mu_0 = 0$; $\Delta\mu_L = 0$ where $L$ is the thickness of the device.

The solutions for the electrochemical potential, and hence the total potential drop over the device, are obtained for parallel and anti-parallel configurations, from which one can derive the magnetoresistance ratio. In the limit of infinite SDL, the drift-diffusion model reduces to the well-known two-current model. Worth mentioning here is that boundary condition of (b) is an approximation since the zero spin-accumulation condition is strictly valid only at the ends of ferromagnetic layers of infinite thickness. However, one can show that the approximation is approached in a practical spin valve device by taking into account the presence of long non-magnetic conductive leads adjacent to the ferromagnetic layers, and applying zero spin-accumulation condition at the ends of the leads (instead of at the ferromagnetic layers). The resulting magnetoresistance at high lead resistance will be significantly lower, due to the non-spin-dependent scattering within the non-magnetic leads.

It has been pointed[85,86] that applying the zero spin accumulation condition at the device terminals and not that of the leads is consistent with assuming a metal contact with infinite conductivity. The requisite high lead conductance can be achieved by having a lead made of Cu, and increasing its cross-sectional area $A$ by the same factor, e.g. a hundred times compared with the active spin valve region. This requirement is usually realized in a physical device, where the leads have a much larger $A$ than the submicron-sized spin valve cross section area. For instance, reasonable cross-sectional areas of 0.01 $\mu m^2$ for the spin valve trilayer and 1 $\mu m^2$ for the contact leads could have achieved the required hundred factor. Thus, it is a reasonable assumption to neglect the effect of the leads in our magnetoresistance calculations and to apply the zero spin accumulation boundary conditions at the terminals. Table 5.10 summarizes the boundary conditions required to solve for the spin accumulation and spin current across the device. The inclusion of interfacial[87] resistance and interfacial spin flip is important for practical devices and their effects on boundary conditions are also provided in Table 5.10.

## 5.9 Spin drift diffusion for non-collinear spin valve

Spin transfer torque was first studied, independently, by L. Berger and J. C. Slonczewski.[88] Spin drift diffusion in the previous section considers only the longitudinal component of the spin accumulation, since the spin of the carriers is either parallel/anti-parallel to the local magnetization direction **M** in spin valve in which the ferromagnetic layers are either parallel or anti-parallel to one another. However, it is necessary to extend the collinear spin transport model to the general

*Table 5.10* Phenomenological solutions for spin accumulation across a multilayer device (e.g. the spin valve)

| | To find parameters: | To find parameters: |
|---|---|---|
| | $\boldsymbol{A, B, C, D, \beta, j^s, r^s}$ | $\boldsymbol{A, B, C, D, \beta_i, \beta_{i+1}, j_i^s, j_{i+1}^s, r^s}$ |
| | Given parameters $\boldsymbol{j}$, $\mathbf{R_0}$, $\gamma$<br>**$R_0$ = interfacial resistance** | Given parameters $\boldsymbol{j}$, $\boldsymbol{R_F}$, $\gamma$<br>**$R_F$ = interfacial spin flip** |
| 1. | Terminal boundary conditions<br>$\Delta\mu_0 = \Delta\mu_L = 0$ | Terminal boundary conditions<br>$\Delta\mu_0 = \Delta\mu_L = 0$ |
| 2. | Continuity of spin accumulation<br>$\Delta\mu_i = \Delta\mu_{i+1}$ | Interfacial spin flip breaks continuity of spin accumulation |
| | Asymmetric interfacial resistance breaks continuity of spin accumulation | $-(\Delta\mu_{i+1} - \Delta\mu_i) = \frac{1}{2}\sum_\sigma S_\sigma r_\sigma (j_i^\sigma + j_{i+1}^\sigma)$ |
| | $-(\Delta\mu_{i+1} - \Delta\mu_i) = r^\uparrow j_i^\uparrow - r^\downarrow j_i^\downarrow$ | |
| 3. | Asymmetry of spin current<br>$j^\uparrow = \beta j \quad j^\downarrow = (1 - \beta) j$ | Asymmetry of spin current<br>$j_i^\uparrow = \beta_i j_i \quad j_i^\downarrow = (1 - \beta_i)\, j_i$ |
| | | Interfacial spin flip breaks continuity of spin current polarization $\beta$ |
| | | $j_{i+1}^\uparrow = \beta_{i+1} j_{i+1} \quad j_{i+1}^\downarrow = (1 - \beta_{i+1})\, j_{i+1}$ |
| 4. | Asymmetry of interfacial resistance<br>$r^\uparrow = \gamma R_0 \quad r^\downarrow = (1 - \gamma) R_0$ | Asymmetry of interfacial resistance<br>$r^\uparrow = \gamma R_0 \quad r^\downarrow = (1 - \gamma) R_0$ |
| | | $\frac{1}{2}(\Delta\mu_{i+1} + \Delta\mu_i) = R_{SF}(\Delta j_i - \Delta j_{i+1})$ |

Note: ***A, B, C*** and ***D*** are the strength of spin accumulation at different layers.

case in which the relative magnetization directions of the ferromagnetic layers are at some arbitrary angle to one another. There are two motivations for this:

1. In practice true parallel/anti-parallel alignment in spin valves can be achieved only at high fields.
2. One needs to calculate the transverse (i.e. perpendicular to the local **M**) spin accumulation ($m_\perp$), which arises when the magnetization alignment is non-collinear, i.e. $\theta \neq 0$. A finite $m_\perp$ is essential to generate a spin transfer torque between the conduction electrons and the local magnetic moments, and thus forms the basis of the current-induced magnetization switching (CIMS) effect.

Since CIMS offers a potentially useful method of magnetization switching, it has been actively investigated in a variety of experimental magnetic nanostructures, including pseudospin valve (PSV) trilayers,[89] exchange-biased spin valves,[90] spin valves with synthetic anti-ferromagnets[91,92] and magnetic tunnel junctions.[93,94]

In this analysis, both the in-plane and out-of-plane components of the transverse accumulation $m_\perp$ are considered; however the former was neglected by many previous models. The effect of mixing between longitudinal $m_\parallel$ and transverse $m_\perp$ spin accumulations due to $s-d$ coupling between the accumulated spins and local moments is also considered. By relating the transverse accumulation $m_\perp$ to the effective torque on the local moments in the free ferromagnetic layer, the non-collinear model becomes particularly suitable for studying spin transfer torque, an increasingly important physical effect in nanoelectronics. One can now imagine that both $m_\parallel$ and $m_\perp$ are generated by the passage of spin-polarized current into the free ferromagnetic layer of a PSV trilayer with non-collinear magnetization alignment. At equilibrium, this generation rate is balanced by spin relaxation processes which proceed at a faster rate for $m_\perp$, compared to $m_\parallel$, as a result of precessional motion. The charge and spin transport are driven along the potential and spin accumulation gradients by $\sigma$ and D, respectively. The coupled equations[95] for the spin and charge currents are given by:

$$j_e = -2\sigma_0\left(\frac{\partial\phi}{\partial x}\right) - 2\Delta D\left(\frac{\partial m_\parallel}{\partial x}\right)$$

$$j_{m_\parallel} = -2\Delta\sigma\left(\frac{\partial\phi}{\partial x}\right) - 2D_0\left(\frac{\partial m_\parallel}{\partial x}\right)$$

$$j_{m_\perp} = -2D_0\left(\frac{\partial m_\perp}{\partial x}\right) \qquad [5.83]$$

where $D_0$ and $\sigma_0$ and are the spin-independent part of the conductivity and diffusivity constants, respectively, and $\Delta D$ and $\Delta\sigma$ are the difference in conductivity/diffusivity experienced by majority and minority spins, respectively. Note that the units of charge '$e$' and spin have for simplicity been taken to be 1. We now apply the non-collinear SDD model on a PSV trilayer structure (Fig. 5.9), consisting of a fixed ferromagnetic (FM1) layer, a non-magnetic (NM) spacer and a free (FM2) layer. The magnetization directions in the PSV are assumed to be non-collinear, i.e. with the magnetization $\mathbf{M}_1$ of the fixed FM1 layer in the vertical ($z-$) direction, while that of FM2 ($\mathbf{M}_2$) is rotated by some arbitrary angle in the $y-z$ plane relative to $\mathbf{M}_1$.

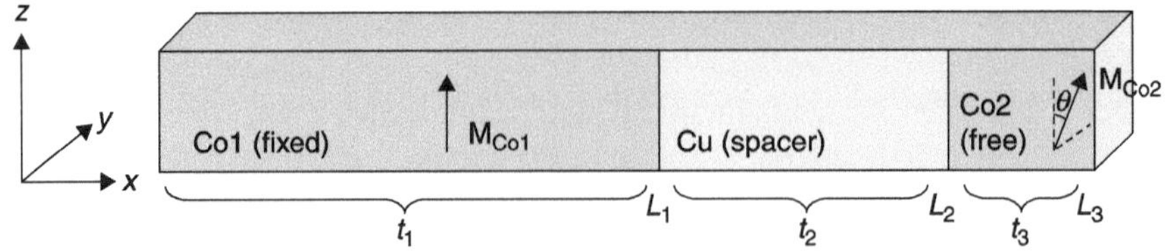


*5.9* Schematic diagram of Co1–Cu–Co2 pseudospin-valve trilayer.

To solve the coupled equations, we express the ansatz for the longitudinal and transverse spin accumulations in the three regions ($k = 1,2,3$, corresponding to the fixed FM1, spacer NM, free FM2 layers, respectively) as:

$$m_{\parallel}^{k}(x) = A_1^k \exp\left[\frac{(x-L_k)}{\lambda_L^k}\right] + A_2^k \exp\left[\frac{-(x-L_k)}{\lambda_L^k}\right]$$
$$m_{\perp}^{k}(x) = B_1^k \exp\left[\frac{(x-L_k)}{\lambda_T^k}\right] + B_2^k \exp\left[\frac{-(x-L_k)}{\lambda_T^k}\right]. \qquad [5.84]$$

In the above, $L_{1,2,3}$ refer to the position of the right boundary of the three layers, while $\lambda_L^k$ and $\lambda_T^k$ are the spin relaxation lengths for the transverse and longitudinal spins, respectively. For the NM spacer, the two lengths are identical, i.e. $\lambda_L^2 = \lambda_T^2$, but they assume very different values in the FM1 and FM2 layers. In these layers, the SDL are summarized in Table 5.11, where $\lambda_{sf}$ is the spin-flip length.

The SDL $\lambda_T$ for Co is much shorter since the relaxation process involves both spin flipping and precessional motion of $m_\perp$ about the local magnetization **M**. Notation $J_{sd}$ is the *s–d* coupling energy between the conduction electrons and the local moments, and is of the order of $10^{-20}$ J, so that $\lambda_T$ is typically 1–2 nm. To solve for the coefficients $A_{1,2}^{(\lambda)}$ and $B_{1,2}^{(\lambda)}$, we need to apply the boundary conditions. The first set of four boundary conditions are applied at the terminals of the PSV, i.e. at $x = 0$ and $x = L_3$, where both longitudinal and transverse spin accumulations vanish, i.e.

$$m_{\parallel}^{1}(0) = m_{\perp}^{1}(0) = 0$$
$$m_{\parallel}^{3}(L_3) = m_{\perp}^{3}(L_3) = 0. \qquad [5.85]$$

*Table 5.11* Spin diffusion length for Co and Cu

| | SDL of different materials[95,96] | |
|---|---|---|
| 1. | Longitudinal spin diffusion length<br>$\lambda_L = \lambda_{sf}\sqrt{1-(\Delta\sigma/\sigma_0)(\Delta D/D_0)} \sim \lambda_{sf}$ | $\lambda_L^{Co} = 60$ nm<br>$\lambda_L^{Cu} = 140$ nm |
| 2. | Transverse spin diffusion length<br>$\lambda_T = \sqrt{2hD_0/J_{sd}}$ | $\lambda_T^{Co} = 2$ nm(ref. 95)<br>$\lambda_T^{Cu} = 140$ nm |
| 3. | Bulk spin polarization of conductivity ($\Delta\sigma/\sigma_0$) of Co is taken to be 38%.<br>Bulk diffusivity ($\Delta D/D_0$) of Co is taken to be 38%. | |

Notes: For calculation, other required numerical values are: Cu and Co conductivity $\sigma_{Co} = 1.6 \times 10^7\ \Omega^{-1}\text{m}^{-1}$ and $\sigma_{Cu} = 5.96 \times 10^7\ \Omega^{-1}\text{m}^{-1}$. Angular deviation of $\theta = \pi/4$ is assumed for $M_{Co1}$ and $M_{Co2}$.

The above ensures that all conduction electrons, regardless of spin orientation, experience the same potential drop across the device. Another set of four boundary conditions is obtained by considering the continuity of the transverse and longitudinal spin accumulations across the FM1–NM and NM–FM2 interfaces, i.e.

$$
\begin{aligned}
&m_{\parallel}^{1}(L_1) = m_{\parallel}^{2}(L_1),\\
&m_{\perp}^{1}(L_1) = m_{\perp}^{2}(L_1),\\
&m_{\parallel}^{2}(L_2) = \left[m_{\parallel}^{3}(L_2)\right]\cos\theta - \left[m_{\perp}^{3}(L_2)\right]\sin\theta,\\
&m_{\perp}^{2}(L_2) = \left[m_{\parallel}^{3}(L_2)\right]\sin\theta + \left[m_{\perp}^{3}(L_2)\right]\cos\theta.
\end{aligned}
\tag{5.86}
$$

In the above the reference spin orientation for the NM layer is chosen arbitrarily to lie along the $z$-direction, similar to that in the FM1 layer. Since the local magnetization in the FM2 is pointing in the oblique direction, we have resolved the components of the spin accumulation in FM2 along the $y$- and $z$-direction to obtain the last two boundary conditions corresponding to the NM–FM2 interface. We have disregarded, for the moment, any interfacial spin-selective scattering events.

The final set of four boundary conditions is derived based on the continuity of the spin currents $j_{m\perp}$ and $j_{m_{\parallel}}$ across the two FM–NM interfaces. As charge current conserves, and by eliminating $\frac{\partial \phi}{\partial x}$ from the equations for $j_{m\perp}$, the longitudinal spin current can then be expressed as:

$$
j_{m_{\parallel}} = \frac{\Delta\sigma}{\sigma_0}\left[j_e + 2\Delta D\left(\frac{\partial m_{\parallel}}{\partial x}\right)\right] - 2D_0\left(\frac{\partial m_{\parallel}}{\partial x}\right). \tag{5.87}
$$

The continuity relations for the longitudinal and transverse spin currents at the two FM–NM interfaces are analogous to those for spin accumulations, i.e.

$$
\begin{aligned}
&j_{m_{\parallel}}^{1}(L_1) = j_{m_{\parallel}}^{2}(L_1),\\
&j_{m_{\perp}}^{1}(L_1) = j_{m_{\perp}}^{2}(L_1),\\
&j_{m_{\parallel}}^{2}(L_2) = \left[j_{m_{\parallel}}^{3}(L_2)\right]\cos\theta - \left[j_{m_{\perp}}^{3}(L_2)\right]\sin\theta,\\
&j_{m_{\perp}}^{2}(L_2) = \left[j_{m_{\parallel}}^{3}(L_2)\right]\sin\theta + \left[j_{m_{\perp}}^{3}(L_2)\right]\cos\theta,
\end{aligned}
\tag{5.88}
$$

where we have again accounted for the rotation in the spin reference axis in the last two relations. In the above continuity relations, we have also assumed no spin flipping at the NM–FM interfaces. However, in practice a finite amount of spin flip may occur at the interfaces, but this can be readily incorporated into the current equations by introducing an interfacial discontinuity in $j_{m\perp}$ and $j_{m\parallel}$. The spin accumulations $m_{\perp}$ and $m_{\parallel}$ are solved by substituting the ansatz of Eq. 5.84 into Eq. 5.88 and determining the coefficients $A_{1,2}^{\lambda}$ and $B_{1,2}^{\lambda}$. One can then use the solutions for $m_{\perp}$ and $m_{\parallel}$ to find the various spin current.

## 5.10 References


[1] Munekata H, Ohno H, von Molnar S, Segmüller A, Chang L L *et al.* (1989), 'Diluted magnetic III–V semiconductors', *Phys. Rev. Lett.* 63, 1849.

[2] Ohno H (1998), 'Making nonmagnetic semiconductors ferromagnetic', *Science* 281, 951.

[3] Kikkawa J M and Awschalom D D (1998), 'Resonant spin amplification in n-type GaAs', *Phys. Rev. Lett.* 80, 4313.

[4] Binasch G, Grünberg P, Saurenbach F and Zinn W (1989), 'Enhanced magnetoresistance in layered magnetic structures with antiferromagnetic interlayer exchange', *Phys. Rev. B*, 39, 4828.

[5] Baibich M N, Broto J M, Fert A, Nguyen Van Dau F, Petroff F *et al.* (1988), 'Giant magnetoresistance of (001)Fe/(001)Cr magnetic superlattices', *Phys. Rev. Lett.* 61, 2472.

[6] Parkin S S P, More N and Roche K P (1990), 'Oscillations in exchange coupling and magnetoresistance in metallic superlattice structures: Co/Ru, Co/Cr, and Fe/Cr', *Phys. Rev. Lett.* 64, 2304.

[7] Ohno H, Chiba D, Matsukura F, Omiya T, Abe E *et al.* (2000), 'Electric-field control of ferromagnetism', *Nature* 408, 944.

[8] Chiba D, Yamanouchi M, Matsukura F and Ohno H (2003), 'Electrical manipulation of magnetization reversal in a ferromagnetic semiconductor', *Science* 301, 943.

[9] Fetter A L and Walecka J D (2003), *Quantum Theory of Many-Particle Systems*, Dover Publications, Mineola, NY.

[10] Craik D (1995), *Magnetism (Principles and Applications)*, John Wiley & Sons, New York.

[11] Majumdar A (1996), 'Effects of intrinsic spin on electronic transport through magnetic barriers', *Phys. Rev. B* 54, 11911; Guo Y, Gu B, Duan W and Zhang Y (1997), 'Transport in asymmetric multiple-barrier magnetic nanostructures', *Phys. Rev. B* 55, 9314.

[12] Papp G and Peeters F M (2001), 'Spin filtering in a magnetic–electric barrier structure', *Appl. Phys. Lett.* 78, 2184; Jiang Y, Jalil M B A and Low T S (2002), 'Comment on "spin filtering in a magnetic–electric barrier structure" [*Appl. Phys. Lett.* 78, 2184 (2001)]', *Appl. Phys. Lett.* 80, 1673.

[13] Xu H Z and Okada Y (2001), 'Does a magnetic barrier or a magnetic-electric barrier structure possess any spin polarization and spin filtering under zero bias?', *Appl. Phys. Lett.* 79, 3119.

[14] Jalil M B A, Tan S G, Liew T, Teo K L and Chong T C (2004), 'High spin filtering using multiple magnetoelectric barriers', *J. Appl. Phys.* 95, 7321; Seo K C, Jhm G, Ahn K-H and Lee S J (2004), 'Spin filtering in an electromagnetic structure', *J. Appl. Phys.* 95, 7252.

[15] Kim J W, Kim N, Lee S J and Kang T W (2006), 'Spin currents modulated by magnetic barriers in semiconductor nanowires', *Semicond. Sci. Technol.* 21, 647.

[16] Lu J D, Shao L, Hou Y-L and Dai H-M (2007), 'The electron transport characters in a nanostructure with the periodic magnetic-electric barriers', *Chin. Phys.* 16, 3080.

[17] Kumar S B, Tan S G, Jalil M B A and Liang G-C (2009), 'High and tunable spin current induced by magnetic–electric fields in a single-mode spintronic device', *Nanotechnology* 20, 365204.

[18] Monzon F G, Johnson M and Roukes M L (1997), 'Strong Hall voltage modulation in hybrid ferromagnet/semiconductor microstructures', *Appl. Phys. Lett.* 71, 3087.

[19] Vancura T, Ihn T, Broderick S, Ensslin K, Wegscheider W *et al.* (2000), 'Electron transport in a two-dimensional electron gas with magnetic barriers', *Phys. Rev. B* 62, 5074; Kubrak V, Neumann A, Gallagher B L, Main P C, Henini M *et al.* (2000), 'Magnetoresistance and Hall magnetometry of single submicron ferromagnetic structures', *J. Appl. Phys.* 87, 5986; Hong J, Kubrak V, Edmonds K W, Neumann A C, Gallagher B L *et al.* (2002), 'Quasi-ballistic transport of 2D electrons through magnetic barriers', *Physica E* 12, 229.
[20] Cerchez M, Hugger S, Heinzel T and Schulz N (2007), 'Effect of edge transmission and elastic scattering on the resistance of magnetic barriers: Experiment and theory', *Phys. Rev. B* 75, 035341.
[21] Datta S and Das B (1989), 'Electronic analog of the electro-optic modulator', *Appl. Phys. Lett.* 56, 665; Bandyopadhyay S and Cahay M (2004), 'Reexamination of some spintronic field-effect device concepts', *Appl. Phys. Lett.* 85, 1433.
[22] Vasko F T (1979), 'Spin splitting in the spectrum of two-dimensional electrons due to the surface potential', *Pis 'ma Zh. Eksp. Teor. Fiz.* 30, 574 [*JETP Lett.* 30, 541]; Bychkov Y A and Rashba E I (1984), 'Properties of a 2D electron gas with lifted spectral degeneracy', *Pis 'ma Zh. Eksp. Teor. Fiz.* 39, 66 [*JETP Lett.* 39, 78].
[23] Bychkov Y A and Rashba E I (1984), 'Oscillatory effects and the magnetic susceptibility of carriers in inversion layers', *J. Phys. C: Solid State Phys.* 17, 6039.
[24] Nitta J, Akazaki T, Takayanagi H and Enoki T (1997), 'Gate control of spin–orbit interaction in an inverted $In_{0.53}Ga_{0.47}As/In_{0.52}Al_{0.48}As$ heterostructure', *Phys. Rev. Lett.* 78, 1335; Koga T, Nitta J, Akazaki T and Takayanagi H (2002), 'Rashba spin–orbit coupling probed by the weak antilocalization analysis in InAlAs/InGaAs/InAlAs quantum wells as a function of quantum well asymmetry', *Phys. Rev. Lett.* 89, 046801.
[25] Meier L, Salis G, Shorubalko I, Gini E, Silke Schön S *et al.* (2007), 'Measurement of Rashba and Dresselhaus spin–orbit magnetic fields', *Nat. Phys.* 3, 650.
[26] Scheid M, Kohda M, Kunihashi Y, Richter K and Nitta J (2008), 'All-electrical detection of the relative strength of Rashba and Dresselhaus spin–orbit interaction in quantum wires', *Phys. Rev. Lett.* 101, 266401.
[27] Nitta J, Bergsten T, Kunihashi Y and Kohda M (2009), 'Electrical manipulation of spins in the Rashba two dimensional electron gas systems', *J. Appl. Phys.* 105, 122402.
[28] Schmidt G, Ferrand D, Molenkamp L W, Filip A T and van Wees B J (2000), 'Fundamental obstacle for electrical spin injection from a ferromagnetic metal into a diffusive semiconductor', *Phys. Rev. B* 62, R4790.
[29] Rashba E I (2000), 'Theory of electrical spin injection: Tunnel contacts as a solution of the conductivity mismatch problem', *Phys. Rev. B* 62, R16267.
[30] Takahashi S and Maekawa S (2003), 'Spin injection and detection in magnetic nanostructures', *Phys. Rev. B* 67, 052409.
[31] Zhu H J, Ramsteiner M, Kostial H, Wassermeier M, Schönherr H-P *et al.* (2001), 'Room-temperature spin injection from Fe into GaAs', *Phys. Rev. Lett.* 87, 016601.
[32] Sun Q F, Guo H and Wang J (2003), 'A spin cell for spin current', *Phys. Rev. Lett.* 90, 258301.
[33] Jalil M B A and Tan S G (2005), 'Spin transfer and current-induced switching in a ferromagnetic single-electron transistor', *Phys. Rev. B* 72, 214417; Seneor P, Bernand-Mantel A and Petroff F (2007), 'Nanospintronics: When spintronics meets single electron physics', *J. Phys.: Condens. Matter* 19, 165222.
[34] Valenzuela S O and Tinkham M (2006), 'Direct electronic measurement of the spin Hall effect', *Nature* 442, 176.

[35] Harii K, Ando K, Inoue H Y, Sasage K and Saitoh E (2008), 'Inverse spin-Hall effect and spin pumping in metallic films (invited)', *J. Appl. Phys.* 103, 07F311.

[36] Dresselhaus G (1955), 'Spin–orbit coupling effects in zinc blende structures', *Phys. Rev.* 100, 580.

[37] Schliemann J and Loss D (2003), 'Anisotropic transport in a two-dimensional electron gas in the presence of spin-orbit coupling', *Phys. Rev. B* 68, 165311.

[38] Perel V I, Tarasenko S A, Yassievich I N, Ganichev S D, Bel'kov V V *et al.* (2003), 'Spin-dependent tunneling through a symmetric semiconductor barrier', *Phys. Rev. B* 67, 201304; Fujita T, Jalil M B A and Tan S G (2009), 'Topological spin-Hall current in waveguided zinc-blende semiconductors with Dresselhaus spin-orbit coupling', *Ann. Phys.* 324, 2265.

[39] Fujita T, Jalil M B A and Tan S G (2008), 'Multi-channel spintronic transistor design based on magnetoelectric barriers and spin–orbital effects', *J. Phys.: Condens. Matter* 20, 045205.

[40] Dyakonov M I and Perel V I (1971), 'Spin orientation of electrons associated with interband absorption of light in semiconductors', *Zh. Eksp. Teor. Fiz.* 60, 1954 [*Sov. Phys. JETP* 38, 1053].

[41] Aronov A G, Pikus G E and Titkov A N (1983), 'Spin relaxation of conduction electrons in p-type III-V compounds', *Zh. Eksp. Teor. Fiz.* 84, 1170 [*Sov. Phys. JETP*, 57, 680].

[42] Elliott R J (1954), 'Theory of the effect of spin–orbit coupling on magnetic resonance in some semiconductors', *Phys. Rev.* 96, 266.

[43] Zutic I, Fabian J and Das Sarma S (2004), 'Spintronics: Fundamentals and applications', *Rev. Mod. Phys.* 76, 323.

[44] Wang X F and Vasilopoulos P (2003), 'Magnetotransport in a two-dimensional electron gas in the presence of spin-orbit interaction', *Phys. Rev. B* 67, 085313.

[45] Zhang D (2006), 'Exact Landau levels in two-dimensional electron systems with Rashba and Dresselhaus spin–orbit interactions in a perpendicular magnetic field', *J. Phys. A: Math. Gen.* 39, L477; Tan S G, Jalil M B A, Teo K L and Liew T (2005), 'Spin polarization of electrons in Landau spin orbit coupling semiconductor system', *J. Appl. Phys.* 97, 10A716.

[46] Guo B-Y, Shen J and Xu C-L (2003), 'Spectral and pseudospectral approximations using hermite functions: Application to the Dirac equation', *Adv. Comput. Math.* 19, 35.

[47] Fok J C M, Guo B and Tang T (2001), 'Combined hermite spectral-finite difference method for the Fokker–Planck equation', *Math. Comput.* 71, 1497.

[48] Dyakonov M I and Perel V I (1971), 'Current-induced spin orientation of electrons in semiconductors', *Phys. Lett. A* 35, 459; Dyakonov M I and Perel V I (1971), 'Possibility of orienting electron spins with current', *Pis'ma Zh. Eksp. Teor. Fiz.* 13, 657 [*JETP Lett.* 13, 467]; Averkiev N S and Dyakonov M I (1983), 'Current due to inhomogeneity of the spin orientation of electrons in a semiconductors', *Fiz. Tekh. Poluprovodn.* 17, 629 [*Sov. Phys. Semicond.* 17, 393].

[49] Bakun A A, Zakharchenya B P, Rogachev A A, Tkachuk M N and Fleisher V G (1984), 'Observation of a surface photocurrent caused by optical orientation of electrons in a semiconductor', *Pis'ma Zh. Eksp. Teor. Fiz.* 40, 464 [*Sov. Phys. JETP Lett.* 40, 1293]; Tkachuk M N, Zakharchenya B P and Fleisher V G (1986), 'Resonant photovoltaic effect in the NMR of nuclei in a semiconductor lattice', *Pis'ma Zh. Eksp. Teor. Fiz.* 44, 47 [*Sov. Phys. JETP Lett.* 44, 59 (1986)].

[50] Hirsch J E (1999), 'Spin Hall effect', *Phys. Rev. Lett.* 83, 1834.

[51] Kato Y K, Myers R C, Gossard A C and Awschalom D D (2004), 'Observation of the spin Hall effect in semiconductors', *Science* 306, 1910.

[52] Inoue J and Ohno H (2005), 'Taking the Hall effect for a spin', *Science* 309, 2004; Kimura T, Otani Y, Sato T, Takahashi S and Maekawa S *et al.* (2007), 'Room-temperature reversible spin Hall effect', *Phys. Rev. Lett.* 98, 156601.
[53] Onoda M, Murakami S and Nagaosa N (2004), 'Hall effect of light', *Phys. Rev. Lett.* 93, 083901.
[54] Bliokh K Y and Bliokh Y P (2004), 'Topological spin transport of photons: The optical Magnus effect and Berry phase', *Phys. Lett. A* 333, 181.
[55] Bliokh K Y and Freilikher V D (2006), 'Polarization transport of transverse acoustic waves: Berry phase and spin Hall effect of phonons', *Phys. Rev. B* 74, 174302.
[56] Yao W and Niu Q (2008), 'Berry phase effect on the exciton transport and on the exciton Bose–Einstein condensate', *Phys. Rev. Lett.* 101, 106401.
[57] Sinitsyn N A, Hill J E, Min H, Sinova J and MacDonald A H (2006), 'Charge and spin Hall conductivity in metallic graphene', *Phys. Rev. Lett.* 97, 106804.
[58] Zhang S (2000), 'Spin Hall effect in the presence of spin diffusion', *Phys. Rev. Lett.* 85, 393; Saitoh E, Ueda M, Miyajima H and Tatara G (2006), 'Conversion of spin current into charge current at room temperature: Inverse spin-Hall effect', *Appl. Phys. Lett.* 88, 182509; Kimura T, Otani Y, Sato T, Takahashi S and Maekawa S (2007), 'Room-temperature reversible spin Hall effect', *Phys. Rev. Lett.* 98, 156601.
[59] Sinova J, Culcer D, Niu Q, Sinitsyn N A, Jungwirth T *et al.* (2004), 'Universal intrinsic spin Hall effect', *Phys. Rev. Lett.* 92, 126603.
[60] Murakami S, Nagaosa N and Zhang S C (2003), 'Dissipationless quantum spin current at room temperature', *Science* 301, 1348; Fujita T, Jalil M B A and Tan S G (2009), 'Topological spin-Hall current in waveguided zinc-blende semiconductors with Dresselhaus spin–orbit coupling', *Ann. Phys.* 324, 2265.
[61] Bernevig B A and Zhang S-C (2004), 'Intrinsic spin Hall effect in n-doped Bulk GaAs', cond-mat/0412550.
[62] Bernevig B A and Zhang S C (2005), 'Intrinsic spin Hall effect in the two-dimensional hole gas', *Phys. Rev. Lett.* 95, 016801.
[63] Fujita T, Jalil M B A and Tan S G (2009), 'Unified model of intrinsic spin Hall effect in spintronic, optical, and graphene systems', *J. Phys. Soc. Jpn.* 78, 104714; Fujita T, Jalil M B A and Tan S G (2010), 'Unified description of intrinsic spin-Hall effect mechanisms', *New J. Phys.* 12, 013016.
[64] Shen S-Q (2005), 'Spin transverse force on spin current in an electric field', *Phys. Rev. Lett.* 95, 187203.
[65] Inoue J, Bauer G E W and Molenkamp L W (2003), 'Diffuse transport and spin accumulation in a Rashba two-dimensional electron gas', *Phys. Rev. B* 67, 033104.
[66] Zyuzin V A, Silvestrov P G and Mishchenko E G (2007), 'Spin Hall edge spin polarization in a ballistic 2D electron system', *Phys. Rev. Lett.* 99, 106601.
[67] Nikolic B K, Souma S, Zarbo L P and Sinova J (2005), 'Nonequilibrium spin Hall accumulation in ballistic semiconductor nanostructures', *Phys. Rev. Lett.* 95, 046601.
[68] Usaj G and Balseiro C A (2005), 'Spin accumulation and equilibrium currents at the edge of 2DEGs with spin–orbit coupling', *Europhys. Lett.* 72, 631.
[69] Hankiewicz E M and Vignale G (2008), 'Phase diagram of the spin Hall effect, *Phys. Rev. Lett.* 100, 026602.
[70] Aharonov Y and Stern A (1992), 'Origin of the geometric forces accompanying Berry's geometric potentials', *Phys. Rev. Lett.* 69, 3593.
[71] Hatano N, Shirasaki R and Nakamura H (2007), 'Non-Abelian gauge field theory of the spin-orbit interaction and a perfect spin filter', *Phys. Rev. A* 75, 032107.

[72] Yang J-S, He X-G, Chen S-H and Chang C-R (2008), 'Spin precession due to a non-Abelian spin-orbit gauge field', *Phys. Rev. B* 78, 085312.

[73] Aharonov Y and Casher A (1984), 'Topological quantum effects for neutral particles', *Phys. Rev. Lett.* 53, 319.

[74] Mathur H (1991), 'Thomas precession, spin-orbit interaction, and Berry's phase', *Phys. Rev. Lett.* 67, 3325.

[75] Lee T and Ryu C-M (1994), 'SU(2) gauge symmetry of a non-relativistic spin 1/2 particle', *Phys. Lett. A* 194, 310.

[76] Jin P Q, Li Y Q and Zhang F C (2006), 'SU(2) × U(1) unified theory for charge, orbit and spin currents', *J. Phys. A: Math. Gen.* 39 7115.

[77] Tan S G, Jalil M B A, Liu X-J and Fujita T (2008), 'Spin transverse separation in a two-dimensional electron-gas using an external magnetic field with a topological chirality', *Phys. Rev. B* 78, 245321.

[78] Berard A and Mohrbach H (2006), 'Spin Hall effect and Berry phase of spinning particles', *Phys. Lett. A* 352, 190.

[79] Basu B and Bandyopadhyay P (2008), 'Spin orbit gauge and quantum spin Hall effect', *Phys. Lett. A* 373, 148.

[80] Zhou B, Ren L and Shen S-Q (2006), 'Spin transverse force and intrinsic quantum transverse transport', *Phys. Rev. B* 73, 165303.

[81] Camley R E and Barnas J (1989), 'Theory of giant magnetoresistance effects in magnetic layered structures with antiferromagnetic coupling', *Phys. Rev. Lett.* 63, 664.

[82] Valet T and Fert A (1993), 'Theory of the perpendicular magnetoresistance in magnetic multilayers', *Phys. Rev. B* 48, 7099.

[83] van Son P C, van Kempen H and Wyder P (1987), 'Boundary Resistance of the Ferromagnetic-Nonferromagnetic Metal Interface', *Phys. Rev. Lett.* 58, 2271.

[84] Johnson M and Silsbee R H (1988), 'Ferromagnet-nonferromagnet interface resistance', *Phys. Rev. Lett.* 60, 377.

[85] Khaetskii A, Egues J C, Loss D, Gould C, Schmidt G, Molenkamp L W (2005), 'Spin injection across magnetic/nonmagnetic interfaces with finite magnetic layers', *Phys. Rev. B* 71, 235327.

[86] Jalil M B A, Tan S G, Kumar S B and Bae S (2006), 'Spin drift diffusion studies of magnetoresistance effects in current-perpendicular-to-plane spin valves with half-metallic insertions', *Phys. Rev. B* 73, 134417; Tan S G, Jalil M B A, Kumar S B, Li K, Zheng Y *et al.* (2007), 'Theoretical modeling of half-metallic current-perpendicular-to-plane spin valves', *J. Appl. Phys.* 101, 09J502.

[87] Fert A and Lee S-F (1996), 'Theory of the bipolar spin switch', *Phys. Rev. B* 53, 6554; Fert A and Jaffrès H(2001), 'Conditions for efficient spin injection from a ferromagnetic metal into a semiconductor', *Phys. Rev. B* 64, 184420.

[88] Berger L (1996), 'Emission of spin waves by a magnetic multilayer traversed by a current', *Phys. Rev. B* 54, 9353; Slonczewski J C (1996),'Current-driven excitation of magnetic multilayers', J. *Magn. Magn. Mater.* 159, L1.

[89] Myers E B, Ralph D C, Katine J A, Louie R N and Buhrman R A (1999), 'Current-induced switching of domains in magnetic multilayer devices', *Science* 285, 867.

[90] Jiang Y, Abe S, Ochiai T, Nozaki T, Hirohata A *et al.* (2004), 'Effective reduction of critical current for current-induced magnetization switching by a Ru layer insertion in an exchange-biased spin valve', *Phys. Rev. Lett.* 92, 167204.

[91] Emley N C, Albert F J, Ryan E M, Krivorotov I N, Ralph D C *et al.* (2004), 'Reduction of spin transfer by synthetic antiferromagnets', *Appl. Phys. Lett.* 84, 4257.

[92] Ochiai T, Jiang Y, Hirohata A, Tezuka N, Sugimoto S *et al.* (2005), 'Distinctive current-induced magnetization switching in a current-perpendicular-to-plane giant-magnetoresistance nanopillar with a synthetic antiferromagnet free layer', *Appl. Phys. Lett.* 86, 242506.
[93] Huai Y, Albert F, Nguyen P, Pakala M and Valet T (2004), 'Observation of spin-transfer switching in deep submicron-sized and low-resistance magnetic tunnel junctions', *Appl. Phys. Lett.* 84, 3118.
[94] Meng H and Wang J P (2005), 'Spin transfer effect in magnetic tunnel junction with a nano-current-channel Layer in free layer', *IEEE Trans. Magn.* 41, 2612.
[95] Zhang S, Levy P M and Fert A (2002), 'Mechanisms of spin-polarized current-driven magnetization switching', *Phys. Rev. Lett.* 88, 236601; Shpiro A, Levy P M and Zhang S (2003), 'Self-consistent treatment of nonequilibrium spin torques in magnetic multilayers', *Phys. Rev. B* 67, 104430.
[96] Vouille C, Barthélémy A, Elokan Mpondo F, Fert A, Schroeder P A *et al.* (1999), 'Microscopic mechanisms of giant magnetoresistance', *Phys. Rev. B* 60, 6710.

## Appendix 5.A Spin current under magnetic fields and spin orbit coupling

This appendix contains the steps required to derive the wavefunctions under the effect of magnetic fields and spin orbit coupling. As described earlier, the purpose here is to find the eigensolutions of $H_C$ so that we can use these solutions as basis solutions for the total Hamiltonian that comprises cross electric magnetic and SOC terms. One sees that on gauge (− *By*, 0, 0):

$$H_C = \frac{-\hbar^2}{2m}\varphi'' + \left(\frac{\hbar^2 k_x^2}{2m} - \hbar\omega k_x y + \frac{1}{2}m\omega^2 y^2 + eyE_y\right)\varphi = \frac{-\hbar^2}{2m}\varphi'' + \frac{1}{2}m\omega^2 (y-y_0)^2\varphi + eyE_y\varphi$$
$$= \frac{-\hbar^2}{2m}\varphi'' + \frac{1}{2}m\omega^2(y-y_a)^2\varphi + V\varphi \qquad [5.A.1]$$

where $V = \frac{1}{2}m\omega^2(y_0^2 - y_a^2)$, $\omega = \frac{eB}{m}$ is the cyclotron frequency, $y_0 = \frac{\hbar k_x}{eB}$ is the cyclotron center and $y_a = y_0 - eE_y/m\omega^2$ is the cyclotron center in the presence of the cross electric field. Note that the cross electric field has been consistently denoted by $E_y$, while the energy is denoted by $E$. Table 5.A.1 summarizes the Hamiltonian at different levels of sophistication and their solutions.

*Table 5.A.1* Hamiltonians with their corresponding solutions and harmonic potentials

| | Hamiltonian | Ansatz/solution | Harmonic potential |
|---|---|---|---|
| 1. | $H_L = \frac{(p_x + eA_x)^2}{2m} + \frac{p_y^2}{2m}$ | $\varphi_n(y-y_0)$ | $U_L(y) = \frac{1}{2}m\omega^2(y-y_0)^2$ |
| 2. | $H_C = H_L + eyE_y$ | $\varphi_n(y-y_a)$ | $U_C(y) = \frac{1}{2}m\omega^2(y-y_a)^2$ |
| 3. | $H = H_C + H_Z + H_{SO}$ | $\psi = \sum_{n=1}\varphi_n(y-y_a)\begin{pmatrix} a_n \\ b_n \end{pmatrix}$ | $U_C(y) = \frac{1}{2}m\omega^2(y-y_a)^2$ |

Multiplying the above equations yields a column vector equation:

$$\begin{bmatrix} (H_C+g\mu B-E)\sum_{n=1}\varphi_n a_n+\left[i\alpha\left(\frac{\partial}{\partial y}-\left(k_x+\frac{e}{\hbar}A_x\right)\right)-\beta\left(\frac{\partial}{\partial y}+\left(k_x+\frac{e}{\hbar}A_x\right)\right)\right]\sum_{n=1}\varphi_n b_n \\ \left[i\alpha\left(\frac{\partial}{\partial y}-\left(k_x+\frac{e}{\hbar}A_x\right)\right)+\beta\left(\frac{\partial}{\partial y}-\left(k_x+\frac{e}{\hbar}A_x\right)\right)\right]\sum_{n=1}\varphi_n a_n+(H_C+g\mu B-E)\sum_{n=1}\varphi_n b_n \end{bmatrix}=0$$

[5.A.2]

where $E = \varepsilon - E_S$. Using the substitutions of $k_x = \frac{y_0 eB}{\hbar}$ and $k_x = \frac{y_0}{r^2}$ where one defines $r^2 = \frac{\hbar}{eB'}$ one has the following:

$$\begin{bmatrix} (H_C+g\mu B-E)\sum_{n=1}\varphi_n a_n+\left[-i\alpha\left(\frac{\partial}{\partial y}+\left(\frac{y-y_0}{r^2}\right)\right)-\beta\left(\frac{\partial}{\partial y}-\left(\frac{y-y_0}{r^2}\right)\right)\right]\sum_{n=1}\varphi_n b_n \\ \left[-i\alpha\left(\frac{\partial}{\partial y}-\left(\frac{y-y_0}{r^2}\right)\right)+\beta\left(\frac{\partial}{\partial y}-\left(\frac{y-y_0}{r^2}\right)\right)\right]\sum_{n=1}\varphi_n a_n+(H_C-g\mu B-E)\sum_{n=1}\varphi_n b_n \end{bmatrix}=0$$

[5.A.3]

where $A_x = -By$. The elements of $\left\{H_n\left(\frac{(y-y_a)}{r}\right) \quad n=1,2,....\right\}$ are mutually orthogonal in a weighted Sobolev space, i.e.

$$\int_\nabla H_n(x)H_m(x)\exp(-x^2)\,dx = c_n\delta_{nm} \rightarrow \int_\nabla\left(\frac{H_n(x)}{\sqrt{c_n}}\exp\left(\frac{-x^2}{2}\right)\right)\left(\frac{H_m(x)}{\sqrt{c_m}}\exp\left(\frac{-x^2}{2}\right)\right)dx = \delta_{nm}$$

[5.A.4]

where $x = \frac{(y-y_a)}{r}$ and $\nabla = \{x| -\infty < x < +\infty\}$, $c_n = \sqrt{\pi}2^n n!$ It is worth recalling that the solutions for $H_C$ or $H_L$ are given below; the content of $x$ depends on the harmonic potential:

$$\varphi_n(x) = \frac{H_n(x)}{\sqrt{\sqrt{\pi}2^n n! r}}\exp\left(\frac{-x^2}{2}\right); \; E_n = \frac{(n+1)}{2}\hbar\omega. \qquad [5.A.5]$$

With the Hamiltonian written in a column vector:

$$\begin{bmatrix} (H_C+g\mu B-E)\sum_{n=1}\varphi_n a_n+\left[-i\frac{\alpha}{r}\left(\frac{\partial}{\partial x}+x\right)-\frac{\beta}{r}\left(\frac{\partial}{\partial x}-x\right)\right]\sum_{n=1}\varphi_n b_n \\ \left[-i\frac{\alpha}{r}\left(\frac{\partial}{\partial x}-x\right)+\frac{\beta}{r}\left(\frac{\partial}{\partial x}+x\right)\right]\sum_{n=1}\varphi_n a_n+(H_C-g\mu B-E)\sum_{n=1}\varphi_n b_n \end{bmatrix}=M=0. \qquad [5.A.6]$$

One can next apply the transform:

$$\int \varphi_m(x)M\,dx = 0. \qquad [5.A.7]$$

*Table 5.A.2* Hermite spectral identities and the integral equations used to derive the eigenfunctions

| | |
|---|---|
| | Useful Hermite identities |
| 1. | $2xH_n(x) = H_{n+1}(x) + 2nH_{n-1}(x)$ |
| 2 | $\partial_x H_n = nH_{n-1} - \frac{1}{2}H_{n+1}$ |
| | Integral equations |
| 1. | $\int \varphi_m(x)(\partial_x + x)\sum_{n=1}^{N}\varphi_n a_n\, dx = \sqrt{2(m+1)}\, a_{m+1}$ |
| 2. | $\int \varphi_m(x)(\partial_x - x)\sum_{n=1}^{N}\varphi_n a_n\, dx = -\sqrt{2m}\, a_{m-1}$ |

With the help of the spectral Hermite identities in Table 5.A.2, one can derive two important integral equations (refer to Exercise 5.2) which are also shown in the table.

Making use of the integral equations and ignoring the Dresselhaus effect, ($\beta = 0$), one now has:

$$\begin{bmatrix} \left[\left(m+\frac{1}{2}\right)\hbar\omega + g\mu B - E'\right] a_m - i\frac{\alpha}{r}\sqrt{2(m+1)}\; b_{m+1} \\ i\frac{\alpha}{r}\sqrt{2m}\; a_{m-1} + \left[\left(m+\frac{1}{2}\right)\hbar\omega - g\mu B - E\right] b_m \end{bmatrix} = 0. \qquad [5.A.8]$$

Substitution of $m \to s - 1$ and $m \to s$ for the upper and lower row simply means selecting, respectively, the $(s-1)^{th}$ term of the upper and the $s^{th}$ term of the lower row. One has the freedom to select because both upper and lower rows contain infinite sets of equations. In the presence of transverse electric field, one can introduce $V$ which is explicitly $V(E_y)$, resulting in:

$$\begin{bmatrix} \left(s-\frac{1}{2}\right)\hbar\omega + V + g\mu B & \frac{-i\alpha}{r}\sqrt{2s} \\ \frac{i\alpha}{r}\sqrt{2s} & \left(s+\frac{1}{2}\right)\hbar\omega + V - g\mu B \end{bmatrix} \begin{pmatrix} a_{s-1} \\ b_s \end{pmatrix} = E \begin{pmatrix} a_{s-1} \\ b_s \end{pmatrix}. \qquad [5.A.9]$$

Let us begin with $s = 0$, and $a_{s-1} = 0$. One thus obtains:

$$\left(\frac{1}{2}\hbar\omega + V - g\mu B\right) b_0 = E_0\; b_0. \qquad [5.A.10]$$

The first energy band has thus been obtained. Its corresponding wavefunction is:

$$\psi_0 = \varphi_0(y - y_a)\begin{pmatrix} 0 \\ b_0 \end{pmatrix}. \qquad [5.A.11]$$

One can further deduce that for $s = 1,2,3,\ldots\ldots\ldots,$ the energy band will be:

$$E_s^{\pm} = s\hbar\omega + V \pm \sqrt{\xi_0^2 + 2s\left(\frac{\alpha}{r}\right)^2} \qquad [5.A.12]$$

where $E_0 = \xi_0 + V$ and $\xi_0 = \frac{1}{2}\hbar\omega - g\mu B$. The above shows that for each value of $s$, there are two energy levels: the (+) branch and the (−) branch. With the eigenvalue equations found, the eigenfunctions corresponding to $E_s^{\pm}$ can be found:

$$\begin{pmatrix} a_{s-1} \\ b_s \end{pmatrix}^{+} = N^{+}\begin{pmatrix} D_s^{+} \\ 1 \end{pmatrix}; \begin{pmatrix} a_{s-1} \\ b_s \end{pmatrix}^{-} = N^{-}\begin{pmatrix} 1 \\ D_s^{-} \end{pmatrix}. \qquad [5.A.13]$$

The normalization constant for both spin branches are:

$$N^{\sigma} = \frac{1}{\sqrt{1+(D_s^{\sigma})^2}} \qquad [5.A.14]$$

where:

$$D_s^{+} = D_s^{-} = \frac{-i\alpha\sqrt{2s}/r}{\left(\xi_0 + \sqrt{\xi_0^2 + 2s\left(\frac{\alpha}{r}\right)^2}\right)}. \qquad [5.A.15]$$

The normalization constant is found from:

$$\int (\psi^{+})^{*}\,\psi^{+}\,dy = 1 \rightarrow (N^{\sigma})^2(1+(D_s^{\sigma})^2) = 1. \qquad [5.A.16]$$

# 6
# Graphene and carbon nanostructures for nanoelectronics

**Abstract:** This chapter introduces electronics based on carbon materials like carbon nanotubes and graphene. Monolayer and bilayer graphene are discussed and the forms of their Hamiltonian analyzed. The concept of gauge potential induced by strain or graphene deformation is introduced. Particular attention is paid to the extra degrees of freedom in graphene, namely, pseudospin and valley spin, which both share the spinor physics of electron spin. Novel graphene behaviors like localization, Klein tunneling, minimal conductivity are also discussed. The integer quantum Hall effect (IQHE) in graphene is discussed in relation to the semiconductor IQHE.

**Key words:** monolayer, bilayer, graphene, pseudospin, valley spin, Klein tunneling, minimal conductivity, strain, integer quantum Hall effect, IQHE.

## 6.1 Introduction to carbon electronics

Carbon, beyond the traditional roles it plays in the steel and chemical industries, powering rail and naval transportation and generating electricity, is in this century set to shed its past grimy image and take on the gleam of high-technology electronics and integrated circuits. Silicon made possible the early human quest into nanoscience, and with metal–semiconductor field-effect transistor (MOSFET) technologies, set us forth on our long march since the 1960s down the scale of electronic channel length. Silicon has given humanity fast and nanoscale electronics, the netbook and the cell phone which can store large colorful photographs. Human needs are unfortunately also expanded by their very fulfillment. With electronic channel length approaching 20 nm and smaller, the post-silicon era becomes ever more real.

The spotlight is on carbon, a close cousin of silicon in the periodic table, but one which exists in various allotropic forms, e.g. $sp^2$ (fullerenes, carbon nanotube and graphene) and $sp^3$ (diamond) bonding. The focus on carbon follows from the failure of molecular electronics to live up to the large-scale fabrication stability required of mass production for integrated circuits. Contact resistance as well as contact effect on channel property remains the major hurdle. Carbon nanotube (CNT), considered a form of molecular electronics based on carbon, possesses quantum confinement and superb electronic properties[1] to succeed silicon (see Table 6.1). However, like molecular electronics, contact resistance as well as patterning process difficulty remains a difficult challenge to stable, large-scale production. Table 6.1 provides a summary of the advantages of CNT over silicon-based devices. MOSFET devices that have been designed and fabricated out of CNT include:

1. Back gate CNT field-effect transistor (FET).
2. Front gate CNT FET.[2]
3. CNT ballistic FET.[3]
4. CNT vertical FET.[4]

Possible examples of spintronic devices based on CNT are CNT spinFET[5] or CNT spin valve.[6] Other applications include CNT superconductivity, which has led to the concept of CNT supercurrent FET.[7] In CNT optoelectronics, a CNT light-emitting diode (LED) has been demonstrated.[8]

The experimental discovery of graphene in 2003 marked the dawn of a new era in condensed matter physics, in which scientists could probe the behavior of massless Dirac fermions in a laboratory setting. Graphene inherits most superior macroscopic electronic features from CNT but largely eludes the patterning and contact problems faced by molecular and CNT electronics on the nanoscale. It also shows large ON current without the need for a large array of parallel CNT. Graphene's electronic capability is further supported by experiments which show that the charge carriers can be tuned continuously between electrons and holes by simply adjusting the gate voltage in concentration of $n = 10^{13}$ cm$^{-2}$ with mobility exceeding 15 000 cm$^2$ V$^{-1}$ s$^{-1}$ at room temperature. Semiconductors like InSb exhibit mobility as high as 77 000 cm$^2$ V$^{-1}$ s$^{-1}$, but such high mobility is meaningless if this agility is possible only with a sparse system with a handful

*Table 6.1* Summary of the superb electronic features of carbon nanotube which have led to hopes of CNT replacing silicon in the future

| | Electronic advantages of a carbon nanotube |
|---|---|
| 1. | It is a one-dimensional nanostructure and has the properties of:<br>(a) high speed due to less scattering<br>(b) ballistic transport – several μm for metal CNT, hundreds of nm for semiconducting CNT<br>(c) low resistance<br>(d) low power consumption. |
| 2. | Due to its small diameter (1–2 nm), CNT FET enhances gate control of the channel, particularly when gates are designed to wrap around the channel. |
| 3. | One crucial advantage of a CNT FET over silicon MOSFET is its much lower capacitance values (0.05 aF nm$^{-1}$) which allows fast switching. |
| 4. | Because of its covalent bonding – it has large current carrying capacity (10$^9$ A cm$^{-2}$) which promises CNT spintronics, e.g. spin transfer switching requires large current density. Its geometrical properties are no dangling bonds, cylindrical shape and smooth surfaces. |
| 5. | CNT possesses small spin orbit coupling, which might translate to large spin diffusion length (SDL). However, CNT will be a weak spin polarizer. It shows large magnetoresistance, e.g. in $La_{2/3}Sr_{1/3}MnO_3$ (LSMO) CNT and LSMO devices. |

electrons. In graphene, high mobility is maintained even for $n > 10^{12}$ cm$^{-2}$, thus yielding long mean free path (MFP) approaching 300 nm even at room temperature. Recent work even showed a mobility of 230 000 cm$^2$ V$^{-1}$ s$^{-1}$ for a suspended graphene layer.[9]

In addition, graphene has shown interesting quantum properties such as phase coherence,[10] integer quantum Hall effects (IQHE) (albeit at high magnetic field)[11] and spin polarization,[12] all at room temperature. Recent progress with graphene on silicon (GOS)[13] and graphene on insulator (GOI)[14] is set to further speed up the industrialization of graphene fabrication. Beside superb macroscopic and quantum electronic properties, graphene also shows novel quantum behavior due to its unique electronic dispersion. The novel features include Klein tunneling, pair production, minimal conductivity, topological zero modes, pseudospin and valley spin polarization. The novel quantum features of graphene represent a completely uncharted territory for nanoelectronics, promising a plethora of romantic nanoscientific ventures that could potentially revolutionize conventional electronics. However, these novel features also pose engineering difficulties in realizing conventional graphene-based electronics. Minimal conductivity results in a relatively poor ON/OFF ratio, a feature not favorable for transistor operation. The lack of quantum confinement in graphene compared to CNT arises due to Klein tunneling.

The lack of bandgap in monolayer graphene has been a major obstacle to realizing graphene electronics. Systematic closing of the graphene gap[15] can, however, be accomplished by increasing sample thickness. The principle of gap opening is based on broken symmetry. Bilayer graphene was epitaxially grown on SiC substrate, producing a gap of approximately 0.26 eV. The interaction between the carbon atoms and the underlying substrate results in a graphene layer configured in such a way that one of the carbon atoms in each unit cell has a neighboring atom in the atomic layer below but the other one does not, thus breaking the symmetry. Bandgap engineering has also been investigated[16] where graphene was patterned into a quasi one-dimensional (1D) structure, also known as the graphene nanoribbon (GNR). It was found that lateral confinement gives rise to an energy gap near the neutrality point. GNR shows a drop of conductivity with decreasing temperature at a much greater magnitude than ordinary graphene, suggesting that lateral confinement could have opened a bandgap. In monolayer graphene, applying the gate voltage could alter the position of the Fermi energy with respect to the Dirac point or the normal Fermi point. However, a practical difficulty still exists because of transport sensitivity to the edge profile and difficulty in fabricating edge nanostructures.

---

**Exercise 6.1**

(1) In MOSFET electronics, carriers can exist in the form of negative charge or positive charge carriers, which correspond to electron or

hole current, respectively. Comment in terms of carrier forms/types on what novel types of current one can expect to detect in the field of spintronics and graphene.

(2) Graphene and CNT have similar electronic properties perceived to be superior to those of silicon in MOSFET devices which, at 20 nm gate length now, face great fabrication challenges in the near future. Describe briefly the shared electronic advantages of graphene and CNT over silicon. Explain why graphene is more promising than CNT in terms of substituting silicon as a future channel material for nanoscale devices.

(3) Despite the many superior properties for electronic applications, CNT devices remain largely in the laboratory. In contrast, metal spintronics has been more successful in terms of technological applications, e.g. spin valve recording head and magnetic tunnel junction.

   (a) Explain in terms of carrier scattering and mobility why CNTs are promising candidates for future electronic devices.
   (b) In your view, what is still holding back the development of CNT electronics?
   (c) What is the prospect of hybrid devices based on CNT spintronics and optics?

## 6.2 Monolayer graphene

The periodic potential of the monolayer graphene honeycomb lattice is defined by two sub-lattices shifted relative to one another. A monolayer graphene is held together by two types of electronic bonding. The in-plane $\sigma$ bonds are hybridization results of the $2s$, $2p_x$, $2p_y$ orbitals in a $sp^2$ configuration; these bonds give rise to its rigid structure. The other type of so-called $\pi$ bonds is formed from the $2p_z$ orbital, which is perpendicular to the graphene plane. The $\pi$ bonds give rise to the valence and conduction band.

### 6.2.1 Electronic structure

Graphene[17,18] is a monolayer of carbon atoms packed into a two-dimensional (2D) honeycomb lattice. Given its thickness of just one atomic layer, it is the first truly 2D system and it is not hard to conceive that such a monolayer 2D system is the basic building block of all carbon-based nanostructures in all dimensions. Table 6.2 provides a summary of some useful and well-known carbon-based nanostructures.

We will study the electronic structure of graphene by first examining its honeycomb lattice in real space (Fig. 6.1a). The reciprocal space lattice can be

*Table 6.2* Graphene sheet can be wrapped and stacked into various carbon nanostructures

| | Geometry | Carbon nanostructure |
|---|---|---|
| 1. | Wrapped into 0D | Fullerene |
| 2. | Rolled into 1D | Carbon nanotube |
| 3. | Stacked into quasi-2D | Bilayer or trilayer graphene |
| 4. | Stacked into thin film, quasi-2D | Few layers (3–10) graphene |
| 5. | Stacked into bulk 3D | Graphite |

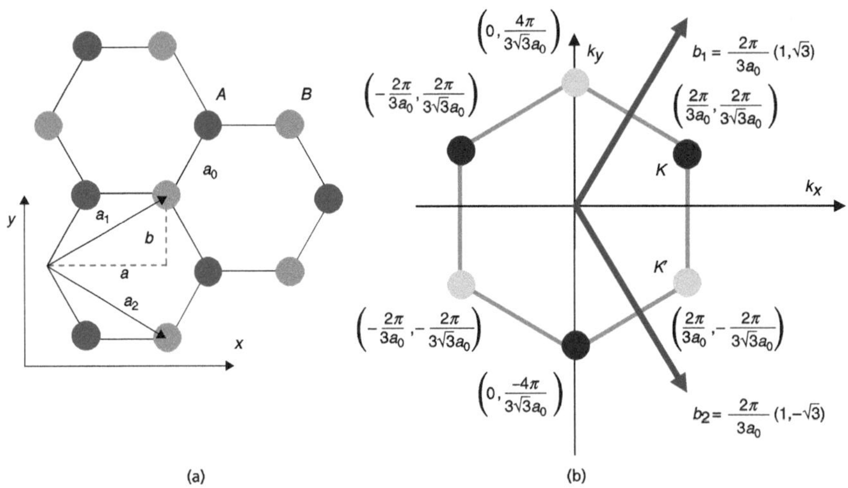


*6.1* (a) The honeycomb lattice of graphene comprises two interpenetrating triangular sub-lattices A and B. (b) The Brillouin zone of graphene is also hexagonal and comprises two non-equivalent corners or valleys (*K* and *K′*) at which the energy spectrum is degenerate.

derived following standard procedures[19] as shown in Fig. 6.1b. The real space and momentum space unit lattice vectors are:

$$\boldsymbol{a}_1 = \frac{a_0}{2}\left(3, \sqrt{3}\right); \boldsymbol{a}_2 = \frac{a_0}{2}\left(3, -\sqrt{3}\right); \boldsymbol{b}_1 = \frac{2\pi}{3a_0}\left(1, \sqrt{3}\right); \boldsymbol{b}_2 = \frac{2\pi}{3a_0}\left(1, -\sqrt{3}\right) \quad [6.1]$$

where $a = \frac{3a_0}{2}, b = \frac{\sqrt{3}a_0}{2}$ and $a_0 \sim 1.42\ A$. The tight binding Hamiltonian that describes the monolayer graphene energies in real space is:

$$H = -t\sum_{ij,\sigma} a^{\dagger}_{\sigma,i} b_{\sigma,j} - t'\sum_{ij,\sigma} a^{\dagger}_{\sigma,i} a_{\sigma,j} + b^{\dagger}_{\sigma,i} b_{\sigma,j} + h.c. \quad [6.2]$$

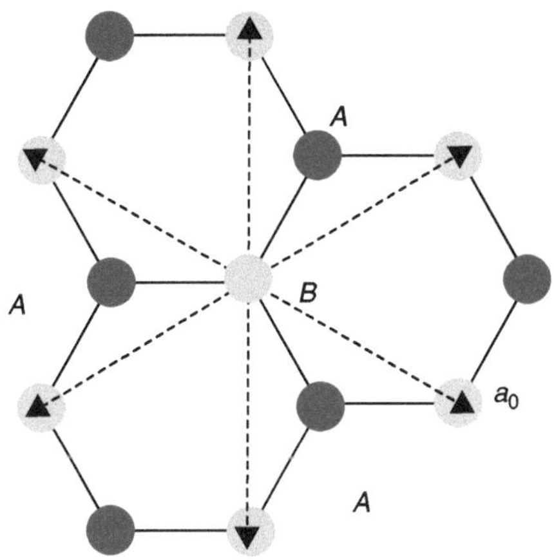


*6.2* Nearest and second nearest neighbors in the graphene honeycomb lattice.

where $t \sim 2.8$ eV is the hopping energy for the three nearest neighbors, while $t' \sim 0.01 - 0.1$ eV is for the six second-nearest neighbors (see Fig. 6.2).

The eigenenergy solution is:

$$E(k) = \pm t\sqrt{3 + g(k)} - t'\, g(k) \qquad [6.3]$$

where $g(k) = 2\cos\left(\sqrt{3}k_y\, a_0\right) + 4\cos\left(\frac{\sqrt{3}}{2}k_y a_0\right)\cos\left(\frac{3}{2}k_x a_0\right)$. Substituting $k = K + p$, with $|p| \ll |K|$ into the above for Dirac points $K$ and $K'$ (also known as valleys), one obtains the respective energy dispersion of $E = \pm v_F\,|p|$, where $v_F = \frac{3ta_0}{2} \sim 10^6\ \mathrm{ms^{-1}}$. The nearest neighbor tight binding model can be used to describe the lattice. By measuring from the valley point $K$, one can arrive at the Hamiltonian of:

$$H^K\begin{pmatrix} \psi_A^K \\ \psi_B^K \end{pmatrix} = \frac{3\gamma_0 d}{2}\begin{pmatrix} 0 & k_x - ik_y \\ k_x + ik_y & 0 \end{pmatrix}\begin{pmatrix} \psi_A^K \\ \psi_B^K \end{pmatrix} \qquad [6.4]$$

where $\gamma_0 \approx 2.7$ eV; a Fermi velocity $v_F = \frac{3\gamma_0 d}{2\hbar} \approx 10^6\ \mathrm{ms^{-1}}$ is introduced. With this, one works out the Hamiltonian for the $K$ and $K'$ valleys to be

$$H^K = v_F\begin{pmatrix} 0 & p_x - ip_y \\ p_x + ip_y & 0 \end{pmatrix}; H^{K'} = v_F\begin{pmatrix} 0 & -p_x - ip_y \\ -p_x + ip_y & 0 \end{pmatrix} \qquad [6.5]$$

or, in short, the Hamiltonian for the $K$ and $K'$ valleys are, respectively:

$$H^K = v_F\boldsymbol{\sigma}.\boldsymbol{p};\ H^{K'} = v_F\boldsymbol{\sigma}^f.\boldsymbol{p} \qquad [6.6]$$

where $\boldsymbol{\sigma} = (\sigma_x, \sigma_y)$; $\boldsymbol{\sigma}^f = (-\sigma_x, \sigma_y)$. The Dirac matrix Hamiltonian combining both valleys is:

$$H_D = v_F\left(\tau_z \otimes \sigma_x p_x + \text{II} \otimes \sigma_y p_y\right) = v_F\begin{pmatrix} \sigma_x p_x + \sigma_y p_y & 0 \\ 0 & -\sigma_x p_x + \sigma_y p_y \end{pmatrix} \quad [6.7]$$

where $\tau_z$ is the Pauli matrix for the valley degree of freedom. The solution for the Dirac matrix above is:

$$\psi_D = \left(\psi_A^K, \psi_B^K, \psi_A^{K'}, \psi_B^{K'}\right)^T. \quad [6.8]$$

The solution is thus regarded as the pseudospin for its resemblance to the spinor solution of the high-energy vacuum Dirac Hamiltonian. Equation 6.8 clearly depicts the additional degree of freedoms in monolayer graphene.

For our studies of the pseudospin effect, only one valley is considered unless otherwise stated. From the above, one can deduce the wavefunction to be:

$$\psi_\pm = \begin{pmatrix} \varphi_A \\ \varphi_B \end{pmatrix} = \begin{pmatrix} \pm e^{\mp i\phi} \\ 1 \end{pmatrix} \quad [6.9]$$

where $\phi = arctan\ ky / kx$, whose spinor is also known as the pseudospin because of its similar mathematical structure to the real electron spin. The $\mp$ in the exponential refers to the $K$ and $K'$ valleys, respectively. Mathematically, $\psi_\pm$ can be viewed to be rotating in the $x$–$y$ circle plane of the special unitary group of degree 2 (SU(2)) Bloch sphere. Physically, such rotation describes the distribution of electrons on sub-lattice sites, $A$ and $B$.

It would not be possible to fully comprehend particle behavior or motion in nanoscale graphene systems or devices without first understanding the relativistic quantum mechanics that governs the behavior of these particles. Thus before going into more qualitative descriptions of the dynamic effects in graphene such as conductivity and confinement, we refer readers to Appendix 6.A for a brief review of relativistic quantum mechanics. As this is not a formal course in physics, the formal interpretations of relativistic physics are not discussed. The purpose of this section is to facilitate understanding of the more elaborate descriptions of graphene particle dynamics in the later parts of this chapter. For comparison, the high-energy massless Dirac Hamiltonian is of the form:

$$\begin{pmatrix} -\boldsymbol{\sigma} \cdot \boldsymbol{p} & 0 \\ 0 & \boldsymbol{\sigma} \cdot \boldsymbol{p} \end{pmatrix}. \quad [6.10]$$

Let us expand our examination of the monolayer graphene Hamiltonian to both valleys: $H_K = \upsilon_F\, \boldsymbol{\sigma}\cdot\boldsymbol{p}$, $H_{K'} = \upsilon_F \boldsymbol{\sigma}^f\cdot\boldsymbol{p}$, where $\boldsymbol{\sigma} = (\sigma_x, \sigma_y)$ and $\boldsymbol{\sigma}^f = (-\sigma_x, \sigma_y)$. One realizes instantly their analogies to the Weyl equations where,, in the massless limit, particles possess handedness. It is espoused by the high energy physics community that electron exhibits right or left helicity because of mass-related mixing. To preserve Lorentz invariance,[20] the electron loses its helicity. Decomposition of the right and left helicity would be possible when particles become massless. For comparison, the Hamiltonian for massless neutrino can be decomposed into $\chi_R$ and $\chi_L$ to result in the Weyl equations of:

$$(i\partial_0 - \boldsymbol{\sigma} \cdot \boldsymbol{p})\chi_R = 0$$
$$(-i\partial_0 - \boldsymbol{\sigma} \cdot \boldsymbol{p})\chi_L = 0 \quad [6.11]$$

which accounts for the positive-energy massless particle and anti-particle, respectively. The massless particle acquires a unique helicity, i.e. its spin projection is constantly parallel to its momentum direction. A comparison is given in Table 6.3. The above is merely for reference; a more mathematical description of the physics of helicity is given in Appendix 6.B. It is hoped that useful applications can be discovered by understanding the graphene analogies to the massless neutrino.

## 6.2.2 Pseudospin and valley spin

*Pseudospin*

As shown above, carriers in monolayer graphene mimic the massless Dirac fermions rather than the usual conduction electrons.[21,22] This is because of the peculiar linear bandstructure at the Dirac cones which subjects particles at the Dirac points to behave like relativistic particles. One wonders whether the monolayer graphene electron should be addressed as an electron after all. But one can regard the graphene 'electron' as a neutrino which has acquired a charge $e$ or an electron that has lost its rest mass.

For simplicity we will focus on the $K$ valley where $H_K = v_F \boldsymbol{\sigma}\cdot\boldsymbol{p}$. The solution to this equation is a column vector spinor $(\psi_A\ \psi_B)^T$ which describes the sub-lattice degree of freedom (DOF). Figure 6.1a shows that graphene's hexagonal lattice is the superposition of two triangular sub-lattices which distinguishes $A$ atoms from $B$ atoms. In exact analogy to the real spin, electrons can reside in one of the two sub-lattices (corresponding to up-spin or down-spin) or a superposition of $A$ and $B$ sites, and are represented by the familiar two-component spinor form. In quantum mechanics, the state vectors of $|\uparrow\rangle$ and $|\downarrow$ represent the full distribution of electrons in sub-lattices $A$ and $B$, respectively. The hopping of electrons between $A$ and $B$ can be mathematically described by the rotation of a pseudospin

*Table 6.3* Analogy of monolayer graphene to the Weyl equations of the neutrino

| | Monolayer graphene | Weyl equations |
|---|---|---|
| Valley $K$ | $H_K = v_F(\sigma_x p_x + \sigma_y p_y)$ | $H_R = c(\sigma_x p_x + \sigma_y p_y)$ |
| Valley $K'$ | $H_K' = v_F(\sigma_x p_x - \sigma_y p_y)$ | $H_L = c(-\sigma_x p_x - \sigma_y p_y)$ |

vector on a Bloch sphere in exact analogy to the SU(2) physics of the electron's real spin. In fact, the rotation of the pseudospin vector can be regarded as a direct result of the rotation of an effective magnetic field induced by the electron momentum measured around the Brillouin zone corners, i.e. $p$ in our notation.

The pseudospin indexes the two non-equivalent lattice sites of graphene on which electrons can reside. The pseudospin of massless Dirac fermions in graphene possesses an analogous property to the electron spin in metal or semiconductors. One interesting fact to note here is that the pseudospin of massless particles is irrevocably tied to the momentum direction. Therefore, if the momentum of massless particles is confined or constrained in motion, such that all massless particles have a unique $p$, total pseudospin polarization reaching full polarity could then be achieved with pseudospin oriented parallel to momentum. One may thus conjecture that such a system could be used to supply highly pseudospin-polarized current for yet another analogous effect, e.g. pseudospin transfer.

*Valleyspin*

The remarkable bandstructure of graphene is due to its unique honeycomb lattice structure (Fig. 6.1a). From the real lattice, one can readily derive the reciprocal space lattice which is also periodic and hexagonal. One can show (e.g. using the tight binding approach) that the conduction and valence bands of graphene touch each other at Dirac points at the corners of the first Brillouin zone. The Brillouin zone of graphene is hexagonal and the energy gap closes at the six corners (see Fig. 6.1b). The corners are of two non-equivalent and degenerate types, also known as valleys, which are labeled $K$ and $K'$. Due to its two-valued nature, and the fact that the valley DOF is quite robust (inter-valley scattering can typically be ignored – for low energy regimes), valleys have been likened to the electron spin and numerous proposals of devices which exhibit valley-dependent transport have quickly emerged. Since inter-valley scattering is normally well suppressed in graphene, the valley DOF might be able to emulate many device functionalities in spintronics rather conveniently in the two current fashions. The physics of valley-based electronics[23–25] – so called 'valleytronics' – in graphene has thus been studied intensively in recent times. Indeed, the term 'valleytronics' was coined just several years after the discovery of graphene to collectively describe the plethora of physical effects and potential devices that make use of graphene's valleys. More recently, the idea of valley-qubits in quantum computing applications has also been proposed.[26]

### Exercise 6.2

Bandgap engineering of graphene is an essential task for building graphene-based nanoelectronic devices. As mentioned, one way of inducing a bandgap in graphene is to grow it on a substrate (e.g. SiC) such that the two sub-lattices of the graphene interact differently with the substrate. This breaks the sub-lattice symmetry, which is modeled by an electrostatic potential $U_A$ ($U_B$) on the $A$ ($B$) sub-lattice, where $U_A \neq U_B$. Thus, we can write the effective Hamiltonian as ($K$ valley):

$$H_K = v_F \boldsymbol{\sigma} \cdot \boldsymbol{p} + \begin{pmatrix} U_A & 0 \\ 0 & U_B \end{pmatrix}$$

(1) Show that the Hamiltonian above can be written as:

$$H_K = \upsilon_F \boldsymbol{\sigma} \cdot \boldsymbol{p} + \Delta\sigma_z + \gamma I,$$

where:

$$\Delta = \frac{U_A - U_B}{2} \quad \text{and} \quad \gamma = \frac{U_A + U_B}{2}$$

(2) Derive the eigenvalues for the Hamiltonian and conclude the existence of an energy gap of magnitude $2|\Delta|$. Show that the same result occurs at the $K'$ valley.

(3) Compute the Berry curvature of the Hamiltonian, and show that it is of equal magnitude but opposite signs in the two valleys.
(Refer to Chapter 7 for more details on the concept of Berry curvature in nanoelectronics.)

## 6.3 Carbon nanostructures

The electronic structure[27,28] of carbon nanotubes (CNT) is similar to monolayer graphene with additional sub-bands. We will look at the momentum space diagram of a monolayer graphene with the two valleys as labeled. A CNT is formed by rolling up a sheet of monolayer graphene. For an arbitrary direction along the circumferential vector $\boldsymbol{c}$ (see Fig. 6.3), a quantized sub-band is formed due to constraints (periodicity) imposed along that direction. Electron momenta take on quantized values along $\boldsymbol{c}$ and continuous values along the lines perpendicular to $c$. The CNT has thus become a one-dimensional electronic system where every line corresponds to an integer value $\upsilon$ which denotes the number of complete stationary wavelength that an electron forms along $\boldsymbol{c}$. A dispersion relation can be written for every integer value $\upsilon$.

Vector $\boldsymbol{c}$ is also known as the circumferential vector, denoted by $\boldsymbol{c} = m\boldsymbol{a}_1 + n\boldsymbol{a}_2$. Recalling in Fig. 6.1 that $\boldsymbol{a}_1 = a\boldsymbol{i} + b\boldsymbol{j}$, $\boldsymbol{a}_2 = a\boldsymbol{i} - b\boldsymbol{j}$, one has:

$$\boldsymbol{c} = m\boldsymbol{a}_1 + n\boldsymbol{a}_2; \qquad \boldsymbol{c} = (m+n)\, a\boldsymbol{i} + (m-n)b\boldsymbol{j}. \tag{6.12}$$

Figure 6.4 shows the honeycomb lattice of a graphene sheet. The chiral vector which defines the folding direction to form a CNT is denoted by $\boldsymbol{c}$. The periodic boundary condition requires that momentum along the circumferential vector can only exist in the multiple of $2\pi$, i.e.

$$\boldsymbol{k.c} = (m+n)\, a\, k_x + (m-n) b k_y = 2\pi v. \tag{6.13}$$

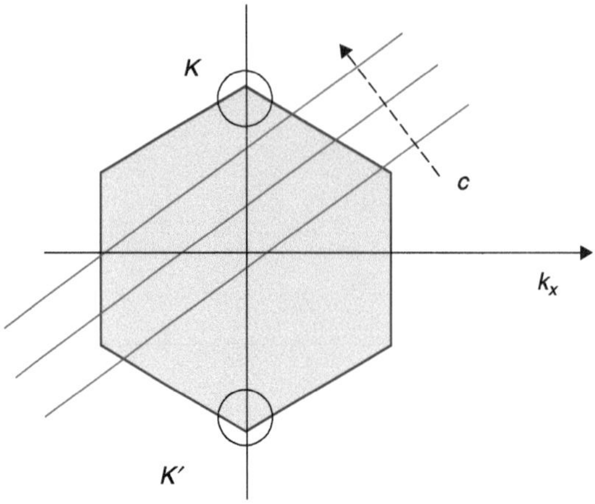


*6.3* Graphene sheet is wrapped along the circumferential vector and momentum becomes quantized along that direction. The electron is only free to move along the diagonal lines, exhibiting one-dimensional behavior.

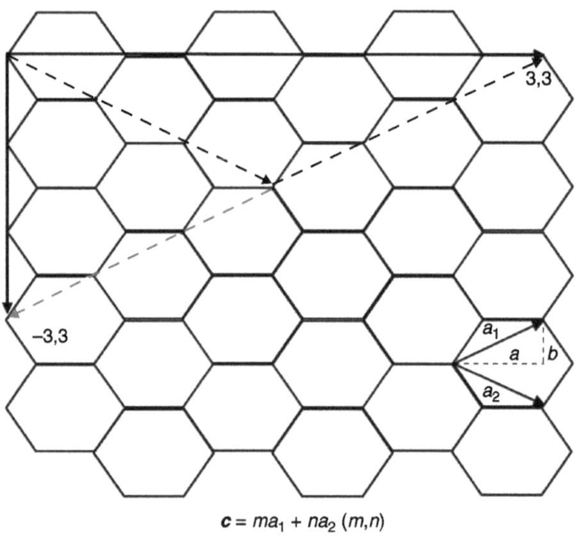


*6.4* Honeycomb lattice of a monolayer graphene before rolling into various carbon nanostructures.

If these lines pass through one of the valleys $(k_x a, k_y b) = (0, \pm 2\pi/3)$, energy $E = 0$ at the valley coincides with the lines, and thus there will be no bandgap in these CNTs. One can see that this can only happen for:

$$|m - n| = 3v. \quad [6.14]$$

The graphene sheet shown in Fig. 6.5a can be wrapped along the $x$ axis, such that side $A$ meets side $B$ to form a CNT structure known as the 'armchair', reflecting the top and bottom circumferential edges of the cylindrical structure. Similarly, Fig. 6.5b when wrapped around in the same manner (along the $y$ axis) yields a cylindrical tube structure with zigzag edges. The allowed momenta for the armchair CNT lie along a set of lines parallel to the $y$ axis, while those for the zigzag CNT lie on lines parallel to the $x$ axis as shown in Fig. 6.6.

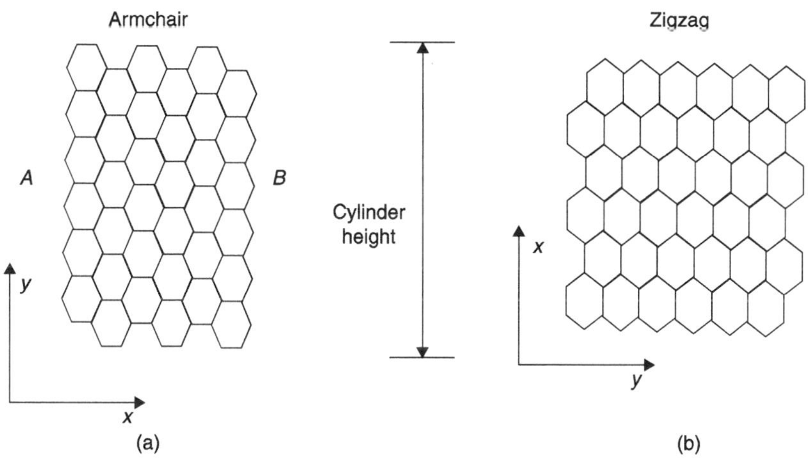


*6.5* (a) Armchair CNT with circumferential vector $\boldsymbol{c} = 2am\boldsymbol{i}$. (b) Zigzag CNT with circumferential vector $\boldsymbol{c} = 2bm\boldsymbol{j}$.

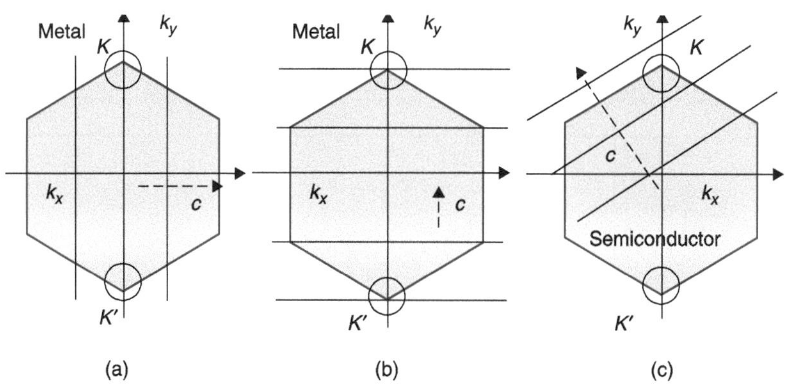


*6.6* (a) Armchair CNT with circumferential vector $\boldsymbol{c} = 2am\boldsymbol{i}$ showing metallic bandstructure at the valleys. (b) Zigzag CNT with circumferential vector $\boldsymbol{c} = 2bm\boldsymbol{j}$ can be metallic or semiconductor depending on the values of satisfying $2m = 3v$. (c) An arbitrary circumferential vector can yield metallic or semiconducting CNT for any which satisfies $|m - n| = 3v$.

**Exercise 6.3**

Deduce using $\boldsymbol{c} = (m + n)a\boldsymbol{i} + (m - n)b\boldsymbol{j}$ that the circumferential vector for the armchair and zigzag CNTs are $\boldsymbol{c} = 2am\boldsymbol{i}$, $\boldsymbol{c} = 2bm\boldsymbol{j}$, respectively. Explain with Brillouin zones and sub-band drawings why the armchair graphene is always metallic regardless of the value of $m$.

***Solution***

In armchair CNT, $m = n$, and the chiral vector is $\boldsymbol{c} = 2am\boldsymbol{i}$. One has $2mak_x = 2\pi v$, or $k_x = v\left(\frac{2\pi}{2ma}\right)$ and $L = 2ma$ is the length of the CNT cylindrical circumference. The wavelength of the wavevector is $k_x = v\left(\frac{2\pi}{L}\right)$.

The sub-band with $v = 0$ will always pass through $K = (k_x a,\ k_y b) = \left(0, \pm\frac{2\pi}{3}\right)$.

## 6.4 Bilayer graphene

Much has been said about the novelty of graphene which promises new electronics with applications wild and aplenty. We begin this section with a brief introduction to the full bilayer graphene Hamiltonian leading to its reduced form. As is well-known, the bilayer graphene comprises two monolayer graphene stacked vertically and has a more complicated energy structure. The bilayer graphene Hamiltonian[29] has been expressed as:

$$H_1\psi_1 = \begin{pmatrix} -u/2 & v_3\pi & 0 & c\pi^\dagger \\ v_3\pi^\dagger & u/2 & c\pi & 0 \\ 0 & c\pi^\dagger & u/2 & t \\ c\pi & 0 & t & -u/2 \end{pmatrix}\begin{pmatrix} \varphi_{A1} \\ \varphi_{B2} \\ \varphi_{A2} \\ \varphi_{B1} \end{pmatrix} \qquad [6.15]$$

where $\pi = p_x + ip_y$, $t$ is the interlayer coupling and $\psi_1$ is a four-component spinor wavefunction. In graphene it is known that interlayer coupling between $A_1$ and $B_2$ is strong. Subscripts $A$, $B$ refer to the sub-lattice index, while 1, 2 refer to the layer index. Alternatively there are different versions of the Hamiltonian found in literatures. The differences arise mainly from symmetry as well as variations due to approximations.

### 6.4.1 Bilayer graphene Hamiltonian

The spinor wavefunction of (1) in Table 6.4 is different from those in (2) and (3). A transformation can be performed as follows:

$$\psi_1 = \begin{pmatrix} \varphi_{A1} \\ \varphi_{B2} \\ \varphi_{A2} \\ \varphi_{B1} \end{pmatrix} \rightarrow \psi' = U\psi_1 = \begin{pmatrix} \varphi_{A1} \\ \varphi_{B1} \\ \varphi_{A2} \\ \varphi_{B2} \end{pmatrix} \quad where\ U = \begin{pmatrix} 1 & 0 & 0 & 0 \\ 0 & 0 & 0 & 1 \\ 0 & 0 & 1 & 0 \\ 0 & 1 & 0 & 0 \end{pmatrix} = U^{\dagger}. \qquad [6.16]$$

Matrix $U$ defined above has the property of hermicity (i.e. $U = U^{\dagger}$) and unitary (i.e. $UU^{\dagger} = 1$). The Hamiltonian can then be transformed as $H_3 = UHU^{\dagger}$.

*Table 6.4* Hamiltonian of the bilayer graphene resembles the massive Dirac Hamiltonian

| | Bilayer graphene Hamiltonian | Reference |
|---|---|---|
| 1. | $H_1\psi_1 = \begin{pmatrix} -u/2 & v_3\pi & 0 & c\pi^{\dagger} \\ v_3\pi^{\dagger} & u/2 & c\pi & 0 \\ 0 & c\pi^{\dagger} & u/2 & t \\ c\pi & 0 & t & -u/2 \end{pmatrix} \begin{pmatrix} \varphi_{A1} \\ \varphi_{B2} \\ \varphi_{A2} \\ \varphi_{B1} \end{pmatrix}$ | 29 |
| 2. | $H_2\psi_2 = \begin{pmatrix} -u/2 & c\pi^{\dagger} & 0 & 0 \\ c\pi & -u/2 & t & 0 \\ 0 & t & u/2 & c\pi^{\dagger} \\ 0 & 0 & c\pi & -u/2 \end{pmatrix} \begin{pmatrix} \varphi_{A1} \\ \varphi_{B1} \\ \varphi_{A2} \\ \varphi_{B2} \end{pmatrix}$ | 30 |
| 3. | $H_3\psi_3 = \begin{pmatrix} -u/2 & c\pi^{\dagger} & 0 & v_3\pi \\ c\pi & -u/2 & t & 0 \\ 0 & t & u/2 & c\pi^{\dagger} \\ v_3\pi^{\dagger} & 0 & c\pi & -u/2 \end{pmatrix} \begin{pmatrix} \varphi_{A1} \\ \varphi_{B1} \\ \varphi_{A2} \\ \varphi_{B2} \end{pmatrix}$ | |

**Exercise 6.4**

Show that after transformation, the Hamiltonian of $H_1$ is identical to $H_3$, and is very close to that of $H_2$ except that $v_3\pi$ and $\upsilon_3\pi^{\dagger}$ have been dropped in $H_2$.

***Solution***

$$H_1\psi_1 = \begin{pmatrix} -u/2 & v_3\pi & 0 & c\pi^\dagger \\ v_3\pi^\dagger & u/2 & c\pi & 0 \\ 0 & c\pi^\dagger & u/2 & t \\ c\pi & 0 & t & -u/2 \end{pmatrix} \begin{pmatrix} \varphi_{A1} \\ \varphi_{B2} \\ \varphi_{A2} \\ \varphi_{B1} \end{pmatrix} \rightarrow H_3\psi_3 = UH_1U^\dagger U\psi_1$$

$$= \begin{pmatrix} -u/2 & c\pi^\dagger & 0 & v_3\pi \\ c\pi & -u/2 & t & 0 \\ 0 & t & u/2 & c\pi^\dagger \\ v_3\pi^\dagger & 0 & c\pi & u/2 \end{pmatrix} \begin{pmatrix} \varphi_{A1} \\ \varphi_{B1} \\ \varphi_{A2} \\ \varphi_{B2} \end{pmatrix}$$

---

It is interesting to examine whether the above Hamiltonian fits the Dirac's equation. In terms of the explicitly covariant form of the Dirac equation, $(i\hbar\gamma^\mu \partial_\mu - mc)\,\psi = 0$, one has:

$$\begin{pmatrix} E - mc & -\boldsymbol{\sigma}.\boldsymbol{p} \\ \boldsymbol{\sigma}.\boldsymbol{p} & -E - mc \end{pmatrix} \psi = 0 \qquad [6.17]$$

where:

$$\gamma^0 = \begin{pmatrix} 1 & 0 \\ 0 & -1 \end{pmatrix}, \qquad \gamma^j = \begin{pmatrix} 0 & \sigma^j \\ -\sigma^j & 0 \end{pmatrix}. \qquad [6.18]$$

Equation 6.17 can be transformed to:

$$\begin{pmatrix} E - mc & -\boldsymbol{\sigma}.\boldsymbol{p} \\ -\boldsymbol{\sigma}.\boldsymbol{p} & E + mc \end{pmatrix} \psi = 0. \qquad [6.19]$$

Equation 6.19 seems a closer fit with $(E - H_1)\,\psi = 0$ which explicitly is:

$$\begin{pmatrix} E - \begin{pmatrix} 0 & v_3\pi \\ v_3\pi^\dagger & 0 \end{pmatrix} & -\boldsymbol{\sigma}.\boldsymbol{p}\,c \\ -\boldsymbol{\sigma}.\boldsymbol{p}\,c & E - \begin{pmatrix} 0 & t \\ t & 0 \end{pmatrix} \end{pmatrix} \psi = 0. \qquad [6.20]$$

In the above, magnetic vector and electric scalar potential has been removed. The bilayer graphene mimics the massive Dirac systems to some extent. Future nanofabrication technologies might be able to produce a graphene-like system which mimics the massive Dirac system with less approximation needed.

### 6.4.2 Reduction of the Hamiltonian

In bilayer graphene, the 4×4 Hamiltonian is cumbersome and one often uses a 2×2 simplified form. In this section we transform the larger bilayer graphene Hamiltonian into a simplified form. In Table 6.4, the Hamiltonian (3) can be written as:

$$H_3\psi_3 = \begin{pmatrix} -u/2 & c\pi^\dagger & 0 & v_3\pi \\ c\pi & -u/2 & t & 0 \\ 0 & t & u/2 & c\pi^+ \\ v_3\pi^\dagger & 0 & c\pi & u/2 \end{pmatrix}\begin{pmatrix} \varphi_{A1} \\ \varphi_{B1} \\ \varphi_{A2} \\ \varphi_{B2} \end{pmatrix} = \begin{pmatrix} -\frac{u}{2}\varphi_{A1} + c\pi^\dagger\varphi_{B1} + v_3\pi\varphi_{B2} \\ c\pi\varphi_{A1} - \varphi_{B1}\frac{u}{2} + t\varphi_{A2} \\ t\varphi_{B1} + \varphi_{A2}\frac{u}{2} + c\pi^\dagger\varphi_{B2} \\ v_3\pi^\dagger\varphi_{A1} + c\pi\varphi_{A2} + \varphi_{B2}\frac{u}{2} \end{pmatrix} \qquad [6.21]$$

where $u$ is the applied electrical voltage between the two layers and $t$ is the interlayer coupling. Subscripts $A$, $B$ refer to the sub-lattice index, while subscripts 1, 2 refer to the layer index. Written in this form, one has chosen to show the interlayer coupling between $A2$ and $B1$. One assumes the limit of strong interlayer coupling, i.e. $t \gg u$.

---

**Exercise 6.5**

Show that in the limit $E \ll t$, the Hamiltonian $H_3$ can be reduced as follows:

$$H_3 = \begin{pmatrix} -u/2 & c\pi^\dagger & 0 & v_3\pi \\ c\pi & -u/2 & t & 0 \\ 0 & t & u/2 & c\pi^\dagger \\ v_3\pi^\dagger & 0 & c\pi & u/2 \end{pmatrix} \rightarrow H_3^{eff} = \begin{pmatrix} -\frac{u}{2}\left(1-\frac{c^2p^2}{t^2}\right) & \left(-\frac{c^2\pi^{\dagger 2}}{t} + v_3\pi\right) \\ \left(\frac{-c^2\pi^2}{t} + v_3\pi^\dagger\right) & \frac{u}{2}\left(1-\frac{c^2p^2}{t^2}\right) \end{pmatrix}.$$

***Solution***

In the limit $u \ll t$ and $E \ll t$, one can obtain an effective Hamiltonian for the lowest energy bands. We can reorder the wave functions according to $(A_1, B_1, A_2, B_2) \rightarrow (A_1, B_2, A_2, B_1)$, and in the new basis, the Hamiltonian becomes:

$$H_3' = \begin{pmatrix} -\frac{u}{2} & v_3\pi & 0 & c\pi^\dagger \\ v_3\pi^\dagger & \frac{u}{2} & c\pi & 0 \\ 0 & c\pi^\dagger & \frac{u}{2} & t \\ c\pi & 0 & t & -\frac{u}{2} \end{pmatrix} = \begin{pmatrix} H_{11} & H_{12} \\ H_{21} & H_{22} \end{pmatrix}$$

where $H_{ij}$ is a 2 × 2 block. Use the identity $det\,(H_3' - E) = det\,(H_{11} - H_{12}\,(H_{22} - E)^{-1} H_{21} - E)\,det\,(H_{22} - E)$. For, $E \ll t$, the substitution $H_{22} - E \rightarrow H_{22}$ reduces the computation of the lowest-energy bands to the diagonalization of the 2 × 2 effective Hamiltonian:

$$H_3^{eff} = H_{11} - H_{12} H_{22}^{-1} H_{21}$$

$$= \begin{pmatrix} -\frac{u}{2} & v_3\pi \\ v_3\pi^{+} & \frac{u}{2} \end{pmatrix} - \frac{1}{t^2 + \frac{u^2}{4}} \begin{pmatrix} 0 & c\pi^{\dagger} \\ c\pi & 0 \end{pmatrix} \begin{pmatrix} \frac{u}{2} & t \\ t & \frac{-u}{2} \end{pmatrix} \begin{pmatrix} 0 & c\pi^{\dagger} \\ c\pi & 0 \end{pmatrix}$$

$$\approx \begin{pmatrix} -\frac{u}{2}\left(1 - \frac{c^2p^2}{t^2}\right) & \left(-\frac{c^2\pi^{\dagger 2}}{t} + v_3\pi\right) \\ \left(\frac{-c^2\pi^2}{t} + v_3\pi^{\dagger}\right) & \frac{u}{2}\left(1 - \frac{c^2p^2}{t^2}\right) \end{pmatrix}.$$

---

Taking the limit of $u \ll t$ and $p = \sqrt{p_x^2 + p_y^2}$, the effective Hamiltonian can now be expressed in the basis $(A_1, B_2)$. Thus the Hamiltonian of the reduced bilayer graphene is:

$$H_3^{eff} = \frac{-c^2}{t}\begin{pmatrix} 0 & \pi^{+2} \\ \pi^2 & 0 \end{pmatrix} - \frac{u}{2}\left(1 - \frac{c^2p^2}{t^2}\right)\begin{pmatrix} 1 & 0 \\ 0 & -1 \end{pmatrix} + v_3\begin{pmatrix} 0 & \pi \\ \pi^{\dagger} & 0 \end{pmatrix}. \qquad [6.22]$$

This Hamiltonian has been used widely as an approximation type of Hamiltonian for bilayer graphene. Future advances in technologies might allow us to engineer the various parameters and new approximations might be possible. It will be interesting if one can pre-conceive a useful Hamiltonian and use technologies to achieve the desired Hamiltonian. The present Hamiltonian above has been used to understand the topological properties[31,32] of graphene as well as the Berry's phase,[33–35] which have great implications to the physics of electron dynamics in graphene.

## 6.5 Deformation-induced gauge potential

It has been understood that free-standing graphene is not flat. In fact it is soft and membrane-like, and conforms to the substrate, creating ripples which are static distortion of the 2D lattice.[36] Because of its flatness, graphene sheet reacts to thermal fluctuation and thermal stress and becomes warped or floppy as a result. Free-standing graphene also generates flexural (out-of-plane) phonons which are simply quantized mode of vibration in the crystal (dynamic distortion). Thus graphene sheets can move in the third dimension – in other words behave like a 'cantilever'. Lattice deformation in graphene modifies the nearest neighbor hopping strength as well as the on-site potential.

We will show that asymmetric hopping strength or site potentials can in fact induce a gauge potential[37] in the graphene sheet. We begin with the nearest neighbor Hamiltonian of a graphene sheet system:

$$H = -t \sum_{i \in A \;\; j=1,2,3} a_i^\dagger b_j + b_j^\dagger a_i \qquad [6.23]$$

where $i$ runs over all sub-lattice sites $A$ and $j$ = 1,2,3 are the three nearest neighbors surrounding each site $i$. Here, we focus on the hopping strength effect. Lattice deformation modifies $t$ as $t \to t + \delta t_j$, generating a perturbation Hamiltonian of:

$$H_I = -\delta t_j(r_i) \sum_{i \in A \; j=1,2,3} a_i^\dagger b_j + b_j^\dagger a_i. \qquad [6.24]$$

Using the Bloch wavefunction for each lattice site:

$$\left|\psi_A^k\right\rangle = \frac{1}{\sqrt{N}} \sum_{i \in A} e^{-i\boldsymbol{k}.\boldsymbol{r}_i} a_i^\dagger \left|0\right\rangle$$

$$\left|\psi_B^k\right\rangle = \frac{1}{\sqrt{N}} \sum_{i \in B} e^{-i\boldsymbol{k}.\boldsymbol{r}_i} b_i^\dagger \left|0\right\rangle$$

[6.25]

where $N$ is the number of unit cells. Considering intra-valley scattering within valley $K$, one finds the matrix elements:

$$\begin{bmatrix} \left\langle \psi_A^{k_F+k+\delta k} \middle| H_I \middle| \psi_A^{k_F+k} \right\rangle & \left\langle \psi_A^{k_F+k+\delta k} \middle| H_I \middle| \psi_B^{k_F+k} \right\rangle \\ \left\langle \psi_B^{k_F+k+\delta k} \middle| H_I \middle| \psi_A^{k_F+k} \right\rangle & \left\langle \psi_B^{k_F+k+\delta k} \middle| H_I \middle| \psi_B^{k_F+k} \right\rangle \end{bmatrix}$$

$$= \begin{bmatrix} 0 & e^{-i\delta\boldsymbol{k}.\boldsymbol{r}_i} \dfrac{v_F}{N} \sum_i A_x(r_i) - iA_y(r_i) \\ e^{-i\delta\boldsymbol{k}.\boldsymbol{r}_i} \dfrac{v_F}{N} \sum_i A_x(r_i) + iA_y(r_i) & 0 \end{bmatrix} \qquad [6.26]$$

where:

$$v_F A_x = \delta t_1 - \frac{1}{2}(\delta t_2 + \delta t_3); \quad v_F A_y = \frac{\sqrt{3}}{2}(\delta t_2 - \delta t_3) \qquad [6.27]$$

and it is understood that $A_x$ and $\delta t_j$ are functions of sub-lattice A site $i$. The same can be worked out for valley $K'$ using similar processes. The full Hamiltonian for valleys $K$ and $K'$ due to lattice deformation modifying the hopping integral are:

$$H_0^K + H_I^K = v_F\, \boldsymbol{\sigma}.(\boldsymbol{p} + \boldsymbol{A}(\boldsymbol{r}))$$

$$H_0^{K'} + H_I^{K'} = v_F\, \boldsymbol{\sigma}^f.(\boldsymbol{p} - \boldsymbol{A}(\boldsymbol{r})).$$

[6.28]

Inter-valley scattering in graphene is well suppressed and we do not discuss its effect here. The on-site potential modification due to lattice deformation or external imposition is also not discussed. As far as electron motion is concerned, it becomes clear that applying strain to graphene is equivalent to applying an external magnetic field.

## 6.6 Application of graphene spin

Spintronics is one of the richest topics in nanoscale electronics. Spintronics in graphene has, however, remained a fairly quiet area of research. This is because spin-based effects in graphene, such as spin orbit coupling (SOC) and Zeeman interactions, are negligibly small, typically of the order 0.1 K.[38] As an example, in 2005, Kane and Mele[39] proposed the existence of the remarkable quantum spin-Hall state in graphene (which led to the launched of a separate and vigorous investigation into a new state of matter known as 'topological insulators'). However, due to the invariably small SOC, graphene was soon ruled out as a likely candidate for the experimental realization of the effect. We note though that recent studies suggest evidence for strong spin orbit coupling in bilayer graphene.

One of the most remarkable things about graphene is the immensely rich and elegant physics that can be drawn from such a simple structure. Despite the lack of interesting spin-based effects, the symmetry of pristine graphene affords us two additional spin-like quantities, namely pseudospin and valley spin. Like the real spin, the pseudo and valley spins in graphene are two-valued, and in the effective Hamiltonian, can be expressed in terms of Pauli matrix-like quantities.

### 6.6.1 Pseudospin electronics

Unlike the real spin, the pseudospin cannot be polarized by a real magnetic field. This is because the polarizing effect in opposite valleys cancel each another out.[37] To induce a Zeeman-like term which actually polarizes the pseudospin, one needs to break the symmetry between the two sub-lattices. This could be realized by applying different electrostatic potentials to the A and B sub-lattice. In the case of monolayer graphene, this can be achieved by epitaxial growth on, for example, SiC substrates. It is particularly simple for the case of bilayer graphene where one need only apply different gate voltages to the two coupled graphene layers (in the low energy regime). This opens up a gap in the electronic spectrum which is proportional to the energy difference on the two sub-lattices, effectively transforming bilayer graphene into a tunable semiconductor.

*Berry's phase*

The pseudospin of graphene has deep and interesting connections with many physical effects, e.g. the half/full integer shift in integer quantum Hall effects

*Table 6.5* Summary of Berry's phase in monolayer and bilayer graphene and the associated gauge potential

| | Monolayer graphene | Bilayer graphene |
|---|---|---|
| Wavefunction | $\psi_{\pm}=\begin{pmatrix}\varphi_A\\ \varphi_B\end{pmatrix}=\frac{1}{\sqrt{2}}\begin{pmatrix}\pm e^{-\varphi}\\ 1\end{pmatrix}$ | $\psi_{\pm}=\begin{pmatrix}\varphi_{A1}\\ \varphi_{B2}\end{pmatrix}=\frac{1}{\sqrt{2}}\begin{pmatrix}\pm e^{-2i\varphi}\\ 1\end{pmatrix}$ |
| Berry's phase | $\phi_{Berry}=\int d\boldsymbol{k}\cdot\boldsymbol{A}=\pm\pi$ | $\phi_{Berry}=\int d\boldsymbol{k}\cdot\boldsymbol{A}=\pm 2\pi$ |
| Local gauge | $\boldsymbol{A}(k)=\pm\frac{\hbar}{2ek^2}\begin{pmatrix}-k_y\\ k_x\\ 0\end{pmatrix}$ | $\boldsymbol{A}(k)=-i\langle\psi_{\pm}\lvert\nabla_k\rvert\psi_{\pm}\rangle-\perp\frac{\hbar}{2ek^2}\begin{pmatrix}-2k_y\\ 2k_x\\ 0\end{pmatrix}$ |

(IQHE), pseudospin analogies of spin Hall and spin torque, particle localization and anti-localization, Klein tunneling, minimum conductivity, zitterbewegung and pair production. It is, however, not always apparent how pseudospin is linked to these phenomena. But if one understands the connection between the pseudospin and the Berry's phase, the link becomes clearer; one needs only to understand the link between the Berry's phase and the physical phenomena. In Chapter 7, we discuss the adiabatic evolution of the pseudospin vector and how it gives rise to the gauge curvature and to Berry's phase, which is a topological invariant. Prior to that, readers are required to accept that Berry's phase associated with pseudospin of graphene can be derived as summarized in Table 6.5.

### *Pseudospin analogies of spin Hall and spin torque*

Spintronics – particularly in semiconductors and magnetic multilayers – has represented an engaging field of study and has become all important for applications in magnetic storage. In semiconductors, the SOC effect, in which carrier spins feel an effective momentum-dependent magnetic field, plays a crucial role in how spins are manipulated. For example, it leads to the remarkable spin Hall effect (SHE) in which dissipationless transverse spin currents flow in response to an applied electric field as discussed in Chapter 5. Since SHE can be linked to the Berry's phase associated with spin, we remark that a similar SHE for the pseudospin might arise in connection with the Berry's phase associated with the pseudospin. Likewise, we comment generally that other spintronic phenomena related to spin orbit gauge potential or Berry's phase (e.g. the spin torque and spin current) might have their respective counterparts in the pseudospin.

*Pseudospin orbit coupling*

Using the Dirac equations, SOC can be derived in vacuum with applied electric fields. One can thus visualize that a similar effect should arise for the pseudospin of graphene particles governed by the Dirac equations. Below, we derive the pseudo SOC for a bilayer graphene with the Hamiltonian $H_1$ (see Table 6.4):

$$H_3\psi_3 = \begin{pmatrix} & & 0 & v_3\pi \\ c\boldsymbol{\sigma}.\boldsymbol{p} & & t & 0 \\ 0 & t & c & \boldsymbol{\sigma}.\boldsymbol{p} \\ v_3\pi^\dagger & 0 & & \end{pmatrix}\begin{pmatrix} A_1 \\ B_1 \\ A_2 \\ B_2 \end{pmatrix} \rightarrow H_1\psi_1 = \begin{pmatrix} 0 & v_3\pi & & \\ v_3\pi^\dagger & 0 & c & \boldsymbol{\sigma}.\boldsymbol{p} \\ c\boldsymbol{\sigma}.\boldsymbol{p} & & 0 & t \\ & & t & 0 \end{pmatrix}\begin{pmatrix} A_1 \\ B_2 \\ A_2 \\ B_1 \end{pmatrix}. \quad [6.29]$$

For brevity, we write $(A_1\ B_1\ A_2\ B_2)$ to denote $\varphi_{A1}\ \varphi_{B1}\ \varphi_{A2}\ \varphi_{B2})$. The transformation performed by $U$ determines the type of pseudospin to be investigated; for example, by the above method, which results in $\chi = (A_1\ B_2)^T$, pseudospin is defined between site $A_1$ and $B_2$. In application to particles which mimic Dirac fermions, it would be instructive to replace the coupling mass term of $mc^2$ for particles in vacuum with a coupling term $\Delta$ which arises due to material bandstructure but plays the same role as the mass term as far as the Dirac matrix is concerned. The coupling term $\Delta$ gives rise to the energy dispersion where the effective mass of particles in the materials can be derived. For monolayer graphene, $\Delta$ vanishes and one deduces from the energy dispersion relation that particles behave like massless Dirac fermions. Following $H_1$ in Table 6.4, we write the graphene Hamiltonian:

$$\begin{pmatrix} E-\Delta_a & -\boldsymbol{\sigma}\cdot\boldsymbol{p}\,c \\ -\boldsymbol{\sigma}\cdot\boldsymbol{p}\,c & E+\Delta_b \end{pmatrix}\cdot\begin{pmatrix} \chi \\ \theta \end{pmatrix} = 0 \quad [6.30]$$

such that it is comparable to Eq. 6.19. In this specific case, $\Delta_a = \begin{pmatrix} 0 & v_3\pi \\ v_3\pi^\dagger & 0 \end{pmatrix}$ and $\Delta_b = \begin{pmatrix} 0 & -t \\ -t & 0 \end{pmatrix}$. Multiplying by $\gamma^0$ to the left, one obtains the graphene Hamiltonian:

$$\begin{pmatrix} (E-e\phi)-\Delta_a & -\boldsymbol{\sigma}\cdot\boldsymbol{p}c \\ \boldsymbol{\sigma}\cdot\boldsymbol{p}c & -(E-e\phi)-\Delta_b \end{pmatrix}\cdot\begin{pmatrix} \chi \\ \theta \end{pmatrix} = 0 \quad [6.31]$$

which is comparable to Eq. 6.17. Note in Eq. 6.31, the electric field has been incorporated via $e\phi$. Equation 6.31 is now written as:

$$i\hbar\gamma^\mu D_\mu - R = 0 \quad [6.32]$$

where

$$R = \begin{pmatrix} \Delta_a & 0 \\ 0 & \Delta_b \end{pmatrix} \text{ and } D_\mu = \partial_\mu - \frac{ie}{\hbar}A_\mu.$$

One can now apply the electric field $\varepsilon$, with electric potential $\phi$. Here we merely wish to illustrate the close connection between the bilayer graphene and the vacuum Dirac Hamiltonian such that useful analogies between pseudo SOC and vacuum SOC can be drawn. Thus, to simplify matters, we temporarily disregard the fact that $\Delta_a = \begin{pmatrix} 0 & v_3\pi \\ v_3\pi^\dagger & 0 \end{pmatrix}$ and $\Delta_b = \begin{pmatrix} 0 & t \\ t & 0 \end{pmatrix}$. Rather, we modify as follows:

$$\Delta_a^2 = \begin{pmatrix} (v_3 p)^2 & 0 \\ 0 & (v_3 p)^2 \end{pmatrix} \to \Delta_a^2 = \begin{pmatrix} (mc^2)^2 & 0 \\ 0 & (mc^2)^2 \end{pmatrix} \qquad [6.33]$$

so that one can write $\Delta_a^2 = \Delta^2 I$, where $\Delta = mc^2$. We make an even more drastic assumption that one can write:

$$\Delta_b = \begin{pmatrix} mc^2 & 0 \\ 0 & mc^2 \end{pmatrix} = \Delta I. \qquad [6.34]$$

Multiplying Eq. 6.32 to the left with $\gamma^v D_v + R$ thus yields:

$$\begin{pmatrix} [E-e\phi]^2 - p^2c^2 - \Delta_a^2 & -ie\hbar c\,\boldsymbol{\sigma}\cdot\boldsymbol{\varepsilon} \\ -ie\hbar c\,\boldsymbol{\sigma}\cdot\boldsymbol{\varepsilon} & [E-e\phi]^2 - p^2c^2 - \Delta_b^2 \end{pmatrix} \cdot \begin{pmatrix} \chi \\ \theta \end{pmatrix} = 0. \qquad [6.35]$$

Equation 6.31 therefore can be reduced to:

$$\theta = \frac{\boldsymbol{\sigma}\cdot\boldsymbol{p}c}{(E-e\phi)-[\Delta_b]}\chi \qquad [6.36]$$

where $[\Delta_b] = \sqrt{[\Delta_a^2]} = \Delta$. The relativistic energy equation, which could be used to describe the analogous effect of pseudo SOC (i.e. the coupling of pseudospin to particle momentum in the presence of electric fields) for Dirac fermions in graphene-like material systems is thus:

$$\left[(E-e\phi)^2 - \Delta_a^2 - p^2c^2 - \frac{ie\hbar c^2}{E-e\phi-[\Delta_b]}\boldsymbol{\varepsilon}\cdot\boldsymbol{p} - \frac{e\hbar c^2}{E-e\phi-[\Delta_b]}\boldsymbol{\sigma}\cdot(\boldsymbol{\varepsilon}\times\boldsymbol{p})\right]\chi = 0. \qquad [6.37]$$

We note again that approximations have been taken to avoid dealing with excessive material science, i.e. we have assumed bilayer graphene material could satisfy $\Delta^2_a = \Delta^2 I$; $\Delta_b = \Delta I$.

## 6.6.2 Valleyspin electronics

The first valleytronics work[24] proposed a valley filtering effect where electrons are transmitted through graphene quantum point contacts. In such contacts, right and left movers reside in opposite valleys. It was suggested that two such valley filters could be connected in cascade to form a valley valve; see reference 40 for further insights into the physics of the valley valve effect. A valley filter in strain-engineered graphene has been proposed.[41,42] In bilayer graphene, there has been a proposal for a valley filter[43] based on irradiation by high intensity light. We have mentioned that inducing an asymmetry in the graphene sub-lattices opens up a gap due to the Zeeman like splitting of the pseudospin. Moreover, it can be shown that this mechanism results in a finite Berry curvature in momentum ($k$) space. This Berry curvature has opposite signs for the two valley spins and thus affects carrier motion in a valley-dependent manner. In fact it leads to the valley Hall effect[25] in direct analogy with SHE in p-doped semiconductors.[44] Such an effect also arises in the case of inhomogeneous sub-lattice asymmetry, where the effective Hamiltonian contains a valley orbit coupling term.[45]

In fact, the analogy with the electron spin goes much deeper. It has been shown[25] that, in graphene with broken inversion symmetry, there is an intrinsic magnetic moment associated with the valley DOF exactly like the electron spin. The requirement for spatial inversion symmetry breaking comes about from symmetry considerations of the effective 'Bohr magneton'. The (low-energy) magnetic moment characterizing monolayer graphene's valleys is:

$$M_v = \xi\mu_B^* \qquad \mu_B^* = \frac{e\hbar}{2m^*} \qquad [6.38]$$

where $\xi = \pm 1$ for the two valleys and the effective mass $m^* = \dfrac{2\Delta\hbar^2}{3a^2t^2}$. Here, $\Delta$ is the size of the inversion asymmetry. Unlike the real spin, the magnetic moment associated with the valley DOF is an orbital magnetic moment and originates from the self-rotation of the wavefunctions about the valley centers. It is found that, for typical values, $\mu^*_B$ is an order of magnitude larger than the spin-Bohr magneton, implying that in the low-energy regime, the response to a magnetic field is dominated by the valley magnetic moment rather than the actual spin. In the limit of high magnetic fields, quantum Landau levels are formed, and several authors have experimentally found that the lowest Landau level ($n = 0$) exhibits a broken valley degeneracy.[46] This was later explained as an intrinsic property of the symmetry of graphene's lattice.[47] The above discussion motivates a serious discussion about the relevance of the valley spin in graphene-based nanotechnology. With this, we proceed to discuss some valleytronics devices proposed in the literature, including valley filters, valley valves and devices which make use of valley-dependent Hall effects.

### *Valley filtering*

In the context of spintronics, one could conceive of a spin filter, in which an unpolarized input current gives rise to a spin-polarized output current. In direct analogy with this concept, valley filtering in graphene has been proposed by several authors. The first proposal[24] was based on valley-dependent transmission through graphene nanoconstrictions. This idea works on the principle that, in narrow graphene nanoribbons with zigzag edges, the lowest energy mode is chiral in the sense that left movers and right movers occupy separate valleys. One caveat is that the filtering property relies on the graphene samples that have well defined zigzag edges, which are difficult to control in experiments. Upon application of a rightward moving electric current to the device, only carriers in one valley are able to be transmitted, while those in the other valley will be totally reflected (there are simply no states in the central narrow region supporting the other valley). Numerical calculations of the transmitted current through such devices reveal that a near perfect valley polarization approaching 100% can be achieved. The polarity of the filter can be changed by applying a sufficiently large electrostatic potential $U > E_F$ to the central nanoconstriction via gating, which pulls up the energy bands and switches transport from the conduction to the valence band. When two valley filters with tunable polarities are connected in series, a valley valve effect can be realized in which the electronic conductance can be modulated depending on the parallel/anti-parallel configuration of the two filters. We note that the valley filter seemingly requires zigzag edges in the nanoconstriction to operate perfectly. However, it was subsequently shown that there is some robustness of the valley polarization with respect to imperfect edge profiles.

Valley filtering in strained graphene was proposed by Fujita *et al.*[41] Graphene is unique in that its electronic properties can be modulated simply by applying mechanical strain to the bonds. The operation of the valley filter can be understood most simply by realizing that strain induces a valley-dependent magnetic vector potential in the effective Hamiltonian:

$$H = \boldsymbol{\sigma} \cdot (\boldsymbol{p} + \xi \boldsymbol{A}) \qquad [6.39]$$

where $\xi$ is the valley index. It is well-known that, when carriers are transported across regions of differing vector potentials (e.g. due to magnetic field barriers), wavevector filtering takes place; namely, incident carriers with momenta $\boldsymbol{k}$ are either transmitted or reflected depending on details such as the magnetic field strength and carrier Fermi energy. Since the strain-induced vector potential is valley-dependent, so is the wavevector filtering and it is possible to tune the strain against the Fermi level of carriers such that exactly half of the filtered states belong to the $K$ valley, whilst the other half belong to the $K'$ valley. Clearly, the transmission of the two valleys is well separated in $\phi$ space. One now only needs to employ a standard magnetic barrier to filter out either the $K$ or $K'$ carriers. Numerically, perfect valley polarization has been predicted for experimentally realizable strain and magnetic field values. Subsequent works have also studied other methods of using the valley-dependent magnetic field for valley filtering.

A valley filter making use of the valley-asymmetric trigonal warping in monolayer graphene was also proposed.[48] Trigonal warping measures the deviation away from the conical Dirac spectrum of the carriers in graphene as the energy increases beyond the neutral point and introduces a small valley degeneracy. It was predicted that a valley polarization of ~20% is achievable from this scheme, although the large energy scales involved would be likely to excite inter-valley scattering processes which would degrade filtering performance. On the other hand, there has been a proposal of a valley filter based on the line defects in graphene.[49]

*Berry curvature and valley Hall effect*

The Berry curvature plays a crucial role in spintronics, in which it governs the topological transport effect, most notably the SHE. Graphene offers a natural setting to study such effects. We have seen above that inversion symmetry breaking in graphene allows one to consider the valley DOF as an additional orbital magnetic moment. In graphene we can break the inversion symmetry by introducing a sub-lattice asymmetry; applying an energy different electrostatic potentials between the $A$ and $B$ sub-lattices ($U_A$ and $U_B$, respectively, $U_A \neq U_B$). In the monolayer Hamiltonian, this can be modeled by an additional Zeeman-like term:

$$H = \boldsymbol{\sigma} \cdot \boldsymbol{B}_\xi + U\sigma_z + V \qquad [6.40]$$

where $\boldsymbol{B}_\xi = (p_x, \xi p_y)$ is the effective pseudospin field, $\xi = \pm 1$ indexes the valley $K$ (+1) and $K'$ (−1), and $U = 0.5(U_A - U_B)$, $V = 0.5(U_A + U_B)$ are pseudospin-dependent and independent potentials, respectively. The sub-lattice asymmetry quantified by $U$ opens up an energy gap in the spectrum of magnitude $\Delta = 2|U|$. Furthermore, it induces a finite Berry curvature as shown below. Applying the well-known result for the Berry connection, $\boldsymbol{A}_k = \pm\frac{1}{2}(1-\cos\theta)\nabla_k \phi$, we find that in graphene:

$$\boldsymbol{A}_k(\boldsymbol{k}) = \pm\frac{1}{2}\left(1 - \frac{U}{\sqrt{p_x^2 + p_y^2 + U^2}}\right)\left(\frac{-\xi p_y}{p_x^2 + p_y^2}, \frac{p_x}{p_x^2 + p_y^2}, 0\right) \qquad [6.41]$$

whose curvature $F_{ij} = \varepsilon_{ijk}\,\partial_i A_j$ is calculated to be:

$$F_z(\boldsymbol{k}) = \frac{\xi U}{2\left(p_x^2 + p_y^2 + U^2\right)^{\frac{3}{2}}}. \qquad [6.42]$$

The $k$ space Berry curvature is known to result in an anomalous velocity of carriers. It is clear that in graphene, the Berry curvature is valley dependent, which means that $K$ and $K'$ carriers will be affected in an equal and opposite manner in terms of topological transport.

One consequence is that a valley Hall effect will result in a sample upon application of an electric field (Fig. 6.7). To see this, we examine the full equations of motion of a carrier in the presence of Berry's curvature is:[50]

$$\dot{\boldsymbol{k}} = -\frac{e}{\hbar}\boldsymbol{E}$$

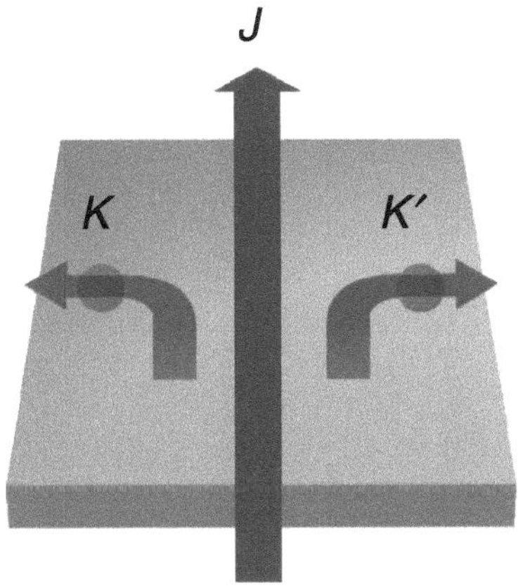


*6.7* Illustration of valley Hall effect in which carriers residing in opposite valleys are deflected in opposite transverse directions. The deflection results from the finite *k*-space Berry curvature in graphene with broken spatial inversion symmetry.

$$\dot{\boldsymbol{r}} = \frac{1}{\hbar}\frac{\partial \varepsilon}{\partial \boldsymbol{k}} \dot{\boldsymbol{k}} \times \boldsymbol{F}(\boldsymbol{k}). \qquad [6.43]$$

The Berry curvature appears as an additional velocity term, which is perpendicular to the applied electric field $\boldsymbol{E}$, and which is equal and opposite for the two valleys. The same mechanism drives the intrinsic SHE in semiconductors, whereby opposite spin species are separated along opposite lateral sides of a sample.

*Quantum valley Hall effect*

Below we discuss an analogue of the quantum SHE in valleytronics, returning our attention to strained graphene.[51] Strain induces a valley-dependent vector potential. The curvature of the vector potential is equivalent to an effective, vertical magnetic field which points in opposite directions for the two valleys. For spatially uniform strains, the effective field strength is zero. For spatially inhomogeneous strain the finite vertical magnetic fields induce a valley resolved quantum Hall effect; namely two copies of the quantum Hall effect for each valley, one chiral and the other anti-chiral, placed on top of one another as shown in Fig. 6.8. The two copies are related by time-reversal symmetry and thus the whole system is time-reversal invariant.

*Valley-orbit coupling*

In spintronics, in particular semiconductor-based spintronics, the notion of SOC is most important as it makes possible the idea of manipulating spins purely by

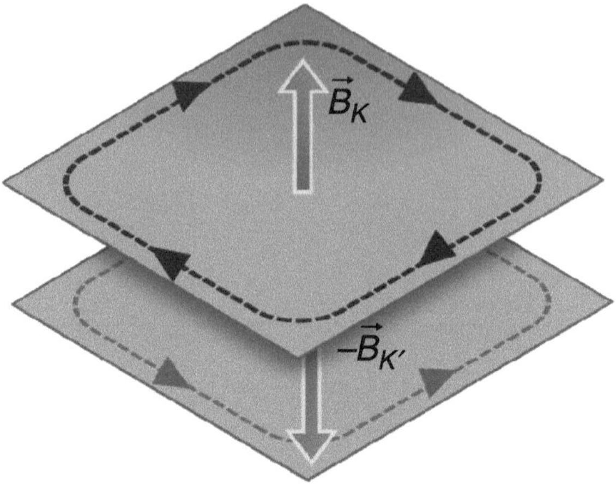


*6.8* The quantum valley Hall effect in which the two valleys experience opposing magnetic fields which induce two copies of the quantum Hall state. The overall time-reversal symmetry of the system is intact.

electric means. It has been proposed that an analogous valley orbit coupling could exist in graphene. In particular, the effective conduction band energy in monolayer graphene in the presence of spatially varying sub-lattice asymmetry, $U = U(\vec{r})$, was found to be:[45]

$$\varepsilon = E - \frac{\xi\hbar}{2E(U+E)} p \times \left[ \left( \frac{2U+E}{E} \right) \nabla_r U - e\nabla_r V \right] \cdot \hat{z} - eV \qquad [6.44]$$

where $E = \sqrt{p_x^2 + p_y^2 + U^2}$. The valley orbit coupling is indicated inside the square brackets of Eq. 6.44. The first term depends on the spatial variation of the sub-lattice asymmetry and suggests that the lattice distortion can be viewed as an effective electric field. The second term is of the same origin as the relativistic spin-orbit coupling in the Dirac equation.

## 6.7 Localization and Klein tunneling

Klein tunneling[52] is a strictly relativistic behavior, which normally occurs in high energy physics. Graphene provides the rare low energy stage in which such high energy behavior can be observed. That Klein tunneling could be seen in graphene introduces new implications, e.g. new confinement methods to modulate device conductance, as well as harnessing the physics of pair production[53] for device applications. In semiconductors one can confine electrons quantum mechanically using electrostatic barriers, e.g. a quantum well confines electrons along one dimension as shown in Fig. 6.9.

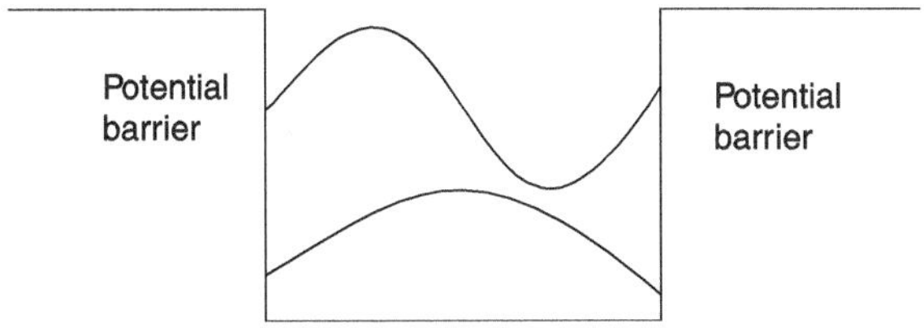


*6.9* A quantum well which normally confines electrons, forming stationary waves within the well.

In graphene, electrostatic barriers cannot confine electrons due to Klein tunneling; high transmission occurs for high and wide barriers. It has been shown for monolayer graphene that normal incidence of a wave on a potential barrier results in unity transmission, while in the case of bilayer graphene, normal incidence results in total reflection. To give readers a foundation of Klein tunneling, we refer to Appendix 6.C, which introduces the general concept of Klein tunneling in relativistic physics.

In the context of nanoscale device, Klein tunneling can be understood by examining the band diagram in Fig. 6.10a, which shows a raised electrostatic barrier where the hole band energy in Region (II) overlaps the energy of the conduction band in Region (I). This is because the bandgap is small compared to the raised electrostatic barrier. Electrons in region I interact with positron (holes) in region II, resulting in the escape of particle. Figure 6.10b shows a large bandgap compared with the raised electrostatic barrier. Figure 6.10b illustrates normal low energy systems, e.g. semiconductors where electron waves encounter a barrier in the conduction band.

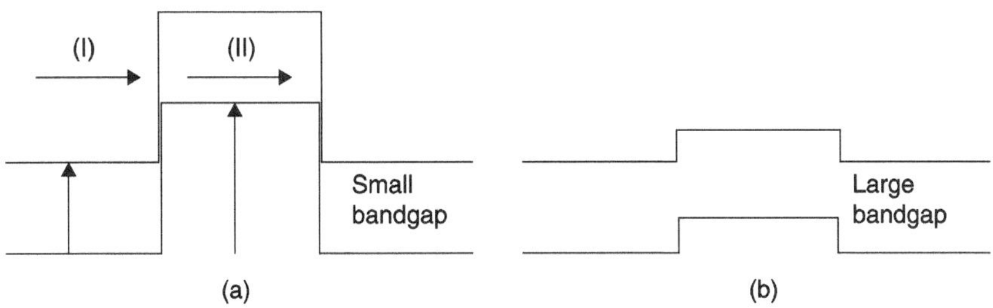


*6.10* (a) Raised barrier is high compared to the small bandgap resulting in an overlap as described above. (b) Raised barrier is small compared to the large bandgap.

We show below the Dirac incident, reflecting and transmitting wavefunctions which explain how flux continuity at the boundaries gives rise to the escape of particles through high and wide barriers:

$$\psi_1(z) = a\begin{bmatrix} \begin{pmatrix} 1 \\ 0 \end{pmatrix} \\ \dfrac{pc}{(E+mc^2)} \begin{pmatrix} 1 \\ 0 \end{pmatrix} \end{bmatrix} e^{\frac{i}{\hbar}pz} = a\begin{bmatrix} 1 \\ 0 \\ \dfrac{pc}{(E+mc^2)} \\ 0 \end{bmatrix} e^{\frac{i}{\hbar}pz}. \quad [6.45]$$

For reflected wave in Region (I), the solution might be:

$$\psi_R(z) = b\begin{bmatrix} \begin{pmatrix} 1 \\ 0 \end{pmatrix} \\ \dfrac{-pc}{(E+mc^2)} \begin{pmatrix} 1 \\ 0 \end{pmatrix} \end{bmatrix} e^{\frac{i}{\hbar}(-pz)} + b'\begin{bmatrix} \begin{pmatrix} 0 \\ 1 \end{pmatrix} \\ \dfrac{pc}{(E+mc^2)} \begin{pmatrix} 0 \\ 1 \end{pmatrix} \end{bmatrix} e^{\frac{i}{\hbar}(-pz)}. \quad [6.46]$$

For Region (II), $E$ is replaced by $E - V$, and $p'c = \sqrt{(E-V)^2 - m^2c^4}$, leading to:

$$\psi_T(z) = d\begin{bmatrix} \begin{pmatrix} 1 \\ 0 \end{pmatrix} \\ \dfrac{p'c}{\left(E-V+mc^2\right)} \begin{pmatrix} 1 \\ 0 \end{pmatrix} \end{bmatrix} e\frac{i}{\hbar}(p'z) + d'\begin{bmatrix} \begin{pmatrix} 0 \\ 1 \end{pmatrix} \\ \dfrac{-p'c}{\left(E-V+mc^2\right)} \begin{pmatrix} 0 \\ 1 \end{pmatrix} \end{bmatrix} e^{\frac{i}{\hbar}(p'z)}. \quad [6.47]$$

## 6.7.1 Interaction, localization and minimal conductivity

In graphene, electron–electron interaction is strong because of the weak screening, the vanishing density of states and the much larger coupling constant of $\alpha = e^2/\hbar v_F$ compared to $\alpha = e^2/c$ in quantum electrodynamics. Moreover electron motion is slow in graphene compared with photons as we know that $v_F \sim \frac{1}{300}c$. The model that describes the interaction in graphene is also not Lorentz invariant due to the great departure of $v_F$ from $c$. The Dirac particles in graphene are individually relativistic particles, but interact in a non-relativistic way because of their slow motion. The particles in graphene are also helical, in the sense that the pseudospins are aligned (anti-aligned) to the momentum in the conduction (valence) band. One deduces from $H = v_F\, \boldsymbol{\sigma}\cdot\boldsymbol{p}$ that electrons in the conduction band have the opposite pseudospin to those in the valence band. Since graphene's Fermi level coincides with the Dirac point, the Fermi sphere shrinks to a point and intra-band electron–hole pair transition might not be possible. Thus in clean, undoped graphene, electron–hole pair is restricted to inter-band excitation. However, as doping pushes the Fermi level above the Dirac point, the intra-band transition is recovered, and the many-body effect of graphene begins to resemble those of two-dimensional electron gas (2DEG).

One interesting but as yet not well-explained observation in graphene is the minimum conductivity.[54] Experiments have shown that graphene exhibits a minimal conductivity of 4 $e^2/h$ at the Dirac point. For ordinary materials, such a low conductivity would normally lead to a metal–insulator transition, but graphene remains conductive with a small conductivity. Since bilayer graphene also exhibits minimum conductivity, one can rule that chirality is perhaps the more important reason than the linear dispersion which might be responsible for this.[22] One other explanation is based on the fact that graphene electrons, being Dirac particles, are not easily confined, as they escape easily from electrostatic barriers through Klein tunneling. Localization can be observed in metals and semiconductors, but in graphene, this well-known phenomenon seems absent.

The lack of localization in graphene remains puzzling, although it is possible that strong electron–electron interactions at the Dirac point could be the reason. One other reason is the short phase relaxation length, within which the effect of localization can manifest. It has been anticipated that, at very low temperature, where the phase relaxation length becomes longer, it might be possible to observe localization in graphene. One could also borrow Mott's physical elucidation of metal conductivity to explain the minimal conductivity of graphene. Under the theory of Boltzmann, increased electron scattering implies the continued shrinking of the average distance known as the mean free path ($l$) between two electrons on colliding course. Mott hypothesized that downscaling of conductivity (shrinking of $l$) cannot continue indefinitely. When $l$ shrinks to the electron's Fermi wavelength ($\lambda$), scattering becomes ineffective. The conductance of an electron shall not be lower than:

$$\sigma = \frac{e^2}{h}\frac{l}{\lambda} = \frac{e^2}{h}. \qquad [6.48]$$

A factor of four arising from valley and spin degeneracy needs to be included. However, a formal derivation based on the Kubo shows a different result:

$$\sigma = \frac{4e^2}{\pi h}. \qquad [6.49]$$

It has been suggested that the difference could be due to sample boundary effects and theoretical approximations of scattering processes that previous derivations have failed to take into account. Overall, graphene has high mobility, high conductivity and weak coupling to phonon scattering. These properties make graphene a promising candidate material for device applications.

## 6.7.2 Pseudospin relation

It may not be apparent how pseudospin could in any way have an effect on the localization of electron. But if one looks from the standpoint of Berry's phase, the connection becomes clear. Take monolayer graphene, for example, which exhibits a $\pi$ shift Berry's phase. The Berry's phase imparts an additional phase to the propagating wave, transforming an otherwise constructive interference to a destructive event. This results in weak anti-localization. However, there are events in graphene which destroys the Berry's phase (e.g. edge scattering), which might restore some localization effects. Therefore in graphene, there is a mix of both localization and anti-localization in competition. The above is also related to the absence of backscattering in graphene by long-range potential impurities.

To obtain a clearer understanding, we rewrite the graphene Hamiltonian for valley $K$:

$$H^K = \hbar v_F \begin{pmatrix} 0 & e^{-i\phi} \\ e^{i\phi} & 0 \end{pmatrix} \qquad [6.50]$$

where $\phi$ is the in-plane angle, $(k_x, k_y) \equiv (k\ cos\phi,\ k\ sin\phi)$. The solution for the conduction band is thus:

$$\psi_c^K = \frac{e^{ik.r}}{\sqrt{25}} \begin{pmatrix} e^{-\frac{i\phi}{2}} \\ e^{\frac{i\phi}{2}} \end{pmatrix}. \qquad [6.51]$$

For impurity with long range potential, scattering strength is given by:

$$\langle -\boldsymbol{k} | H_{imp} | \boldsymbol{k} \rangle = \int \frac{dxdy}{2S} e^{2i\,\boldsymbol{k}\cdot\boldsymbol{r}} \begin{pmatrix} e^{-\frac{i\phi}{2}} \\ e^{\frac{i\phi}{2}} \end{pmatrix}^{\dagger} \begin{pmatrix} V(r) & 0 \\ 0 & V(r) \end{pmatrix} \begin{pmatrix} e^{-\frac{i\phi}{2}} \\ e^{\frac{i\phi}{2}} \end{pmatrix} = 0. \qquad [6.52]$$

A backscattered particle going around an impurity in real space inscribes a $2\pi$ rotation in momentum space around valley $K$. The Berry's phase of $\pi$ shift acquired by the quasiparticle results in $\phi(-\boldsymbol{k}) = \phi(\boldsymbol{k}) \pm \pi$. The matrix element above vanishes. Physically, the pseudospin components of the backscattered wavefunction interact with the oncoming wavefunction such that the probability that an oncoming state transits to the backscattered state is zero.

Klein tunneling is another direct consequence of the relativistic wavefunction of graphene quasiparticles, whose internal dimension is of course the pseudospin. It has been shown for monolayer graphene that the normal incidence of a wave on a potential barrier results in unity transmission, while in the case of bilayer graphene, normal incidence results in total reflection. Klein tunneling has also been attributed to the minimum conductivity of graphene.

## 6.8 Integer quantum Hall effect

Quantum Hall effects in graphene[55,56] have been studied intensively. Generally speaking, the IQHE in graphene has the same underlying mechanism as that in the semiconductor 2DEG. Readers are referred to Chapter 4 for the basic concepts of quantum Hall effects in semiconductors, e.g. Landau levels, cyclotron frequency, degeneracy strength, flux quantum, incompressibility, Shubnikov–de Haas (SdH) oscillations, integer-shift Hall plateau, edge and localized states, impurities effects, and others. Bearing the above in mind, the IQHE in graphene can be understood with some modifications due to its different Hamiltonian. Table 6.6 provides a comparison summarizing the important IQHE physical effects in semiconductors and graphene.

Pseudospin has a well-known physical consequence to IQHEs in graphene. As described earlier, Berry's phase arises as a result of the rotation of the pseudospin in an adiabatic manner. When the graphene quasiparticle's momentum encircles the Dirac point in a closed contour (i.e. $2\pi$), the pseudospin for graphene acquires a Berry's phase of $J\pi$, where:

$$H = \begin{pmatrix} 0 & (p_x - ip_y)^J \\ (p_x + ip_y)^J & 0 \end{pmatrix} \qquad [6.53]$$

and $J$ = 1/2 indicates a monolayer/bilayer graphene, respectively. Thus, for a monolayer graphene, the quasiparticle gains a $\pi$ Berry's phase while for the bilayer graphene it is $2\pi$. Berry's phase affects both the SdH oscillations as well as the shift in the first quantum Hall effect plateau. In monolayer graphene, the Berry's phase contributes to the $\pi$-shift in the SdH oscillations and a half-integer shift in the Hall conductivity plateau as the Fermi energy ($E_F$) crosses the $n = 0$ Landau level. The conductivity shift is $\pm ge^2/2h$ depending on electron/hole, respectively, and $g$ is the degeneracy factor. But as $E_F$ crosses higher Landau

*Table 6.6* Summary of physical quantities relevant to the understanding of IQHE in semiconductors, monolayer and bilayer graphene.

| | Semiconductor 2DEG | Monolayer graphene | Bilayer graphene |
|---|---|---|---|
| Hamiltonian | $H = (\boldsymbol{p} + e\boldsymbol{A})^2$ | $H = v_F(\boldsymbol{p} + e\boldsymbol{A})\cdot\boldsymbol{\sigma}$ | $H = \frac{-c^2}{t}\begin{pmatrix} 0 & \pi^{\dagger 2} \\ \pi^2 & 0 \end{pmatrix}$ |
| Energy | $E_n = \hbar\omega_c\left(n + \frac{1}{2}\right)$ | $E_n = \pm\hbar\omega_c\sqrt{n}$ | $E_n = \pm\hbar\omega_c\sqrt{n(n-1)}$ |
| Cyclotron frequency | $\omega_c = \frac{eB}{m}$ | $\omega_c = v_F\sqrt{eB/\hbar}$ | ? |
| Transverse conductivity | $\sigma_{xy} = \pm(gn)\frac{e^2}{h}$ | $\sigma_{xy} = \pm g\left(n + \frac{1}{2}\right)\frac{e^2}{h}$ | $\sigma_{xy} = \pm(gn)\frac{e^2}{\hbar}$ $n \geq 1$ |
| Energy step for B = 10T | $\omega_c \sim 1\,\text{K}$ | $\omega_c \sim 1500\,\text{K}$ | |

Note: In bilayer graphene $\pi = (p_x + eA_x) + i(p_y + eA_y)$. The factor $g$ denotes the spin and valley degeneracy. In monolayer and bilayer graphene, $g = 4$.

levels, the conductivity shift is $\pm ge^2/h$. Thus, the effect of Berry's phase is to yield the quantization condition of $\sigma_{xy} = \pm g(n + 1/2)e^2/h$. For the bilayer graphene with $J = 2$, one observes a $J\pi$ Berry's phase which can be associated with the $J$-fold degeneracy of the zero-energy Landau level. The double-degenerate zero-energy Landau level explains the integer shift of the Hall conductivity just across the zero energy. The eigenenergies of monolayer and bilayer graphene:

$$E_n = \pm\hbar\omega_c\sqrt{n}; \qquad E_n = \pm\hbar\omega_c\sqrt{n(n-1)} \qquad [6.54]$$

show that a zero energy Landau level exists. For the monolayer graphene, a zero Landau level occurs for $n = 0$ and, for bilayer graphene, a zero Landau level occurs for $n = 0$ and $n = 1$. An inspection of the Hall conductivity at energy just across the zero Landau level shows that it has shifted a half-integer vertically, resulting in the first conductivity step in either direction being half the size of subsequent steps. Figure 6.11 provides a pictorial description of IQHE in graphene for both the monolayer and the bilayer.

The half-integer shift of Hall conductivity can be deduced straightforwardly where Hall conductivity for monolayer graphene is (Table 6.6):

$$\sigma_{xy} = \pm 4\left(n + \frac{1}{2}\right)\frac{e^2}{h}. \qquad [6.55]$$

The degeneracy factor of $g = 4$ arises due to two contributed by valley and two by spin. The underlying physics is related to the particle–hole symmetry and electron–hole degeneracy at the zero energy level. Nonetheless, one can imagine the zero Landau level to consist of both electrons and holes, and thus at energy just across the zero energy in either direction, Hall conductivity due respectively, to electrons and holes will be a 1/2 integer shift compared to conductivity due to the first Landau level. Thus when the Fermi energy surpasses the first Landau level, Hall conductivity contributed by carriers of both zero and first Landau level will give a total of 3/2 shift integer shift. In bilayer graphene where the Hall conductivity is (for $n \geq 1$):

$$\sigma_{xy} = \pm(4n)\frac{e^2}{h}, \qquad [6.56]$$

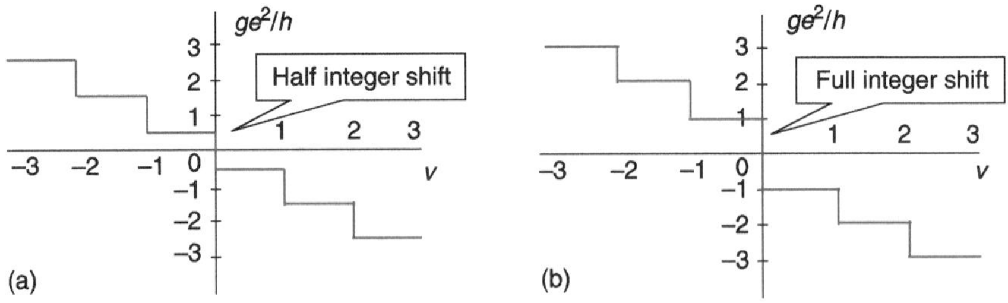


*6.11* (a) IQHE for monolayer graphene showing half integer shift. (b) IQHE for bilayer graphene showing full integer shift.

a full integer shift of conductivity is obtained for $n = 1$. Inspection of $E_n = \pm\hbar\omega_c\sqrt{n(n-1)}$ shows that at $n = 0,1$, energy is zero. The double-degenerate zero energy Landau level explains the full integer shift of the Hall conductivity. Therefore, the main difference between monolayer and bilayer lies in the half shift for monolayer and full shift for bilayer at zero Landau level. But in both monolayer and bilayer, the first Hall plateau appears just across the zero energy. This is the major difference between the IQHE in graphene and conventional semiconductors. On the other hand, IQHE in bilayer graphene resembles the semiconductor 2DEG in that full integer conductivity shift occurs for the Landau level of all *n*. Thus, while the physics of half shift in monolayer is related to electron and hole degeneracy, the full shift in bilayer graphene is due to the doubling of this effect due to the double-degenerate Landau level at zero energy for $n = 0$ and $n = 1$ explained earlier.

## 6.9 References


[1] Bachtold A, Hadley P, Nakanishi T and Dekker C (2001), 'Logic circuits with carbon nanotube transistors', *Science* 294, 1317.

[2] Wind S J, Appenzeller J, Martel R, Derycke V and Avouris P (2002), 'Vertical scaling of carbon nanotube field-effect transistors using top gate electrodes', *Appl. Phys. Lett.* 80, 3817.

[3] Javey A (2003), 'Ballistic carbon nanotube field-effect transistors', *Nature*, 424, 654.

[4] Hoenlein W, Kreupl F, Duesberg G S, Graham A P, Liebau M *et al.* (2003), 'Carbon nanotubes for microelectronics: status and future prospects', *Mater. Sci. Eng. C* 23, 663.

[5] Alphenaar B W, Tsukagoshi K and Wagner M (2001), 'Spin transport in nanotubes (invited)', *J. Appl. Phys.* 89, 6863; Tsukagoshi K, Bruce W A and Hiroki A (1999), 'Coherent transport of electron spin in a ferromagnetically contacted carbon nanotube', *Nature* 401, 572.

[6] Hueso L E, Pruneda J M, Ferrari V, Burnell G, Valdés-Herrera J P *et al.* (2007), 'Transformation of spin information into large electrical signals using carbon nanotubes', *Nature* 445, 410.

[7] Jarillo-Herrero P, van Dam J A and Kouwenhoven L P (2006), 'Quantum supercurrent transistors in carbon nanotubes', *Nature* 439, 953.

[8] Avouris P, Freitag M and Perebeinos V (2008), 'Carbon-nanotube photonics and optoelectronics', *Nat. Photonics* 2, 341.

[9] Bolotin K I, Sikes K J, Hone J, Stormer H L and Kim P (2008), 'Temperature-dependent transport in suspended graphene', *Phys. Rev. Lett.* 101, 096802.

[10] Berger C, Song Z, Li T, Li X, Ogbazghi A Y *et al.* (2004), 'Ultrathin epitaxial graphite: 2D electron gas properties and a route toward graphene-based nanoelectronics', *J Phys. Chem. B* 108, 19912.

[11] Novoselov K S, Jiang Z, Zhang Y, Morozov S V, Stormer H L *et al.* (2007), 'Room-temperature quantum Hall effect in graphene', *Science* 315, 1379.

[12] Son Y W, Cohen M L and Louie S G (2006), 'Half-metallic graphene nanoribbons', *Nature* 444, 347.

[13] Suemitsu M and Fukidome H (2010), 'Epitaxial graphene on silicon substrates', *J. Phys. D: Appl. Phys.* 43, 374012.

[14] Tsukamoto T and Ogino T (2010), ' 'Graphene-on-insulator' fabricated on atomically controlled solid surfaces', *J. Phys. D: Appl. Phys.* 43, 374014.

[15] Zhou S Y, Gweon G-H, Fedorov A V, First P N, de Heer W A *et al.* (2007), 'Substrate-induced bandgap opening in epitaxial graphene', *Nat. Mater.* 6, 770.

[16] Han M Y, Özyilmaz B, Zhang Y and Kim P (2007), 'Energy band-gap engineering of graphene nanoribbons', *Phys. Rev. Lett.* 98, 206805.

[17] Castro Neto A H, Guinea F, Peres N M R, Novoselov K S and Geim A K (2009), 'The electronic properties of graphene', *Rev. Mod. Phys.* 81, 109.

[18] Lozovik Y E, Merkulova S P, Sokolik A A, Morozov S V, Novoselov K S *et al.* (2008), 'Collective electron phenomena and electron transport in graphene. Scientific Session of the Physical Sciences Division of the Russian Academy Of Sciences (27 February 2008)', *Phys.-Usp.* 51, 727.

[19] Ashcroft N W and Mermin N D (1976), *Solid State Physics*, Brooks/Cole, Publishing, Pacific Grove, CA; Kittel C (1953), *Introduction to Solid State Physics*, John Wiley & Sons, New York.

[20] Schweber S S (2005), *An Introduction to Relativistic Quantum Field Theory*, Dover Publications, Mineola, NY.

[21] Novoselov K S, Geim A K, Morozov S V, Jiang D, Katsnelson M I *et al.* (2005), 'Two-dimensional gas of massless Dirac fermions in graphene', *Nature* 438, 197.

[22] Geim A K and Novoselov K S (2007), 'The rise of graphene', *Nat. Mater.* 6, 183.

[23] Rycerz A (2008), 'Non-equilibrium valley polarization in graphene nanoconstrictions', *Phys. Status Solidi A*, 205, 1281.

[24] Rycerz A, Tworzydlo J and Beenakker C W J (2007), 'Valley filter and valley valve in graphene', *Nat. Phys.* 3, 172.

[25] Xiao D, Yao W and Niu Q (2007), 'Valley-contrasting physics in graphene: Magnetic moment and topological transport', *Phys. Rev. Lett.* 99, 236809.

[26] Wu G Y, Lue N-Y and Chang L (2011), 'Graphene quantum dots for valley-based quantum computing: A feasibility study', arXiv: 1104.0443.

[27] Ando T (2006), 'Aharonov–Bohm effect and symmetry crossover in carbon nanotubes', *J. Phys. Soc. Jpn.* 75, 054701.

[28] Datta S (2005), *Quantum Transport – Atom to Transistor*, Cambridge University Press, Cambridge.

[29] McCann E (2006), 'Asymmetry gap in the electronic band structure of bilayer graphene', *Phys. Rev. B* 74, 161403(R).

[30] Martin I, Blanter Y M and Morpurgo A F (2008), 'Topological confinement in bilayer graphene', *Phys. Rev. Lett.* 100, 036804.

[31] Manes J L, Guinea F and Vozmediano M A H (2007), 'Existence and topological stability of Fermi points in multilayered graphene', *Phys. Rev. B* 75, 155424.

[32] Martinez J C, Jalil M B A and Tan S G (2009), 'Robust localized modes in bilayer graphene induced by an antisymmetric kink potential', *Appl. Phys. Lett.* 95, 213106.

[33] Tan S G, Jalil M B A and Fujita T (2010), 'Monopole and topological electron dynamics in adiabatic spintronic and graphene systems', *Ann. Phys.* 325, 1537.

[34] Vozmediano M A H, Katsnelson M I and Guinea F (2010), 'Gauge fields in graphene', *Phys. Rep.* 496, 109.

[35] Englman R and Vértesi T (2008), 'Large Berry phases in layered graphene', *Phys. Rev. B* 78, 205311.

[36] Bunch J S, van der Zande A M, Verbridge S S, Frank I W, Tanenbaum D M *et al.* (2007), 'Electromechanical resonators from graphene sheets', *Science* 315, 490;

Pereira V M and Castro Neto A H (2009), 'Strain engineering of graphene's electronic structure', *Phys. Rev. Lett.* 103, 046801.

[37] Sasaki K I and Saito R (2008), 'Pseudospin and deformation-induced gauge field in graphene', *Prog. Theor. Phys.* Suppl. 176, 253.

[38] Guinea F (2010), 'Spin–orbit coupling in a graphene bilayer and in graphite', *New J. Phys.* 12, 083063.

[39] Kane C L and Mele E J (2005), Quantum spin Hall effect in graphene, *Phys. Rev. Lett.* 95, 226801.

[40] Akhmerov A R, Bardarson J H, Rycerz A and Beenakker C W J (2008), 'Theory of the valley-valve effect in graphene nanoribbons', *Phys. Rev. B* 77, 205416.

[41] Fujita T, Jalil M B A and Tan S G (2010), 'Valley filter in strain engineered graphene', *Appl. Phys. Lett.* 97, 043508.

[42] Chaves A, Covaci L, Rakhimov K Y, Farias G A and Peeters F M (2010), 'Wave-packet dynamics and valley filter in strained graphene', *Phys. Rev. B* 82, 205430; Zhai F, Zhao X, Chang K and Xu H Q (2010), 'Magnetic barrier on strained graphene: A possible valley filter', *Phys. Rev. B* 82, 115442.

[43] Abergel D S L and Chakraborty T (2009), 'Generation of valley polarized current in bilayer graphene', *Appl. Phys. Lett.* 95, 062107.

[44] Murakami S, Nagaosa N and Zhang S-C (2003), 'Dissipationless quantum spin current at room temperature', *Science* 301, 1348.

[45] Gosselin P, Berard A, Mohrbach H and Ghosh S (2009), 'Berry curvature in graphene: A new approach', *Eur. J. Phys. C* 59, 88.

[46] Zhang Y, Jiang Z, Small J P, Purewal M S, Tan Y-W *et al.* (2006), 'Landau-level splitting in graphene in high magnetic fields', *Phys. Rev. Lett.* 96, 136806.

[47] Luk'yanchuk I A and Bratkovsky A M (2008), 'Lattice-induced double-valley degeneracy lifting in graphene by a magnetic field', *Phys. Rev. Lett.*100, 176404.

[48] Pereira Jr J M, F M Peeters, F M, Costa Filho R N and Farias G A (2009), 'Valley polarization due to trigonal warping on tunneling electrons in graphene', *J. Phys.: Condens. Matter* 21, 045301.

[49] Gunlycke D and White C T (2011), 'Graphene valley filter using a line defect', *Phys. Rev. Lett.* 106, 136806.

[50] Sundaram G and Niu Q (1999), 'Wave-packet dynamics in slowly perturbed crystals: Gradient corrections and Berry-phase effects', *Phys. Rev. B* 59, 14915.

[51] Guinea F, Katsnelson M I and Geim A K (2010), 'Energy gaps and a zero-field quantum Hall effect in graphene by strain engineering', *Nat. Phys.* 6, 30.

[52] Katsnelson M I, Novoselov K S and Geim A K (2006), 'Chiral tunnelling and the Klein paradox in graphene', *Nat. Phys.* 2, 620.

[53] Martinez J C, Jalil M B A and Tan S G (2010), 'Klein tunneling and zitterbewegung and the formation of a polarized p-n junction in graphene', *Appl. Phys. Lett.* 97, 062111.

[54] Katsnelson M I (2006), 'Zitterbewegung, chirality, and minimal conductivity in graphene', *Eur. Phys. J. B* 51, 157.

[55] Zhang Y, Tan Y W, Stormer H L and Kim P (2005), 'Experimental observation of the quantum Hall effect and Berry's phase in graphene', *Nature* 438, 201.

[56] Novoselov K S, McCann E, Morozov S V, Fal'ko V I, Katsnelson M I *et al.* (2006), 'Unconventional quantum Hall effect and Berry's phase of 2pi in bilayer graphene', *Nat. Phys.*, 2, 177.

## Appendix 6.A Relativistic quantum mechanics

In relativistic quantum mechanics, Dirac matrices could have different representations, but all obey Dirac algebra. Table 6.A.1 summarizes the various representations.

*Table 6.A.1* Different representations for the Dirac matrices

| | $\gamma^0$ | $\alpha^i$ | $\gamma^i$ | $i\gamma^5 = i\gamma^0\gamma^1\gamma^2\gamma^3$ | $\gamma^5 = \gamma^0\gamma^1\gamma^2\gamma^3$ |
|---|---|---|---|---|---|
| Set A standard representation | $\begin{pmatrix} I & 0 \\ 0 & -I \end{pmatrix}$ | $\begin{pmatrix} 0 & \sigma^i \\ \sigma^i & 0 \end{pmatrix}$ | $\begin{pmatrix} 0 & \sigma^i \\ -\sigma^i & 0 \end{pmatrix}$ | $\begin{pmatrix} 0 & I \\ I & 0 \end{pmatrix}$ | $-i\begin{pmatrix} 0 & I \\ I & 0 \end{pmatrix}$ |
| Set B[20] | $\begin{pmatrix} 0 & I \\ I & 0 \end{pmatrix}$ | $\begin{pmatrix} -\sigma^i & 0 \\ 0 & \sigma^i \end{pmatrix}$ | $\begin{pmatrix} 0 & \sigma^i \\ -\sigma^i & 0 \end{pmatrix}$ | $\begin{pmatrix} -I & 0 \\ 0 & I \end{pmatrix}$ | $i\begin{pmatrix} I & 0 \\ 0 & -I \end{pmatrix}$ |
| Set C | $\begin{pmatrix} 0 & I \\ I & 0 \end{pmatrix}$ | $\begin{pmatrix} \sigma^i & 0 \\ 0 & -\sigma^i \end{pmatrix}$ | $\begin{pmatrix} 0 & -\sigma^i \\ \sigma^i & 0 \end{pmatrix}$ | $\begin{pmatrix} I & 0 \\ 0 & -I \end{pmatrix}$ | |

In the position basis, the Schrödinger equation for a free particle (electron) isolated from all external force fields is:

$$H\psi(x,t) = i\hbar\frac{\partial\psi}{\partial t}(x,t) \qquad [6.A.1]$$

$$i\hbar\frac{\partial\psi}{\partial t} = \left[c\boldsymbol{\alpha}\cdot\boldsymbol{p} + \alpha_0 mc^2\right]\psi \qquad [6.A.2]$$

$$i\hbar\alpha_0\partial_0\psi = \left[c\alpha_0\boldsymbol{\alpha}\cdot\boldsymbol{p} + \alpha_0\alpha_0 mc^2\right]\psi = \left[c\boldsymbol{\gamma}\cdot\boldsymbol{p} + mc^2\right]\psi. \qquad [6.A.3]$$

One could thus surmise that the relativistic energy equation must resemble:

$$\sqrt{\sum_{J=1}^{3}(p_j c)^2 + (mc^2)^2}\,\psi = i\hbar\frac{\partial\psi}{\partial t}. \qquad [6.A.4]$$

The square of Eq. 6.A.4 leads to the Klein–Gordon equation. Dirac wanted to prevent obtaining a second order equation and wanted to give equal treatment to space and time. Thus, he reasoned that, since the right side of the equation contains a first-order derivative in time, the left side should contain equally simple first-order derivatives in space (i.e. in the momentum operators). One way for this to happen is if the quantity in the square root is a perfect square, which also has the form of $\boldsymbol{E}\cdot\boldsymbol{I} = \alpha_0 mc^2 + c\,\Sigma_{j=1}^{3}\,\alpha_j p_j$. One would then obtain the free Dirac equation of:

$$i\hbar\frac{\partial\psi}{\partial t} = \left[c\sum_{j=1}^{3}\alpha_j p_j + \alpha_0 mc^2\right]\psi = H\psi. \qquad [6.A.5]$$

The coefficients $\alpha_0$ and $\alpha_j$ are to be determined. Since:

$$E^2 = \sum_{j=1}^{3}(p_j c)^2 + (mc^2)^2 = \left[c\sum_{j=1}^{3}\alpha_j p_j + \alpha_0 mc^2\right]^2 \qquad [6.A.6]$$

one could easily deduce that:

$$E(E)\psi = \left[c\sum_{j=1}^{3}\alpha_j p_j + \alpha_0 mc^2\right]\left[c\sum_{k=1}^{3}\alpha_k p_k + \alpha_0 mc^2\right]\psi$$

$$= \left[c^2\sum_{j=1}^{3}\sum_{k=1}^{3}\alpha_j\alpha_k p_j p_k + \sum_{j=1}^{3}\alpha_j\alpha_0 p_j mc^2 + \sum_{k=1}^{3}\alpha_0\alpha_k p_k mc^2 + \alpha_0^2 m^2c^4\right]\psi$$

$$= \left[\frac{1}{2}\{\alpha_j,\alpha_k\}c^2\sum_{j=1}^{3}\sum_{k=1}^{3}p_j p_k + \{\alpha_j,\alpha_0\}\sum_{j=1}^{3}p_j mc^2 + \alpha_0^2 m^2 c^4\right]\psi. \qquad [6.A.7]$$

And, by comparing coefficients, every term of Eq. 6.A.7 with every term of the left-hand side of Eq. 6.A.6, one obtains the relations for the coefficients shown in Table 6.A.2.

*Table 6.A.2* Some identities involving the $\alpha$ matrices

| | |
|---|---|
| 1. | $\alpha_0^2 = I$ |
| 2. | $\{\alpha_j, \alpha_0\} = 0$ |
| 3. | $\{\alpha_j, \alpha_k\} = 2\delta_{jk}$ |

Additional properties are that $\alpha_i$ is hermitian since it follows from the hermicity property of the Hamiltonian:

$$\mathrm{Tr}(\alpha_0) = 0. \qquad [6.A.8]$$

From Eq. 6.A.2, $\alpha_j\alpha_0 = -\alpha_0\alpha_j$, it follows that $\alpha_j = -\alpha_0\alpha_j\alpha_0$. This leads one to conclude that:

$$\mathrm{Tr}(\alpha_j) = -\mathrm{Tr}(\alpha_o\alpha_j\alpha_0) = -\mathrm{Tr}(\alpha_j\alpha_0^2) = -\mathrm{Tr}(\alpha_j) \Rightarrow \mathrm{Tr}(\alpha_j) = 0. \qquad [6.A.9]$$

One can now determine the dimensionality of $N$. Using the relation $\alpha_j\alpha_0 = -\alpha_0\alpha_j$ and taking the determinants as follows:

$$\det(\alpha_j\alpha_0) = \det(-\alpha_0\alpha_j) = \det(-\alpha_0)\det(\alpha_j) = (-1)^N\det(\alpha_j)\det(\alpha_0) \Rightarrow N = \text{even}. \qquad [6.A.10]$$

One has thus concluded that the dimensionality ($N$) must be even. In fact one can deduce further that $N \geq 4$ since the Pauli matrices ($\sigma$, $I$) form a complete set of 2×2 matrices, with the unit matrix $I$ always commuting, and they do not satisfy the above anti-commutative form. This leaves $N = 4$ as the lowest possibility. It is convention to let:

$$\alpha_0 = \begin{bmatrix} I & 0 \\ 0 & -I \end{bmatrix} = \begin{bmatrix} 1 & 0 & 0 & 0 \\ 0 & 1 & 0 & 0 \\ 0 & 0 & -1 & 0 \\ 0 & 0 & 0 & -1 \end{bmatrix} \qquad [6.A.11a]$$

$$\alpha_i = \begin{bmatrix} 0 & \sigma_i \\ \sigma_i & 0 \end{bmatrix} \textit{where } \sigma_1 = \begin{bmatrix} 0 & 1 \\ 1 & 0 \end{bmatrix}, \quad \sigma_2 = \begin{bmatrix} 0 & -i \\ i & 0 \end{bmatrix}, \quad \sigma_3 = \begin{bmatrix} 1 & 0 \\ 0 & -1 \end{bmatrix}. \qquad [6.A.11b]$$

## Appendix 6.B Helicity and masslessness

In the massless limit, we will evaluate $i\hbar\alpha_0\partial_0 = [c\alpha_0\boldsymbol{\alpha} \cdot \boldsymbol{p} + \alpha_0\alpha_0 mc^2] = [c\gamma \cdot \boldsymbol{p} + mc^2]$ using the set B convention of Table 6.A.1, which can be written as:

$$\frac{i}{c}\partial_0\psi = (\gamma^5\boldsymbol{\sigma}\cdot\boldsymbol{p} + \gamma^0 mc)\psi. \qquad [6.B.1]$$

Expanding fully:

$$\begin{bmatrix} p_0 - mc & 0 & p_3 & p_1 - ip_2 \\ 0 & p_0 - mc & p_1 + ip_2 & -p_3 \\ p_3 & p_1 - ip_2 & p_0 + mc & 0 \\ p_1 + ip_2 & -p_3 & 0 & p_0 + mc \end{bmatrix}\psi = 0 \qquad [6.B.2]$$

the solution to Eq. 6.B.1 is:

$$\psi = \sqrt{\frac{E + mc^2}{2E}}\begin{bmatrix} 1 \\ 0 \\ \Delta p_3 \\ \Delta(p_1 + ip_2) \end{bmatrix} \qquad [6.B.3]$$

where $\Delta = c/(E + mc^2)$. But $\psi$ is not a solution of the helicity operator. The helicity solution can be found by:

$$\boldsymbol{\sigma}\cdot\boldsymbol{p} = \begin{bmatrix} p_3 & p_1 - ip_2 & 0 & 0 \\ p_1 + ip_2 & -p_3 & 0 & 0 \\ 0 & 0 & p_3 & p_1 - ip_2 \\ 0 & 0 & p_1 + ip_2 & -p_3 \end{bmatrix}; \psi = \sqrt{\frac{E + mc^2}{(2E)2(p_3 + p)}}\begin{bmatrix} p_3 + p \\ p_1 + ip_2 \\ \Delta(p_3 + p) \\ \Delta(p_1 + ip_2) \end{bmatrix} \qquad [6.B.4]$$

where $\Delta = \dfrac{cp}{E + mc^2}$. In the massive limit, a solution to the Dirac equation can also be a solution to the helicity operator only when particle is at rest, i.e. $p = 0$. In the massless limit, a solution to the Dirac equation is also a solution to the helicity operator for the z dimension only, i.e. $\sigma_3 p_3$. Using the following Dirac matrices, $\beta = \begin{pmatrix} 0 & I \\ I & 0 \end{pmatrix}$, $\alpha = \begin{pmatrix} \sigma & 0 \\ 0 & -\sigma \end{pmatrix}$, and letting $\psi = \begin{bmatrix} \theta \\ \chi \end{bmatrix}$, the massless Dirac equation is $E\psi = \boldsymbol{\alpha} \cdot \boldsymbol{p} + \beta m$ leading to:

$$(\boldsymbol{\sigma} \cdot \boldsymbol{p})\, \theta + m\chi = E\theta$$

$$-(\boldsymbol{\sigma} \cdot \boldsymbol{p})\, \chi + m\, \theta = E\chi. \qquad [6.B.5]$$

And since $m = 0$ and $E = |p|$:

$$(\boldsymbol{\sigma} \cdot \boldsymbol{p})\, \theta = |p|\theta$$

$$(\boldsymbol{\sigma} \cdot \boldsymbol{p})\, \chi = -\, |p|\chi. \qquad [6.B.6]$$

With the helicity operator, $h = \dfrac{\boldsymbol{\sigma} \cdot \boldsymbol{p}}{|p|}$, we can then write:

$$\theta = \frac{\boldsymbol{\sigma} \cdot \boldsymbol{p}}{|p|}\theta = h\theta$$

$$\chi = \frac{\boldsymbol{\sigma} \cdot \boldsymbol{p}}{|p|}\chi = h\chi. \qquad [6.B.7]$$

From experimental results, a massless particle (neutrino) has helicity 1 while its anti-particle counterpart has helicity $-1$. Thus, to remove the $\theta$ components, with the matrices of set C, the operator $P = \frac{1}{2}\left(I - \gamma^5\right) = \begin{pmatrix} 0 & 0 \\ 0 & I \end{pmatrix}$ can be used to project the $\theta$ components, leaving the helicity $-1$ components. Thus we get the chirality operators:

$$\psi_L = \frac{1}{2}(I - \gamma^5)\psi$$

$$\psi_R = \frac{1}{2}(I + \gamma^5)\psi. \qquad [6.B.8]$$

The chirality of a particle is right-handed if the direction of its spin is the same as the direction of its motion. It is left-handed if the directions of spin and motion are opposite. By convention for rotation, a standard clock, tossed with its face directed forwards, has left-handed chirality. Mathematically, chirality is the sign of the projection of the spin vector onto the momentum vector: Left is negative, right is positive. Particles with mass have relative chirality, which depends on the observer's reference frame. In the case of these particles, it is possible for an observer to change to a reference frame that overtakes the spinning particle, in which case the particle will then appear to move backwards and its apparent chirality will reverse. Massless particles have absolute chirality (a given massless particle appears to spin in the same direction along its axis of motion regardless of the point of view of the observer). A massless particle moves with the speed of light, so a real observer (who must always travel at less than the speed of light) cannot be in any reference frame to see particle reverse its relative direction, meaning that all real observers see the same chirality. Because of this, the direction of spin of massless particles is not affected by a Lorentz boost (change of viewpoint) in the direction of motion of the particle, and the sign of the projection (chirality) is thus considered to have been fixed for all reference frames, i.e. the chirality is absolute. The solution to $\gamma^5\ \boldsymbol{\sigma}\cdot\boldsymbol{p}$ is:

$$\gamma^5(\boldsymbol{\sigma}\cdot\boldsymbol{p})\psi = \begin{bmatrix} 0 & 0 & p_3 & p_1 - ip_2 \\ 0 & 0 & p_1 + ip_2 & -p_3 \\ p_3 & p_1 - ip_2 & 0 & 0 \\ p_1 + ip_2 & -p_3 & 0 & 0 \end{bmatrix}\psi \rightarrow$$

$$\begin{bmatrix} 0 & 0 & p_3 & p_1 - ip_2 \\ 0 & 0 & p_1 + ip_2 & -p_3 \\ p_3 & p_1 - ip_2 & 0 & 0 \\ p_1 + ip_2 & -p_3 & 0 & 0 \end{bmatrix}\begin{bmatrix} p_3 + p \\ p_1 + ip_2 \\ (p_3 + p) \\ (p_1 + ip_2) \end{bmatrix} = p\begin{bmatrix} p_3 + p \\ p_1 + ip_2 \\ (p_3 + p) \\ (p_1 + ip_2) \end{bmatrix} \qquad [6.B.9]$$

## Appendix 6.C Klein tunneling and paradox

Klein tunneling has significance to graphene devices and confinement. Below we provide a general derivation and interpretation of Klein tunneling. Understanding the physical effects of electron–hole interaction could assist one's understanding of surprised electron escape for very high potential barriers.

In Region I (Fig. 6.C.1), the incident particle acts as a free particle:

$$i\hbar\frac{\partial\psi}{\partial t} = [c\boldsymbol{\alpha}.\boldsymbol{p} + \alpha_0 mc^2]\psi = E\,\psi \qquad [6.C.1]$$

where $\psi(\chi,t) = \psi(\chi)e^{-i\frac{E}{\hbar}t}$. Letting

$\psi = [\psi_1\ \psi_2\ \psi_3\ \psi_4]^T\ e^{ipz/\hbar} = \begin{bmatrix} \theta \\ \hline \chi \end{bmatrix} e^{ipz/\hbar}$, where $\theta = \begin{bmatrix} \psi_1 \\ \psi_2 \end{bmatrix}$ and $\chi = \begin{bmatrix} \psi_3 \\ \psi_4 \end{bmatrix}$, one has:

$$E\begin{bmatrix} \theta \\ \chi \end{bmatrix} e^{ipz/\hbar} = \left[c\boldsymbol{\alpha}\cdot\boldsymbol{p} + \alpha_0 mc^2\right]\begin{bmatrix} \theta \\ \chi \end{bmatrix} e^{ipz/\hbar}. \qquad [6.C.2]$$

Since the electron wave propagates in the z axis, we have $\boldsymbol{p} = \begin{bmatrix} 0 \\ 0 \\ p \end{bmatrix}$ and:

$$E\begin{bmatrix} \theta \\ \chi \end{bmatrix} e^{ipz/\hbar} = \left(c\begin{bmatrix} 0 & \sigma_z \\ \sigma_z & 0 \end{bmatrix}\begin{bmatrix} \theta \\ \chi \end{bmatrix}\frac{-i\hbar\partial}{\partial z} + \begin{bmatrix} I & 0 \\ 0 & -I \end{bmatrix} mc^2 \begin{bmatrix} \theta \\ \chi \end{bmatrix}\right) e^{ipz/\hbar}, \qquad [6.C.3]$$

which leads straightforwardly to:

$$(E - mc^2)\theta - c\sigma_z p\chi = 0 \qquad [6.C.4a]$$

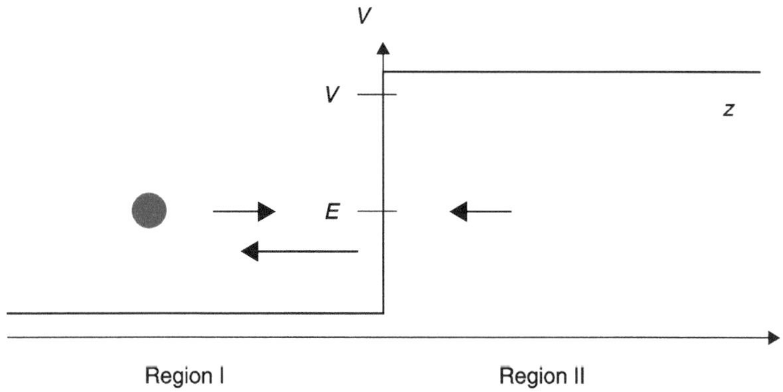


*6.C.1* One-dimensional tunneling of particle with wave-like behavior.

$$-c\sigma_z p\theta + (E + mc^2)\chi = 0. \qquad [6.C.4b]$$

To have non-trivial solutions (linearly dependent), the determinant vanished, i.e. $\begin{vmatrix} E - mc^2 & -c\sigma_z p \\ -c\sigma_z p & (E + mc^2) \end{vmatrix} = 0$ which leads to the energy square equation of:

$$E^2 - m^2c^4 - c^2\sigma_z p\sigma_z p = 0. \qquad [6.C.5]$$

Since $\sigma_z \cdot \sigma_z = I$, one gets $E^2 = m^2c^4 + c^2p^2 \Rightarrow pc = \sqrt{E^2 - m^2c^4}$ and:

$$\chi = \frac{c(\sigma_z p)}{(E + mc^2)}\theta. \qquad [6.C.6]$$

Note that helicity is the projection of the spin onto the direction of the momentum defined by $S_z = \frac{\hbar}{2}\sigma_z$. The general solution is:

$$\psi = \begin{bmatrix} \theta \\ \chi \end{bmatrix} e^{ipz/\hbar} = \begin{bmatrix} \theta \\ \frac{c(\sigma_z p)}{(E + mc^2)}\theta \end{bmatrix} e^{ipz/\hbar}. \qquad [6.C.7]$$

Noting the Pauli $z$ matrix, it is an easy guess that the eigenvectors are $\theta =$ $\theta = \begin{pmatrix} 1 \\ 0 \end{pmatrix}$ and $\begin{pmatrix} 0 \\ 1 \end{pmatrix}$. The more specific solutions in Region I would thus be:

$$\psi_{inc}(z) = a \begin{vmatrix} \begin{pmatrix} 1 \\ 0 \end{pmatrix} \\ \frac{pc}{(E + mc^2)} \begin{pmatrix} 1 \\ 0 \end{pmatrix} \end{vmatrix} e^{\frac{i}{\hbar}pz}. \qquad [6.C.8a]$$

For reflected wave in Region I, the solution might be:

$$\psi_{\mathrm{refl}}(z) = b\begin{vmatrix} \begin{pmatrix} 1 \\ 0 \end{pmatrix} \\ \dfrac{-pc}{(E+mc^2)}\begin{pmatrix} 1 \\ 0 \end{pmatrix} \end{vmatrix} e^{\frac{i}{\hbar}(-pz)} + b'\begin{vmatrix} \begin{pmatrix} 1 \\ 0 \end{pmatrix} \\ \dfrac{pc}{(E+mc^2)}\begin{pmatrix} 1 \\ 0 \end{pmatrix} \end{vmatrix} e^{\frac{i}{\hbar}(-pz)}. \quad [6.C.8b]$$

For Region II (Fig. 6.C.1), $E$ is replaced by $E - V$, and $p'c = \sqrt{(E-V)^2 - m^2c^4}$, leading to:

$$\psi_{\mathrm{trans}}(z) = d\begin{bmatrix} 1 \\ 0 \\ \dfrac{p'c}{(E-V+mc^2)} \\ 0 \end{bmatrix} e^{\frac{i}{\hbar}(p'z)} + d'\begin{bmatrix} 0 \\ 1 \\ 0 \\ \dfrac{-p'c}{(E-V+mc^2)} \end{bmatrix} e^{\frac{i}{\hbar}(p'z)}. \quad [6.C.8c]$$

The paradox involves $V > E + mc^2$, $E > mc^2$. For the wavefunction to be equal at the border, z = 0, $\psi_1 + \psi_R = \psi_T$, and one has two independent relations as below:

$$a + b = d \quad [6.C.9a]$$

$$\frac{pc}{(E+mc^2)}(a-b) = \frac{-p'c}{(V-E-mc^2)}d \quad [6.C.9b]$$

which implies $\Rightarrow a - b = -\dfrac{p'(E+mc^2)}{p(V-E-mc^2)}d$. This would then lead, noting the expression for $p$ and $p'$, to:

$$a - b = -\frac{\sqrt{(V-E)^2 - m^2c^4}\,(E+mc^2)}{\sqrt{E^2 - m^2c^4}\,(V-E-mc^2)}d \quad [6.C.10]$$

which after straightforward steps is:

$$a - b = -\frac{\sqrt{(E+mc^2)(V-E+mc^2)}}{\sqrt{(E-mc^2)(V-E-mc^2)}}d = -rd. \quad [6.C.11]$$

Since $V > E + mc^2$ and $> mc^2$, it is easily seen that $r > 1$. One also observes under the above boundary conditions there are two more independent relations:

$$b' = d' \quad [6.C.12a]$$

$$\frac{pc}{(E+mc^2)}b' = \frac{p'c}{(V-E-mc^2)}d'. \quad [6.C.12b]$$

With $b' = d' = 0$, it can be deduced straightforwardly that $a = \dfrac{(1-r)d}{2}$, $b = \dfrac{(1+r)d}{2}$, or finally:

$$\frac{b}{a} = \frac{1+r}{1-r} \quad and \quad \frac{d}{a} = \frac{2}{1-r}. \qquad [6.C.13]$$

One would now proceed to deduce that the set of current involved in the process of Klein tunneling using the relation $j = c\ \psi^* \alpha_3 \psi$, namely:

$$j_I = a^* a \frac{2pc^2}{E + mc^2}$$
$$j_R = b^* b \frac{-2pc^2}{E + mc^2}$$
$$j_T = d^* d \frac{2p'c^2}{E - V + mc^2}. \qquad [6.C.14]$$

Examining the current carefully, one notes that for $V > E + mc^2$:

$$\frac{j_T}{j_I} = \frac{d^* d(2p'c^2)(E + mc^2)}{a^* a(E - V + mc^2)(2pc^2)} = \frac{-4r}{(1-r)^2} < 0 \qquad [6.C.15]$$

and

$$\left|\frac{j_R}{j_I}\right| = \frac{|-b^* b|}{|a^* a|} = \left(\frac{1+r}{1-r}\right)^2 = 1 + \frac{4r}{(1-r)^2} = 1 - \frac{j_T}{j_I} > 1. \qquad [6.C.16]$$

In fact one notes that when $V \rightarrow \infty, j_T \rightarrow -2c^2$ but not zero, showing that electron cannot be confined in the barrier. The above results can be explained with the aid of Fig. 6.C.2. The 'negative energy solutions' in Region II are larger and lies in the same range as the solutions of positive energy in Region I. The potential in

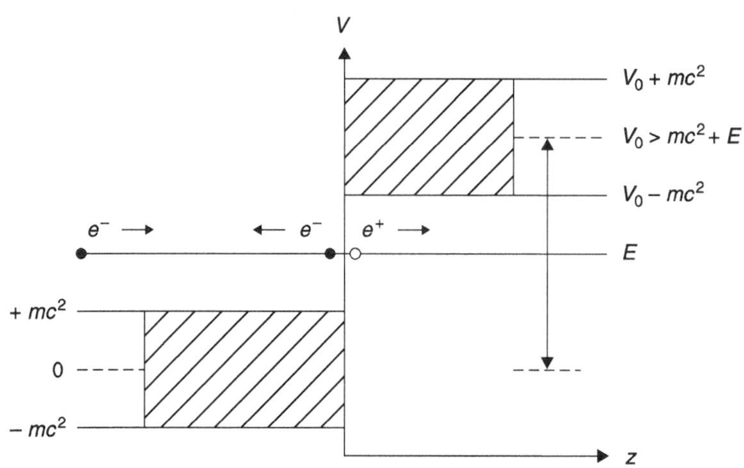


*6.C.2* Schematic illustration of physics of Klein tunneling in relativistic physics.

Region II raises the electron energy in Region II sufficiently until it overlaps with the positive continuum in Region I, resulting in electron–positron pair creation at the potential barrier. There are incident wave packets coming from the right (Region II). Electron and hole interacts in the finite potential region to generate left-moving electron and right-moving current. This explains why the reflecting current in Region I is larger than the incident current. Electron interaction with hole is the physics behind the escape of an electron in a very high potential region.

# 7

# Topological dynamics and gauge potential in nanoelectronics

**Abstract:** This chapter introduces the gauge theory of quantum physics and its application in nanoscale electronics. The concept of effective magnetic fields in the magnetic space and the momentum space is introduced. Exemplary systems are spintronics, graphene and topological insulators where effective magnetic fields in this system contribute to the conductivity, which is topologically invariant. Anomalous Hall effects (AHE) are introduced and their topological versions are studied in both momentum and local space. Spin torque induced by the effects of spin orbit coupling will be discussed using gauge theory. Lastly, mathematical properties such as the Dirac string and magnetic monopole are presented and their physical contribution to the effective magnetic fields is discussed.

**Key words:** topology, gauge theory, topological insulators, anomalous Hall effects, Dirac string, monopole, Berry's phase, Berry's curvature.

## 7.1 Introduction to gauge physics in nanoelectronics

The physics of symmetry and gauge field has traditionally been an integral part of quantum field theory written mainly for high energy physics. The use of gauge theory in condensed matter physics was largely related to superconductivity and quantum Hall effects. In recent times, gauge theoretic methods have become popular in the various fields of nanoelectronics, where curvature of gauge potential and Berry's phase have been related to the physics of electron dynamics in these systems. One example is the spin Hall effects (SHE),[1] where gauge potential related to spin Hall conductivity can be found. In Chapter 5 we discussed SHE in the Rashba two-dimensional electron gas (2DEG) system, and one of the ways to obtain the universal conductivity is via gauge potential in energy space.

In this chapter we give a pedagogical introduction of the mathematics and physics of gauge theoretic applications in nanoscale electronic systems, e.g. spintronics, graphene, magnetic and topological insulator systems. We discuss in particular the application to one important aspect of spintronics, namely the phenomenon of spin transfer torque,[2,3] which is under intense research for the technology of current-induced magnetization switching, so-called CIMS. We also discuss the topological Hall effect or conductivity in magnetic systems, where micromagnetic studies of magnetization configurations have improved the design of magnetic media and read-heads for recording purposes.

We begin our discussion with the time-reversal symmetric system where:

$$THT^{-1} = H \qquad [7.1]$$

where $T$ is the anti-unitary time-reverse operator that satisfies $TPT^{-1} = -P$, $TST^{-1} = -S$, $TQT^{-1} = Q$, where $P$, $S$, $Q$ are the momentum, spin and position operator, respectively. One notable example is the Rashba spin orbit coupling (SOC) system[4] in semiconductor or metal spintronics, and topological insulators. SOC has direct implication to the spin and the momentum dynamics of electrons:

$$H_{SOC} = g\,\boldsymbol{\sigma}\cdot(\boldsymbol{p}\times\boldsymbol{E}) \qquad [7.2]$$

where $\boldsymbol{\sigma}$ is the Pauli matrix, $\boldsymbol{p}$ is the momentum operator and $g$ is the SOC strength. In fact, SOC is highly relevant to spintronics, ranging from the well-known anisotropic magnetoresistance, the anisotropy energy of local moment density, the keenly studied spin Hall and spin current in semiconductor spintronics, to more subtle implications like spin torque, spin dynamics, spin oscillations and zitterbewegung. An SOC system can be viewed as one with an effective Zeeman magnetic ($b$) field which varies in the momentum (K) space, e.g. the momentum dependent $b$ field of the above is $\boldsymbol{b} = a(\boldsymbol{p} \times \boldsymbol{E})$, which is perpendicular to the plane containing both $\boldsymbol{p}$ and $\boldsymbol{E}$. One can thus draw an analogy between an SOC system and one with locally varying $b$ fields. The conduction electron experiences the varying $b$ field in real (R) space.

Let us imagine that electron spin evolves and aligns adiabatically to the $b$ field in their respective K or R spaces. Under the theoretical framework of gauge and symmetry,[5,6] the two systems will be analogous in that electron spin evolving adiabatically 'sees' the gauge potential of a magnetic monopole. In fact the magnetic monopole is the curvature ($\Omega$) of this gauge potential in the magnetic (B) space. In other words, one can now disregard the Zeeman $b$ field and imagine that the electron feels only the impact of the monopole magnetic field. In simple language, electrons traveling in these systems experience forces due to the monopole magnetic field. It is, however, not physically obvious how gauge curvature in B space is related to the dynamics of electrons. For better physical illustration, it might be necessary to transform the B space curvature to one in the more useful spaces of K or R under which equations of motion[7,8] can be constructed to describe the electron's orbital dynamics.

The monopole field can be viewed as a mathematical object that can lead to instructive description of the electron motion.[9–11] Here we present a thorough description of the Dirac gauge potential arising from spinor dynamics (fast alignment with $b$ fields) in the strong field (adiabatic) limit. A strong Zeeman effect in momentum space manifests in SOC systems (e.g. semiconductor spintronics, topological insulators) in which gauge curvature in K space will be derived. In local micromagnetic systems, which have been studied intensively in the magnetic media for hard disk drives, or domain wall spintronics, one needs instead to investigate the gauge curvature in real spaces. A similar system, which resembles the SOC, is the special carbon system of monolayer and bilayer graphene. But the spinor of carriers in these systems does not represent the spin state of conducting carriers in the carbon system. Instead, the spinor describes the pseudospin which consists of a linear combination of waves due to different sub-lattice sites.

## 7.2 Magnetic field in magnetic (B) space – monopole

In a nanoscale electronic system with spinor eigénstates, the adiabatic change of an eigenstate (e.g. spin or pseudospin in graphene) in parameter space (e.g. *b* field) gives rise to a geometric phase also known as Berry's phase.[12,13] By symmetry, the Hamiltonian will be modified by a set of gauge potentials which are Abelian by nature:

$$\boldsymbol{A}^N = \frac{\hbar}{2e}(1-\cos\theta^N)\frac{\partial\phi^N}{\partial\boldsymbol{n}} \qquad [7.3]$$

where $N$ indicates the type of space and $n$ is the parameter in $N$ space. In Eq. 7.3, the spin eigenstate at any point in time shares the same spherical coordinates ($\theta$, $\phi$) as those of the parameters (e.g. $n$). In the event where the parameter is oriented differently than the evolving state (e.g. spin is perpendicular to the parameter of momentum in the SOC at any one time), transformation would be required to find the gauge field in the momentum space. Explicitly, Eq. 7.3 is:

$$\boldsymbol{A}^N = \frac{\hbar}{2e}(1-\frac{n_z}{n})\left(\frac{\partial\phi}{\partial n_x},\frac{\partial\phi}{\partial n_y},0\right). \qquad [7.4]$$

To understand the origin of the geometric phase in an adiabatic spinor system, we first examine the evolution of an eigenstate from initial to final state as described by the path integral of spatial propagators as below:

$$\psi(x_{n+1},t_{n+1}) = \int G(x_{n+1}t_{n+1},x_0t_0)\,\psi(x_0,t_0)dx_0 \qquad [7.5]$$

where $G(x_{n+1}\,t_{n+1}, x_0t_0)$ is the propagator between times $t_0$ and $t$. The propagator in explicit spatial terms is:

$$G(x_{n+1}t_{n+1},x_0t_0) = \int \langle x_{n+1}|U(t_{n+1}t_n)|x_n\rangle\langle x_n|U(t_nt_{n-1})|x_{n-1}\rangle$$
$$\ldots.\langle x_1|U(tt_0)|x_0\rangle dx_1dx_2\,\ldots.\,dx_n \qquad [7.6]$$

where $\langle x_n|U(t_nt_{n-1})|x_{n-1}\rangle = \sqrt{\frac{m}{2\pi i\hbar\Delta t}}e^{\frac{im}{2\hbar}\left(\frac{x_n-x_{n-1}}{\Delta t}\right)^2\Delta t}.e^{-\frac{iV}{\hbar}\Delta t} \equiv \sqrt{\frac{m}{2\pi i\hbar\Delta t}}\gamma_{n-1}$ is the propagator between two spatial points. Substituting this into Eq. 7.6 yields:

$$\psi(x_{n+1},t_{n+1}) = \int\left[\left(\frac{m}{2\pi i\hbar\Delta t}\right)^{\frac{n+1}{2}} e^{\frac{iS(t)}{\hbar}}dx_n\ldots dx_1\right]\psi(x_0t_0)dx_0 \qquad [7.7a]$$

$$\psi(x_{n+1},t_{n+1}) = \int\left(\frac{m}{2\pi i\hbar\Delta t}\right)^{\frac{n+1}{2}}\left[\gamma_n\,\ldots.\,\gamma_0\;dx_n\ldots dx_1\right]\psi(x_0t_0)dx_0 \qquad [7.7b]$$

where the infinitesimal propagators are $\gamma_n \ldots\ldots \gamma_0 = e^{\frac{iS(t)}{\hbar}}$ and $S = i\int_0^T \frac{m}{2}\left(\frac{dx}{dt}\right)^2 - V(x_n)\, dt$ is the action of the system characterizing a spatial propagation. Note that $\left(\frac{m}{2\pi i\hbar\Delta t}\right)^{\frac{n+1}{2}}$ has dimension of $\left(\frac{1}{l}\right)^{n+1}$, which cancels that due to the volume element $dx_n \ldots dx_1 dx_0$. The infinitesimal propagator of $\sqrt{\frac{m}{2\pi i\hbar\Delta t}}\gamma_n = \langle x_{n+1}|U(t_{n+1}t_n)|x_n\rangle$ represents the probability of moving from one spatial point at $t = t_n$ to another at $t = t_{n+1}$; the subscripts 0,1,2, . . . $n,n + 1$ are time stamps. The strength of this probability is higher if $x_n$ is closer to $x_{n+1}$.

The above is a general description. In a dynamic spinor system which evolves with the changing $b$ fields, the infinitesimal propagator corresponding to $\sqrt{\frac{m}{2\pi i\hbar\Delta t}}\gamma_n = \langle x_{n+1}|U(t_{n+1}t_n)|x_n\rangle$ is:

$$C\gamma_n = \left\langle z_{n+1}\left|e^{-\frac{i}{\hbar}\delta t\,\mu \mathbf{b}.\boldsymbol{\sigma}}\right|z_n\right\rangle = \langle z_{n+1}|z_n\rangle e^{-\frac{i}{\hbar}\delta t\,\mu\, b} = e^{-\delta t z^{\dagger}\partial_t z}\, e^{\mp\frac{i}{\hbar}\delta t\,\mu b} \qquad [7.8]$$

where $n$ and $n + 1$ correspond to interval $t$ and $t + \delta t$, respectively; $\boldsymbol{b} = b\hat{n}$ and $\hat{n}$ is the unit vector of the $b$ field. In Eq. 7.8, use has been made of the approximation:

$$\langle z_{n+1}|z_n\rangle = 1 - z_{n+1}^{\dagger}\langle z_{n+1} - z_n\rangle \approx 1 - z_{n+1}^{\dagger}\partial_t z_n \delta t \approx e^{-\delta t z_n^{\dagger}\partial_t z_n} \qquad [7.9]$$

where propagation from $|z_n\rangle$ to $|z_{n+1}\rangle$ depends on whether $|z_n\rangle$ is parallel or anti-parallel to the $b$ field. Letting $\hat{n}.\,\boldsymbol{\sigma} = \sigma_n$, one could deduce that $\sigma_n|z_n\rangle = \pm|z_n\rangle$ for $|z_n\rangle$ parallel/anti-parallel to $b\,\hat{n}$; note that the direction of spin state $|z_n\rangle$ is $(z_n\,|\,\boldsymbol{\sigma}\,|z_n\rangle = \hat{n}$. It can be deduced simply by inspection that the action of the system due to spinor evolution would be:

$$S = \mp\mu bT + i\hbar\int_0^T \langle z_n(t)|\partial_t|z_n(t)\rangle\, dt. \qquad [7.10]$$

Neglecting the dynamic phase of $\mu bT$ and expanding the action leads to:

$$S = \pm\frac{\hbar}{2}\int_0^t (1-\cos\theta)\frac{\partial\phi}{\partial t}\, dt = \pm\frac{\hbar}{2}\int(1-\cos\theta)\frac{\partial\phi}{\partial\boldsymbol{b}}.d\boldsymbol{b}. \qquad [7.11]$$

One can derive the above taking different representation of the spin state. It is worth noting that, depending on $|z_n\rangle$, the respective expressions for the gauge potential in fact correspond to the north and south versions of the potential which we discuss later. For any one $|z_n\rangle$ chosen, the sign ± corresponds respectively to the parallel/anti-parallel spin alignment with the Zeeman field as the spin evolves in time. Here, $(\theta,\ \phi)$ is understood to be $(\theta_B,\ \phi_B)$. Table 7.1 summarizes the descriptions.

The term $\frac{\hbar}{2}(1-\cos\theta)\frac{\partial\phi}{\partial\boldsymbol{b}}$ is a gauge potential defined on the $S_2$ manifold as a regular expression except at $\theta = \pi$. One can thus see that the gauge field in B space

*Table 7.1* Gauge potentials appear differently depending on the choice of the eigenstate representation

| | Spin states | Gauge potential $i\langle z_n(t) \mid \partial_t \mid z_n(t)\rangle$ |
|---|---|---|
| 1. | $\|z_n\rangle = \left(\cos\frac{\theta}{2}\ \sin\frac{\theta}{2}e^{i\phi}\right)^T$ | $-\frac{1}{2}(1-\cos\theta)\partial_t\phi$ |
| 2. | $z_n\rangle = \left(\cos\frac{\theta}{2}e^{-i\phi}\ \sin\frac{\theta}{2}\right)^T$ | $\frac{1}{2}(1+\cos\theta)\partial_t\phi$ |

for any system under consideration is related to its Berry's phase. One could have also seen by now that taking a curvature of the gauge potential does not produce a regular quantity that can be defined everywhere on a $S_2$ manifold parameterized by $(r, \theta, \phi)$. In fact it has a singularity on the $-z$ axis (see Appendix 7.A).

This problem can be resolved[14,15] by conceiving that the gauge over the $S_2$ should be represented by at least two different expressions. Each expression covers all parts of the manifold except its own singularity. We define a north/south expression, which covers all parts of the manifold except the south/north poles. The overlap between the two expressions is the area where transition from one expression to the other must be carried out if one intends to cover the entire manifold with the two expressions. Thus, the total flux $\oint (\nabla_B \times \boldsymbol{A^B}).\boldsymbol{dS}$ cutting through the manifold can be derived using these two expressions. These fluxes can be regarded to have radiated from a monopole charge.

In summary, what has been described above is that the spinor dynamic in B space under the adiabatic approximation has generated a Dirac potential. The curvature of this gauge potential is consequently a Dirac monopole magnetic field. We now focus on $+\frac{\hbar}{2}(1-\cos\theta)\frac{\partial\phi}{\partial b}$, which arises due to the spin assuming only the lower energy eigenstate of the magnetic fields. Let us begin with a two-dimensional (2D) surface defined by the $b$ field. In the adiabatic approximation, the spin aligns closely with $b$, and the spin and $b$ thus share the same spherical coordinates, i.e. $(\theta_S, \phi_S) = (\theta_B, \phi_B)$. The gauge potential in B space is:

$$\boldsymbol{A}^B = \frac{1}{2}(1-\cos\theta_S)\frac{\partial\phi_S}{\partial\boldsymbol{b}} = \frac{1}{2}\left(1-\cos\theta_B\right)\frac{\partial\phi_B}{\partial\boldsymbol{b}} \qquad [7.12a]$$

and in the momentum space, $(\theta_S, \phi_S) \neq (\theta_K, \phi_K)$, the gauge potential in K space is:

$$\boldsymbol{A}^k = \frac{1}{2}(1-\cos\theta_S)\frac{\partial\phi_S}{\partial\boldsymbol{k}} = \frac{1}{2}(1-\cos\theta_S)\frac{\partial\phi_S}{\partial b_v}\frac{\partial b_v}{\partial\boldsymbol{k}} = A_V^B\frac{\partial b_v}{\partial\boldsymbol{k}}. \qquad [7.12b]$$

A pictorial description of this relationship is shown in Fig. 7.1, where the subscript '$s$' indicates the spin spherical coordinates.

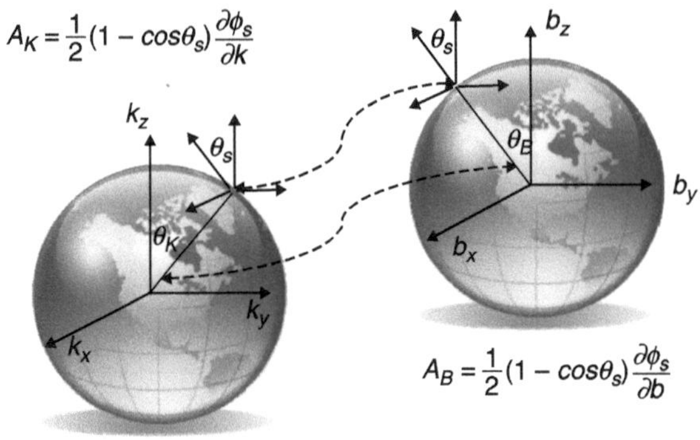


*7.1* Gauge potential in B and K spaces arising due to fast assumption of spinor eigenstates as the Hamiltonian evolves.

In general, a gauge potential in N space can be converted to an arbitrary L space by:

$$A_\mu^L = \frac{\hbar}{2e}(1-\cos\theta_s)\frac{\partial\phi_S}{\partial \boldsymbol{n}}.\frac{\partial \boldsymbol{n}}{\partial l_\mu} = \boldsymbol{A}^N.\frac{\partial \boldsymbol{n}}{\partial l_\mu}. \quad [7.13]$$

We have seen earlier that Eq. 7.11 is a result of adiabatic evolution of a spin up particle tracking the $b$ field. Alternatively, one could start with a Hamiltonian as below:

$$H = \frac{\boldsymbol{p}^2}{2m} + g\boldsymbol{\sigma}.\boldsymbol{b}(\boldsymbol{k}) + V(\boldsymbol{r}) \quad [7.14]$$

where $g$ is the SOC strength, $\boldsymbol{b}(\boldsymbol{k})$ is the momentum-dependent spin orbit magnetic field, and $V(\boldsymbol{r})$ is the potential energy. In the presence of an external electric field $\boldsymbol{E}$, we have $V(\boldsymbol{r}) = -e\boldsymbol{E}.\boldsymbol{r}$. One can diagonalize the expression above with respect to $\boldsymbol{b}(\boldsymbol{k})$ by parameterizing the SOC field in terms of the spherical angles, i.e. $\boldsymbol{b} = b(\sin\theta\cos\phi, \sin\theta\sin\phi, \cos\theta)$, where $\theta$ and $\phi$ are understood to be spherical coordinates of the $b$ field. Diagonalization is achieved through the rotation matrix $U(k)$ where:

$$U = \begin{pmatrix} \cos\frac{\theta}{2} & \sin\frac{\theta}{2}e^{-i\phi} \\ \sin\frac{\theta}{2}e^{i\phi} & -\cos\frac{\theta}{2} \end{pmatrix} = \boldsymbol{\sigma}\cdot\boldsymbol{m}. \quad [7.15]$$

The gauge potential resulting from the above is a 2 × 2 matrix which can be written generally as $\mathcal{A}_\mu(k) = (\boldsymbol{m} \times \partial_\mu \boldsymbol{m})\cdot\boldsymbol{\sigma} = \boldsymbol{A}_\mu\cdot\boldsymbol{\sigma}$ where $\boldsymbol{A}_\mu$ is the scalar gauge potential and $\boldsymbol{m} = \left(\sin\frac{\theta}{2}\cos\phi, \sin\frac{\theta}{2}\sin\phi, \cos\frac{\theta}{2}\right)$. The Hamiltonian under the rotated frame of reference is:

$$H' = U^\dagger H U = \frac{\boldsymbol{p}^2}{2m} + g\sigma_z b - e\boldsymbol{E}.(\boldsymbol{r} + iU^\dagger\partial_k U) \quad [7.16]$$

where $\mathcal{A}_\mu(\boldsymbol{k}) = -iU^\dagger \partial_k U$ is the $2 \times 2$ gauge potential, which is also commonly known as the special unitary group of degree 2 (SU(2)) gauge field in momentum space. The term $H_{SOC} = \gamma \boldsymbol{\sigma} \cdot \boldsymbol{b}(\boldsymbol{k})$ can be written as $g\sigma_\theta b$.

Performing the unitary transformation in the manner $U^\dagger \gamma \sigma_\theta b U = \gamma U^\dagger U \sigma_z U^\dagger U\, b$ results in $H_{SOC} = g\, \sigma_z b$. In fact in a general form, one can view the SU(2) gauge in an arbitrary space of $\mu$. In the example above, $\mu$ is the momentum:

$$\sigma_v A^v_\mu = \begin{pmatrix} i\dfrac{(1-\cos\theta)}{2}\partial_\mu \phi & -i\dfrac{e^{-i\phi}}{2}(\sin\theta\, \partial_\mu\phi + i\partial_\mu\theta) \\ -i\dfrac{e^{-i\phi}}{2}(\sin\theta\,\partial_\mu\phi - i\partial_\mu\theta) & -i\dfrac{(1-\cos\theta)}{2}\partial_\mu\phi \end{pmatrix}. \quad [7.17a]$$

In the usual adiabatic approximation, one considers only the diagonal terms of the matrix, decoupling the spin up and down effects. The top left diagonal term is associated with the anomalous velocity for the spin up while the bottom right diagonal term for the spin down electron:

$$\sigma_v A^v_\mu = \begin{pmatrix} i\dfrac{(1-\cos\theta)}{2}\partial_\mu\phi & 0 \\ 0 & -i\dfrac{(1-\cos\theta)}{2}\partial_\mu\phi \end{pmatrix}. \quad [7.17b]$$

The SU(2) gauge can thus be reduced to two versions of Abelian U(1) gauge potential, or the scalar potential:

$$A_\mu = \frac{1}{2}\begin{pmatrix} -\partial_\mu\theta\sin\phi - \sin\theta\cos\phi\,\partial_\mu\phi \\ \partial_\mu\theta\cos\phi - \sin\theta\sin\phi\,\partial_\mu\phi \\ (1-\cos\theta)\partial_\mu\phi \end{pmatrix}. \quad [7.18]$$

The gauge potential is also known as the connection. The curvature of this connection is the Berry curvature. The curvature is gauge invariant and acts like a magnetic field. It could thus give rise to the modified equations of motion for the electrons. One can thus generalize that a U(1) gauge potential in L space e.g. $\boldsymbol{A}^N \cdot \dfrac{\partial \boldsymbol{n}}{\partial l_\mu}$ constitutes the top left diagonal term of $U\partial_\mu U^\dagger$ where $U$ is the unitary matrix which rotates the laboratory axis to a $n$ field where spin is considered to be aligned to the field in the adiabatic limit. The above can be summarized by $\boldsymbol{A}^N \cdot \dfrac{\partial \boldsymbol{n}}{\partial l_\mu} = diag\left[U\partial_\mu U^\dagger\right] = A^v_{mon}\partial_\mu n^v$, where it is common to write $A^v_{mon} = A(\theta)$ $\partial\phi/\partial n^v$, which is also the gauge potential in N space.

One of the reasons it is legitimate to ask if the Dirac gauge potential in B space has any physical implication to a particle which experiences its presence could be because the Dirac potential is not well-defined everywhere (e.g. at $\theta = \pi$) and its surface integral is vanishing as a consequence. This suggests that the Dirac potential derived through the path integral approach might be mathematically inadequate to represent the monopole field. To make physical sense of this gauge potential, the region described with $\delta(x)\delta(y)\theta(-z)$, which resembles a string, must be avoided. This region is also known as the Dirac string. One needs to find ways to cover the manifold completely, but avoiding the string if one wants to obtain a curvature that resembles a physical magnetic field. This is the method first espoused by T.T. Wu and C. N. Yang.[14] One can thus view the Wu–Yang treatment as having reasonably affirmed that the monopole field can be regarded as a physical magnetic field. It is hence reasonable to accept that the monopole field influences the dynamics of charged particle, e.g. the electrons in the same way that the externally applied magnetic field causes their motion.

The curvature of the gauge potential in B space can be expressed as:

$$\Omega^B_\kappa = \frac{\partial}{\partial b_\mu} A^B_\nu \varepsilon_{\mu\nu\kappa} \tag{7.19}$$

where $\mu$, $\nu$, $\kappa$ are the three space-like dimensions with summation over double indices implied by Einstein's convention. Noting the explicit form of $A^B$, and setting $\hbar/2e$ to 1, one obtains:

$$\Omega^B_\kappa = \left(\left(1-\frac{b_z}{b}\right)\frac{\partial^2\phi}{\partial b_\mu \partial b_\nu} - \left(\frac{\partial}{\partial b_\mu}\frac{b_z}{b}\right)\frac{\partial\phi}{\partial b_\nu}\right)\varepsilon_{\mu\nu\kappa}. \tag{7.20}$$

For clarity, the $z$ component of the curvature can be deduced directly to be:

$$\Omega^B_z = \left(1-\frac{b_z}{b}\right)\left[\frac{\partial}{\partial b_x}, \frac{\partial}{\partial b_y}\right]\phi - \left(\frac{\partial}{\partial b_x}\frac{b_z}{b}\right)\frac{\partial\phi}{\partial b_y} + \left(\frac{\partial}{\partial b_y}\frac{b_z}{b}\right)\frac{\partial\phi}{\partial b_x}. \tag{7.21}$$

The monopole field has been expressed explicitly as a function of the Zeeman $b$ field. In the SU(2) system, it is normally straightforward to deduce an effective Zeeman $b$ field from the Hamiltonian. The Zeeman $b$ field is related to the energy which forms the generator of time translations for the spinor part of the wavefunction. To derive the curvature in B space explicitly, we note that $\cos\theta = b_z/b$ and $\tan\phi = b_y/b_x$; restoring $\hbar/2e$, the curvature reveals their monopole signatures:

$$\Omega^B_x = \frac{\hbar}{2e}\frac{b_x}{b^3}, \qquad \Omega^B_y = \frac{\hbar}{2e}\frac{b_y}{b^3}, \qquad \Omega^B_z = \frac{\hbar}{2e}\frac{b_z}{b^3}. \tag{7.22}$$

Note that in the above expressions, the Dirac string has been deliberately ignored; more discussion of the Dirac string is found in Appendix 7.A. One could easily check that, had the above derivation been carried out in real space (i.e. replacing $b$ with $r$), the expressions of Eq. 7.22 would have the dimension of the magnetic fields. However, in application this is only possible for the specific case of spin aligning with $r$, i.e. spin aligned to the $r$ coordinates with respect to the spherical center. One way of achieving this is by means of technology whereby one creates local magnetic fields or moments with orientations in B space that overlaps the r coordinates in R space, i.e. $\boldsymbol{b} = c\boldsymbol{r}$ where $c$ is a constant. On the other hand there are many natural systems which provide a Zeeman $b$ field in the momentum space, e.g. in the SOC systems of semiconductors and topological insulators, as well as the pseudospin or valley spin systems of graphene.

## 7.3 Magnetic field in momentum (K) space – spintronics, graphene, topological insulators

Above, we have shown the path integral derivation of the gauge potential and its monopole in the B space. As mentioned earlier, the curvature in B space is not useful for heuristically elucidating the motion of a particle which 'sees' these fields. In nature, there exist many systems that provide the mimic of Zeeman $b$ fields, and these fields normally depend on the momentum. It is of interest indeed to find a curvature in the K space, which would be more useful for deriving the equation of motion and hence for elucidating qualitatively and visually the dynamics of electrons which 'see' these fields. The SOC systems, which have been known to exist in atomic physics (hyperfine interaction), semiconductors, metals and topological insulators, are the most conspicuous systems that provide such $k$-dependent Zeeman fields. The SOC effect manifests in the bandstructure, lifting the degeneracy of valence electrons at momentum other than zero. Other analogy of $b(k)$ systems include carbon-based systems such as monolayer and bilayer graphene, and superconducting systems. In the non-relativistic limit:

$$\left(\frac{p^2}{2m}+e\phi+\frac{\hbar e}{4m^2c^2}\boldsymbol{\sigma}\cdot(\boldsymbol{p}\times\boldsymbol{E})+\frac{i\hbar e}{4m^2c^2}\boldsymbol{E}\cdot\boldsymbol{p}\right)\psi=\in\psi. \qquad [7.23]$$

In SOC, graphene or superconducting systems, the parameters of interest lie in the K space. To find the Dirac gauge potential in K space, one merely needs the Zeeman field as a function of momentum. Space conversion for the gauge potential is straightforward, but that for the monopole field is slightly more complex. But we show below that, to transform the monopole field in B space to a gauge curvature in K space, one merely needs $b(k)$. Looking at the $z$ component of the curvature only:

$$\Omega_z^K=\frac{\partial A_y^K}{\partial k_x}-\frac{\partial A_x^K}{\partial k_y}. \qquad [7.24]$$

Using $A_\mu^K \equiv \boldsymbol{A}^B.\frac{\partial \boldsymbol{b}}{\partial k_\mu}$, it is straightforward to obtain:

$$\Omega_z^K=\frac{\partial}{\partial k_x}\left(\boldsymbol{A}^B.\frac{\partial \boldsymbol{b}}{\partial k_y}\right)-\frac{\partial}{\partial k_y}\left(\boldsymbol{A}^B.\frac{\partial \boldsymbol{b}}{\partial k_x}\right)=\left(\frac{\partial \boldsymbol{b}}{\partial k_x}.\partial_b\right)\left(\boldsymbol{A}^B.\frac{\partial \boldsymbol{b}}{\partial k_y}\right)-\left(\frac{\partial \boldsymbol{b}}{\partial k_y}.\partial_b\right)\left(\boldsymbol{A}^B.\frac{\partial \boldsymbol{b}}{\partial k_x}\right) \qquad [7.25]$$

which can be re-expressed in:

$$\Omega_z^K=\left(\frac{\partial b_\nu}{\partial k_x}\frac{\partial A_\mu^B}{\partial b_\nu}\frac{\partial b_\mu}{\partial k_y}\right)-\left(\frac{\partial b_\nu}{\partial k_y}\frac{\partial A_\mu^B}{\partial b_\nu}\frac{\partial b_\mu}{\partial k_x}\right)+A_\mu^B\left(\frac{\partial b_\nu}{\partial k_x}\frac{\partial^2 b_\mu}{\partial b_\nu\partial k_y}-\frac{\partial b_\nu}{\partial k_y}\frac{\partial^2 b_\mu}{\partial b_\nu\partial k_x}\right). \qquad [7.26]$$

The last term written in $A_\mu^B\left(\frac{\partial^2 b_\mu}{\partial k_x \partial k_y}-\frac{\partial^2 b_\mu}{\partial k_y \partial k_x}\right)$ clearly vanishes as $[\partial_x,\partial_y] = 0$ in general:

$$\Omega_z^K=\left(\frac{\partial A_\mu^K}{\partial b_\nu}-\frac{\partial A_\nu^K}{\partial b_\mu}\right)\frac{\partial b_\nu}{\partial k_x}\frac{\partial b_\mu}{\partial k_y}=\frac{\partial A_i^K}{\partial b_j}(\delta_{i\mu}\delta_{j\nu}-\delta_{i\nu}\delta_{j\mu})\left(\frac{\partial b_\nu}{\partial k_x}\frac{\partial b_\mu}{\partial k_y}\right). \qquad [7.27]$$

Using the identity $(\delta_{i\mu}\delta_{j\nu}-\delta_{i\nu}\delta_{j\mu}) = \varepsilon_{ijk}\varepsilon_{\mu\nu k}$ where $\varepsilon_{ijk}$ is the fully anti-symmetric tensor and $\delta_{i\mu}$ is the Kronecker delta, one obtains the $z$ component of the curvature in K space:

$$\Omega_z^K=\frac{\partial A_i^K}{\partial b_j}\varepsilon_{ijk}\left(\frac{\partial b_\nu}{\partial k_x}\frac{\partial b_\mu}{\partial k_y}\right)\varepsilon_{\mu\nu k}. \qquad [7.28]$$

Likewise, the other components of this curvature field can be derived and expressed in the more familiar vector calculus form:

$$\Omega_x^K=\frac{\boldsymbol{b}}{b^3}\cdot\left(\frac{\partial \boldsymbol{b}}{\partial k_z}\times\frac{\partial \boldsymbol{b}}{\partial k_y}\right)$$

$$\Omega_y^K=\frac{\boldsymbol{b}}{b^3}\cdot\left(\frac{\partial \boldsymbol{b}}{\partial k_x}\times\frac{\partial \boldsymbol{b}}{\partial k_z}\right)$$

$$\Omega_z^K=\frac{\boldsymbol{b}}{b^3}\cdot\left(\frac{\partial \boldsymbol{b}}{\partial k_y}\times\frac{\partial \boldsymbol{b}}{\partial k_x}\right) \qquad [7.29]$$

where $\frac{\partial A_i^K}{\partial b_j}\varepsilon_{ijk}=-\frac{b_k}{b^3}$. It has thus been shown that the curvature in K space can be developed by substituting the Zeeman $b$ field into Eq. 7.29.

When the spin of a particle inscribes a path on $S_B^2$, the gauge potential an electron carrying that spin 'sees' at every point on the path is $\boldsymbol{A}^B=\frac{\hbar}{2e}\left(1-\cos\theta^s\right)\frac{\partial \phi^S}{\partial \boldsymbol{b}}$. This potential is directly related to the change of the spin angles with the change of the magnetic field over the manifold. Under the theory of adiabatic alignment, the spin angles coincide identically with the magnetic field angles, resulting in $\boldsymbol{A}^B=\frac{\hbar}{2e}\left(1-\cos\theta^B\right)\frac{\partial \phi^B}{\partial \boldsymbol{b}}$. Likewise, if one were to perform the same over the momentum manifold, i.e. when the electron inscribes a path on $S_K^2$, the potential the electron 'sees' would be $\boldsymbol{A}^K=\frac{\hbar}{2e}\left(1-\cos\theta^S\right)\frac{\partial \phi^S}{\partial \boldsymbol{k}}$. This gauge potential is clearly not equivalent to $\boldsymbol{A}^B$. What one has is:

$$\boldsymbol{A}^K\neq\frac{\hbar}{2e}(1-\cos\theta^K)\frac{\partial \phi^K}{\partial \boldsymbol{k}}. \qquad [7.30]$$

One can simplify the above descriptions by considering that an electron 'sees' $\boldsymbol{A}^B$ or $\boldsymbol{A}^K$ depending on the type of parameter variation that is being considered. It is worth cautioning that while $\nabla_B \times A^B$ is a monopole field, $\nabla_K \times A^K$ is not generally a monopole field.

---

**Exercise 7.1**

Show that for $b = ck$, or $b = cr$, one does not need to perform conversion and that the monopole fields will, respectively, be $\boldsymbol{\Omega}^K \propto \frac{\boldsymbol{k}}{k^3}$; $\boldsymbol{\Omega}^R \propto \frac{\boldsymbol{r}}{r^3}$.

***Solution***

When $\boldsymbol{b} = c\boldsymbol{k}$ or $\boldsymbol{b} = c\boldsymbol{r}$ is obeyed, the spin will share the same polar coordinates with $\boldsymbol{k}$ or $\boldsymbol{r}$.

Hence, $A^K = \frac{\hbar}{2e}(1-\cos\theta^K)\frac{\partial\phi^K}{\partial k}, A^R = \frac{\hbar}{2e}(1-\cos\theta^R)\frac{\partial\phi^R}{\partial r}$.

---

Below, we study the various nanoelectronic fields where the curvature in K space can now be viewed as the spin-dependent Lorentz magnetic fields. Table 7.2 summarizes the curvature fields for the cubic Dresselhaus, the linear Dresselhaus, the Perel-modified Dresselhaus, the linear Rashba and Dresselhaus in quantum well systems, the monolayer Weyl system of massless Dirac fermions in graphene, and topological insulators. First the SOC effects and their corresponding effective $b$ fields are presented. These effective Zeeman $b$ fields can be derived by considering the first of $H = \mu_B b = \frac{eg\hbar}{4m} b$ where $\mu_B$ is the Bohr magneton which has the SI unit of a joule/tesla; note that $\mu_B$ is also equivalent to $IA$ where $I$ is a circulating current and $A$ is the area enclosed by the circulating current. The last column of Table 7.2 summarizes the gauge curvature arising in these various condensed matter or device channel systems. Heuristically, the curvature denoted by $\boldsymbol{\Omega}$ can be regarded as a type of Lorentz magnetic field that provides Lorentz forces to bend electron path. These Lorentz magnetic fields can be derived from the effective Zeeman fields denoted by in Table 7.2.

The fact that the force experienced by the spin up/down particle is opposite in directions arising due to $\pm(\nabla \times A)$ provides the physical picture of spin transverse separation. However, more rigorous quantification of such separation, or the Hall conductivity, is related to the Berry's phase which can be interpreted as a summation of the Lorentz forces over all electron momenta. The Berry's phase in K space, which has important relevance to electron transport, can be derived from $\phi = \frac{e}{\hbar}\int \boldsymbol{A} \cdot d\boldsymbol{k} = \frac{e}{\hbar}\int (\nabla \times A) \cdot d^2\boldsymbol{k}$, where it is worth noting that $(\nabla \times A)$ can be viewed as the magnetic field that the electron 'feels'.

It is not the focus of this paper to conduct an extended analysis of the Berry's phase of all systems. But in view of the modern trend of quantifying system

*Table 7.2* Various SOC and other spinor systems, their Hamiltonian, relativistic effective fields and corresponding K space monopole curvature

| Spin orbit coupling type (material system) | Hamiltonian and effective *b* field | $\boldsymbol{\Omega}$ is the curvature field in K space |
|---|---|---|
| *Semiconductor systems with linear SOC (GaAs, GaSb, InAs, InSb, GaInAs, GaInSb)* | | |
| Linear Dresselhaus | $H = \eta_D (\sigma_x k_x - \sigma_y k_y)$ <br> $\boldsymbol{b} = \frac{\eta_D}{\mu_B}\begin{pmatrix} k_x \\ -k_y \\ 0 \end{pmatrix}$ | $\boldsymbol{\Omega} = \pm \frac{\hbar\pi}{e}\begin{pmatrix} 0 \\ 0 \\ \delta^2(k) \end{pmatrix}$ |
| Linear Rashba | $H = \eta_R (\sigma_x k_y - \sigma_y k_x)$ <br> $\boldsymbol{b} = \frac{\eta_D}{\mu_B}\begin{pmatrix} k_y \\ -k_x \\ 0 \end{pmatrix}$ | $\boldsymbol{\Omega} = \pm \frac{\hbar\pi}{e}\begin{pmatrix} 0 \\ 0 \\ \delta^2(k) \end{pmatrix}$ |
| *Graphene systems* | | |
| Monolayer graphene (massless Weyl) (one valley only) | $H = A(\sigma_x k_x + \sigma_y k_y)$ <br> $\boldsymbol{b} = \frac{A}{\mu_B}\begin{pmatrix} k_x \\ k_y \\ 0 \end{pmatrix}$ | $\boldsymbol{\Omega} = \pm \frac{\hbar\pi}{e}\begin{pmatrix} 0 \\ 0 \\ \delta^2(k) \end{pmatrix}$ |
| Bilayer graphene (one valley only) | $H = \frac{\hbar^2}{2m}\left( \sigma_x \frac{k_+^2 + k_-^2}{2} - \sigma_y \frac{k_-^2 - k_+^2}{2i} \right)$ <br> $\boldsymbol{b} = \frac{1}{2\mu_B m}\begin{pmatrix} k_x^2 - k_y^2 \\ 2k_x k_y \\ 0 \end{pmatrix}$ | $\boldsymbol{\Omega} = \pm \frac{\hbar 2\pi}{e}\begin{pmatrix} 0 \\ 0 \\ \delta^2(k) \end{pmatrix}$ |
| Graphene with sub-lattice asymmetry | $H = \hbar v_F (k_x \sigma_x + k_y \sigma_y) + B_z \sigma_z$ <br> $\boldsymbol{b} = \begin{pmatrix} \hbar v_F k_x \\ \tau\hbar v_F k_y \\ B_z \end{pmatrix}$ | $\Omega_z = \frac{\tau\hbar^2 v_F^2 B_z}{2\left[B_z^2 + \hbar^2 v_F^2 k^2\right]^{3/2}}$ <br> $\tau = \pm$ is the sign representing valley *K/K′* |
| *Semiconductor systems with cubic SOC* | | |
| Dresselhaus cubic (bulk III–V) | $H = \eta_{DC}\left( \sigma_x k_x \left[k_y^2 - k_z^2\right]\right) + c.p.$ <br> $\boldsymbol{b} = \frac{\eta_{DC}}{\mu_B}\begin{pmatrix} k_x\left[k_y^2 - k_z^2\right] \\ k_y\left[k_z^2 - k_x^2\right] \\ k_z\left[k_x^2 - k_y^2\right] \end{pmatrix}$ | $\boldsymbol{\Omega} = \frac{\pm\hbar\left(k_x^2 - k_y^2\right)\left(k_y^2 - k_z^2\right)\left(k_z^2 - k_x^2\right)}{2eb^3}\begin{pmatrix} k_x \\ k_y \\ k_z \end{pmatrix}$ |

| | | |
|---|---|---|
| Dresselhaus (high kinetic, Perel) | $H=\eta_{Dk}\left(-\sigma_x k_x k_z^2+\sigma_y k_y k_z^2+\sigma_z k_z\left[k_x^2-k_y^2\right]\right)$<br>$\boldsymbol{b}=\frac{\eta_{Dk}}{\mu_B}\begin{pmatrix}-k_x k_z^2\\ k_y k_z^2\\ k_z\left[k_x^2-k_y^2\right]\end{pmatrix}$ | $\boldsymbol{\Omega}=\frac{\pm\hbar}{2eb^3}\left(k_y^2-k_x^2\right)\begin{pmatrix}k_z^4 k_x\\ k_z^4 k_y\\ k_z^5\end{pmatrix}$ |
| *Topological insulators* | | |
| Three-dimensional surface topological insulators | $H=-\hbar v_F(k_y\sigma_x-k_x\sigma_y)+(\alpha-\gamma k^2)\sigma_z$<br>$\boldsymbol{b}=\begin{pmatrix}\hbar v_F k_y\\ \hbar v_F k_x\\ (\alpha-\gamma k^2)\end{pmatrix}$ | $\Omega_z=\frac{\pm\hbar^2 v_F^2(\alpha-\gamma k^2)}{2[(\alpha-\gamma k^2)^2+\hbar^2 v_F^2 k^2]^{3/2}}$ |
| *Optics* | | |
| Optical Magnus effect | $H=\frac{1}{2}(p^2-n^2(r))-Q$<br>$A=\pm\left(\frac{p_y p_z}{p\left(p_x^2+p_y^2\right)},\frac{-p_x p_z}{p\left(p_x^2+p_y^2\right)},0\right)$ | $\Omega=c\frac{\boldsymbol{p}}{p^3}$<br>$c=\mp$ is the sign representing right/left circular polarization |

conductivity using the Berry's phase, we merely provide its derivations for common systems only, i.e. linear Rashba, linear Dresselhaus, monolayer and bilayer graphene. To derive the Dirac potential for these systems in the intended space of K, one needs to recall that $A_\mu^k=\frac{\hbar}{2e}(1-\cos\theta)\frac{\partial\phi}{\partial k_\mu}$. To derive $A_\mu^k$ for the systems, one would thus need the effective magnetic fields of each system and keep in mind that in the B space, $\cos\theta=b_z/b$, $\tan\phi=b_y/b_x$. To obtain the Berry's phase, one can then examine the gauge of $\boldsymbol{A}=\pm\frac{\hbar}{2ek^2}(pk_y\ qk_x\ 0)^T$, which is the gauge potential at $\theta=\pi/2$, i.e. in the plane of the K space two-dimensional system.

In monolayer graphene with sub-lattice asymmetry, $B_z(-B_z)$ is the strength of the electrostatic energy at sub-lattice $A(B)$. It can also be treated like a vertical magnetic field acting on the pseudospin of the graphene. Because the effective magnetic field of this graphene system contains an explicitly valley-dependent constant $\tau$, a valley Hall effect could be deduced from the valley-dependent curvature. Careful examination of the curvature shows that, in the limit that $B_z$ vanishes, the delta function curvature of monolayer graphene can be recovered but with an explicit $\tau$ constant. Considering just one valley, the normal pseudospin Hall effect for monolayer graphene can be recovered.

In the optical Magnus system, $p$ is the momentum of photon, while $n(r)$ is the spatially varying refractive index and $Q_{ij}=p_ip_j$ is a $3\times3$ matrix in the spin-1 space. In geometric optics, light propagation in a spatially inhomogeneous medium is governed, to a good approximation, by the Hamiltonian in Table 7.2.

*Table 7.3* Various SOC systems, their gauge and corresponding Berry's phase

| Spin orbit coupling type (material system) | Gauge | $\phi$ is the Berry's phase |
|---|---|---|
| Linear Dresselhaus | $\boldsymbol{A}=\pm\frac{\hbar}{2ek^2}\begin{pmatrix}k_y\\-k_x\\0\end{pmatrix}$ | $n\pi$ |
| Linear Rashba | $\boldsymbol{A}=\pm\frac{\hbar}{2ek^2}\begin{pmatrix}-k_y\\k_x\\0\end{pmatrix}$ | $n\pi$ |
| Massless Weyl (monolayer graphene) | $\boldsymbol{A}=\pm\frac{\hbar}{2ek^2}\begin{pmatrix}-k_y\\k_x\\0\end{pmatrix}$ | $n\pi$ |
| Massive Dirac (bilayer graphene) | $\boldsymbol{A}=\pm\frac{\hbar}{2ek^2}\begin{pmatrix}-2k_y\\2k_x\\0\end{pmatrix}$ | $2n\pi$ |
| Combined linear Rashba and Dresselhaus | $\boldsymbol{A}=\frac{\pm\frac{1}{2}(\alpha^2-\beta^2)}{(\alpha^2+\beta^2-2\alpha\beta\sin 2\phi)k^2}\begin{pmatrix}-k_y\\k_x\\0\end{pmatrix}$ | $\frac{\alpha^2-\beta^2}{\lvert\alpha^2-\beta^2\rvert}\pi$ |

Note: $\phi=\frac{e}{\hbar}\int \boldsymbol{A}.d\boldsymbol{k}=\oint_{SF}\left[\nabla\times\boldsymbol{A}\left(\theta=\frac{\pi}{2}\right)\right].d^2\boldsymbol{k}.$

This model remains remarkably valid for wavelength of light much smaller than the length scales of the spatial inhomogeneity.

---

**Exercise 7.2**

Consider a system in an eigenstate adiabatically transported around a closed path in phase space. The eigenstate returns to its original state with an additional dynamic phase, and a geometrical phase. For a 2D system, the Berry's phase can be derived with:

$$\gamma_n=\oint_C A\left(\theta=\frac{\pi}{2}\right).dk$$

where $C$ represents a closed path. Derive the results in Table 7.3.

---

**Exercise 7.3**

Berry's phase can be represented by a solid angle $\Omega$ traced out as a closed path by the spin vector of an electron. Note that in the adiabatic regime, an electron will align its spin to the magnetic field it experiences. That is to say, for an electron in a changing magnetic field (or an electron moving through micromagnetic patterns), the spin aligns to the changing magnetic field such that a closed path in parameter space is traced out and it acquires a Berry's phase.

(1) With $\nabla_B = \frac{\partial}{\partial b}\hat{\boldsymbol{a}}_b + \frac{1}{b}\frac{\partial}{\partial \theta}\hat{\boldsymbol{a}}_\theta + \frac{1}{b\sin\theta}\frac{\partial}{\partial \phi}\hat{\boldsymbol{a}}_\phi$, show that the gauge potential is:

$$\boldsymbol{A} = \langle z_n | \nabla_B | z_n \rangle = i\frac{1-\cos\theta}{2b\sin\theta}\hat{\boldsymbol{a}}_\phi \quad or \quad -i\frac{1+\cos\theta}{2b\sin\theta}\hat{\boldsymbol{a}}_\phi$$

(2) The curvature of this gauge potential is:

$$\boldsymbol{\Omega} = \nabla_B \times \langle z_n | \nabla_B | z_n \rangle = \frac{i}{2b^2}\hat{\boldsymbol{a}}_b$$

(3) Show, using Stoke's theorem of $\gamma_n = i\oint \boldsymbol{A}\,.d\boldsymbol{b} = i\oint \nabla_B \times \boldsymbol{A}\,.d\boldsymbol{S}_b$, and the relation $\boldsymbol{S}_b = b^2 d\Omega\,\hat{\boldsymbol{a}}_b$, that the geometric phase is a function of the solid angle $\gamma_n = -\frac{1}{2}\Omega$. Note that $\Omega$ denotes the solid angle, not gauge curvature.

***Solution***

Use $|z_n\rangle = \left(\cos\frac{\theta}{2} \;\; \sin\frac{\theta}{2}e^{i\phi}\right)^T$ to obtain $i\frac{1-\cos\theta}{2b\sin\theta}\hat{\boldsymbol{a}}_\phi$

or $|z_n\rangle = \left(\cos\frac{\theta}{2}e^{-i\psi} \;\; \sin\frac{\theta}{2}\right)^T$ to obtain $-i\frac{1+\cos\theta}{2b\sin\theta}\hat{\boldsymbol{a}}_\phi$.

---

We have introduced a general approach which allows the derivation of the monopole field of a specific system in the B space using the Dirac gauge potential defined in the same B space. But the physical significance of the curvature fields depends on the space in which the curvature is taken. The usefulness of the general approach is that it provides a unified underlying picture for the curvature fields in any arbitrary spaces (e.g. in K and R) under a common origin, i.e. the Dirac gauge potential and its monopole field in B space. One merely requires an effective $b$ field of the form $\boldsymbol{\sigma}\,.\,\boldsymbol{b}$ in the Hamiltonian of a specific system. The gauge curvature can be derived in any space outside the B space.

The surface integral of the curvature yields a non-vanishing quantized value, which is invariant under deformation of the surface of integration. The surface integral is hence a topological object. In the context of Dirac monopole, this is associated with the quantization of the electric charge. In SOC or graphene systems, this quantity is associated with quantized magnetic flux or the Berry's phase. It therefore becomes clear that the existence of K space gauge curvature (Lorentz magnetic field) can be related heuristically to particle trajectory, and in the case of SOC system, spin-dependent separation of charges would be resulted from the spin-dependent curvature fields in K space. The summary in Table 7.2 provides a unifying picture for the physics of particle motion in various systems.

## 7.4 Introduction to anomalous Hall effects (AHE)

AHE was observed shortly after the discovery of the classical Hall effect in 1879. In fact, it was found that the Hall resistivity in ferromagnetic materials behaves differently than in paramagnetic and diamagnetic substances. While the transverse Hall voltage in paramagnetic and diamagnetic substances is directly proportional to the $b$ field, the Hall voltages in ferromagnetic materials were found to deviate from this proportion at external magnetic fields, causing the magnetization of the material to saturate.[16] This anomaly is evidence that the resistivity also depends significantly on the magnetization of the material. The expression for transverse resistivity $\rho_H$ in a ferromagnetic material is expressed as a function of both the magnetic induction and the magnetization of the material. The ordinary Hall resistivity is always proportional to the $b$ field, and the magnetization dependence is taken to be linear in M to a first approximation. Hence we have:

$$\rho_H = R_H b + 4\pi R_A M \quad [7.31]$$

where $b$ is the applied magnetic field, $M$ is the magnetization of the material, $R_H$ is the ordinary Hall coefficient and $R_A$ is the anomalous Hall coefficient. AHE requires the presence of non-coplanar local moments. While classical charge Hall causes charge separation, SHE causes both charge and spin separation (Fig. 7.2). It can thus be viewed as a combined presence of both charge and spin Hall.

The origins of AHE have yet to be fully explained. Many models have been proposed in the literature. In 1954, Karplus and Luttinger[17] offered an explanation based on the bandstructure and spin orbit interaction. The theory is based on skew scattering where electrons of different spins are deflected differently to the right and left. The other mechanism given by Berger *et al.*[18] is based on the side jumping

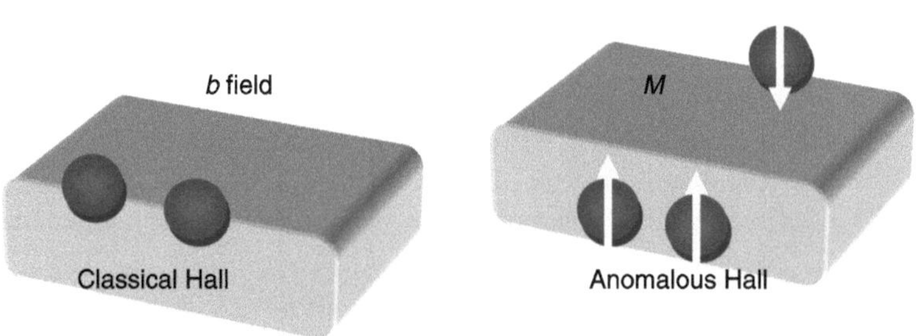


*7.2* The classical Hall effect is a charge Hall effect which builds up transverse charge accumulation. The anomalous Hall effect is a combination of charge and spin Hall which builds up transverse charge and spin accumulation.

of the electron's trajectory close to the scattering center. In the engineering community it is crudely accepted that at low resistivity (e.g. at low temperature in pure metals), the dominant effect is skew scattering. In disordered alloys where resistivity is large, the dominant mechanism is side jumping. AHE can be detected at room temperature. In fact, there has been intense research recently on utilizing AHE for sensing magnetic fields. In the engineering community, AHE is also known as the extraordinary Hall effect (EHE).[19] It has been reported that EHE shows high Hall resistance.

Large Hall resistivity could generate high Hall voltage for a fixed current density for useful device application. Current density is limited by the damaging effect of electromigration to not more than $j = 10^7$ A cm$^{-1}$. But in devices, the applied longitudinal voltage is normally fixed (e.g. to 1 V). Since the same current density is shared by the longitudinal and the Hall arms, a large Hall angle $\rho_{xy}/\rho_{xx}$ would be most favorable. It is somewhat unfortunate that most materials tested so far could not deliver the simultaneous need for high Hall resistivity and Hall angle. For example, material 4 in Table 7.4 shows high Hall resistivity but low Hall angle. To keep the longitudinal voltage to 1 V, a small current is used and this results in low Hall voltage as well. As recently as 2002, Onoda and Nagaosa[20] approached the problem of AHE or EHE with gauge theoretic physics. On the other hand, there have been studies of topological AHE by the same group and others.

*Table 7.4* Material systems with Hall and longitudinal resistivity

| | Material system | $\rho_{xy}$ (μΩ cm) | $\rho_{xx}$ (μΩ cm) | Hall angle $\theta_H = \rho_{xy}/\rho_{xx}$ |
|---|---|---|---|---|
| 1. | Co/Au, Co/Pt, Co/Cu, Co/Pd | 1 | 41 | 2.3% |
| 2. | MgO/FePt(10 nm)/Pt(1.5 nm) | | 39 | 1.5% |
| 3. | MgO/Cr (3 nm)/Pt (60 nm)/ FePt (40 nm)/Pt (2 nm) | | 30 | 1% |
| 4. | Composite materials (e.g. Ni:$SiO_2$, Co:$SiO_2$, CoFe:$Al_2O_3$) | 200 | | 0.1% |

## 7.5 Topological anomalous Hall effects

We now consider a system where the effective Zeeman $b$ field is infinitely strong, whether it is in the momentum space or the real space. In this strong field limit, electron spin relaxes to the field. The alignment of the electron spin to the local field means that the electron assumes the low-energy spin eigenstate of the system, with no admixture from the other spin eigenstate. Such AHE systems correspond to material with an intrinsic spin orbit and exchange energy (magnetic moment) bandstructure. One can take the spin to align along the vector sum of the spin orbit and the exchange magnetic field in momentum space, and derive ($\boldsymbol{A}_k$, $\boldsymbol{\Omega}_k$) as described earlier. If one is interested in the local space effect, one aligns the spin to local moment in real space and derives ($A_r$, $\Omega_r$). In the momentum space, the Kubo conductivity is used; in real space, the semiclassical conductivity is used.

It would be interesting to understand how the two conductivities are related or whether they provide overlapping information. It is, at least, clear that in the event of real space analysis, under specific moment configuration, real space conductivity vanishes. But momentum space conductivity continues to exist. Under this particular circumstance, one may surmise that the two conductivities may not be the same.

### 7.5.1 Topological anomalous Hall effects (momentum space)

Previous sections deal with the gauge physics underlying the magnetic fields in the B and K spaces. Tabulated summaries of the gauge curvature and associated Berry's phase for different nanoelectronic systems are also given. It becomes clear that magnetic fields in K space could be related to electron dynamics via the anomalous velocity method, which is a form of Lorentz force picture in the momentum space. However, falling short of deriving the full expression for conductivity, electron dynamic remains unquantified. It is also not physically clear why the electron dynamic is topological in nature.

Answers to these questions lie in the Kubo Hall conductivity given by:

$$\sigma_{xy} = \int\int \Omega_z (f_+ - f_-)\, dk_x\, dk_y \qquad [7.32]$$

where $f(k)$ is the occupation function for band $\pm$. Since $\int\int \Omega_z\, dk_x\, dk_y$ is a topological invariant, one can deduce that in 2D systems, conductivity will be topological in the energy region where $f_+ - f_- = C$ and $C$ is a constant. Thus in bulk insulator with a bandgap, topological conductivity can be derived in the gap where $(f_+ - f_-) - 1$. In the following, we discuss the derivation of the conductivity which has the physical significance of, upon summing over momentum, the vertical effective magnetic field bending the transverse trajectory of electron in a 2D planar system. The general expression for Kubo conductivity in an equilibrium system is:

$$\sigma_{xy} = \lim_{\substack{E\to 0\\ \eta\to 0}} \frac{i}{E} Q_{xy}(iq_m \to E + i\eta) \qquad [7.33]$$

where:

$$\begin{aligned} Q_{xy}(iq_m) &= \frac{1}{A\beta}\sum_{k,n,s,t} Tr[J_x(\boldsymbol{k})G(\boldsymbol{k}, ik_n + iq_m)J_y(\boldsymbol{k})G(\boldsymbol{k}, ik_n)] \\ &= \frac{1}{A\beta}\sum_{k,n,s,t} Tr\left[\left(\frac{J_x(\boldsymbol{k})}{ik_n + iq_m - E_{ks}}\right)\left(\frac{J_y(\boldsymbol{k})}{ik_n - E_{kt}}\right)\right] \\ &= \frac{1}{A\beta}\sum_{k,n,s,t}\left[\frac{\alpha(\boldsymbol{k})}{\left(ik_n - (E_{ks} - iq_m)\right)\left(ik_n - E_{kt}\right)}\right] = \frac{1}{A\beta}\sum_{k,n,s,t}[g(ik_n)] \end{aligned} \qquad [7.34]$$

and $Q_{xy}(iq_m)$ is also known as the Matsubara sum, where $iq_m$ is the Matsubara bosonic frequency, $ik_n$ the fermionic frequency and $A$ is the area. Note that $\alpha(\boldsymbol{k}) = Tr\left\{ J_x(\boldsymbol{k})\, J_y(\boldsymbol{k}) \frac{1}{2}(I + s\boldsymbol{\sigma}\cdot\boldsymbol{b}) \frac{1}{2}(I + t\boldsymbol{\sigma}\cdot\boldsymbol{b}) \right\}$.

---

**Exercise 7.4**

Show that for a Hamiltonian system which sees a $K$ space magnetic field as given by $H = E(k) + g\,\boldsymbol{\sigma}.\boldsymbol{b}(\boldsymbol{k})$, the Green's function is:

$$G(\boldsymbol{k}, E) = \frac{1}{2}\left( \frac{I + \boldsymbol{\sigma}\cdot\boldsymbol{b}}{iE - E_+} + \frac{I - \boldsymbol{\sigma}\cdot\boldsymbol{b}}{iE - E_-} \right)$$

where $g$ is the strength of the Zeeman field $b$, and $E_+ = E_k + g$, $E_- = E_k - g$.

---

With the Green's function expressed in terms of the Matsubara discrete frequencies, the Matsubara sum can be performed with the prescriptions of:

$$S = \frac{1}{A\beta}\sum_{k,n,}[g(ik_n)] = \frac{1}{A}\sum_{k,j,} Res\,[g(z_j)]\, f(z_j) \qquad [7.35]$$

where $z_j$ are poles of $g(ik_n)$ and $Res\ [g(z_j)]$ are residues at these poles; $f$ is the distribution function of the electrons. The two poles for the functions are $z_1 = E_{kt}$ and $z_2 = E_{ks} - iq_m$.

---

**Exercise 7.5**

Find the residues at poles $z_1 = E_{kt}$ and $z_2 = E_{ks} - iq_m$.

***Solution***

$$Res\,[g(z_1)] = \left[ \frac{\alpha(\boldsymbol{k})\,(ik_n - E_{kt})}{\left(ik_n - (E_{ks} - iq_m)\right)(ik_n - E_{kt})} \right]_{ik_n = z_1} = \frac{\alpha(\boldsymbol{k})}{\left(E_{kt} - (E_{ks} - iq_m)\right)}$$

$$= \frac{\alpha(\boldsymbol{k})}{iq_m - E_{ks} + E_{kt}}$$

$$Res\,[g(z_2)] = \left[ \frac{\alpha(\boldsymbol{k})\left(ik_n - (E_{ks} - iq_m)\right)}{\left(ik_n - (E_{ks} - iq_m)\right)(ik_n - E_{kt})} \right]_{ik_n = z_2} = \frac{\alpha(\boldsymbol{k})}{(E_{ks} - iq_m) - E_{kt}}$$

$$= \frac{-\alpha(\boldsymbol{k})}{iq_m - E_{ks} + E_{kt}}$$

---

With careful summation, one can show that:

$$\sigma_{xy} = \lim_{\substack{E\to 0\\ \eta\to 0}} \frac{i}{E} Q_{xy}(iq_m \to E + i\eta) = \frac{-i}{A} \sum_{s,t\in\{\pm\}} \sum_{k} \frac{\alpha(\boldsymbol{k})\,(f_s - f_t)}{\left(E_s(k) - E_t(k)\right)^2}$$
$$= \frac{-i}{A}\sum_k \frac{r(k)}{\left(E_-(k) - E_+(k)\right)^2}(f_- - f_+) = \frac{-i}{A}\sum_k \frac{r(k)}{4g^2b^2}(f_- - f_+) \qquad [7.36]$$

where $r(k) = TrJ_x(\boldsymbol{k})J_y(\boldsymbol{k})\left\{\frac{1}{2}(I + \boldsymbol{\sigma}\cdot\boldsymbol{b})\frac{1}{2}(I - \boldsymbol{\sigma}\cdot\boldsymbol{b}) - \frac{1}{2}(I - \boldsymbol{\sigma}\cdot\boldsymbol{b})\frac{1}{2}(I + \boldsymbol{\sigma}\cdot\boldsymbol{b})\right\}$. Finally, the Hall conductivity can be shown to be:

$$\sigma_{xy} = \frac{i}{4g^2b^2A}\iint \frac{4}{4\pi^2} r(k)\,(f_- - f_+)dk_x\, dk_y$$
$$= \frac{-1}{8\pi^2}\iint \frac{\boldsymbol{b}}{b^3}\cdot\left(\frac{\partial \boldsymbol{b}}{\partial k_y}\times\frac{\partial \boldsymbol{b}}{\partial k_x}\right)(f_- - f_+)\, dk_x dk_y$$
$$= \frac{-1}{8\pi^2}\iint \Omega_z (f_- - f_+)\, dk_x\, dk_y \qquad [7.37]$$

noting that $b^2 = 1$. It is also worth noting that $\Omega_z$ is the gauge curvature in momentum space. The physical significance of the above would be clear if one refers to our analysis in previous sections of how the gauge curvature arises in B space and its conversion to K space. Here, with the use of Kubo formula, the momentum space gauge curvature arises naturally, supporting the earlier prediction that gauge curvature in momentum space is closely related to electron dynamics. With proper choice of the Fermi surface, e.g. in the region where $(f_- - f_+) = 1$, the Hall conductivity above would be independent of small fluctuation in Fermi distribution, as well as inhomogeneities in spin orbit constant and other material parameters. Such conductivity is said to possess topological property. Topological conductivity has great significance in nanoelectronics because the electric current and voltage of such system will be stable against the device's geometrical non-uniformity, inhomogeneous film property due to mass fabrication tolerance and impurities scattering.

In Chapter 5 we deduced that in some systems the spin Hall conductivities are universal constant. For example, the conductivity of a Rashba 2DEG and a Rashba heavy hole system is not a function of any material parameter (e.g. size, film property). Conductivity shows exact quantized values. The relation between the methods used to derive the expressions in Table 7.5 may have a subtle relation to the method described above. It is believed that, in the case of Rashba, SHE conductivity is due to spin polarization of $p_y$ by a time-dependent field but not the gauge curvature, as gauge curvature in a Rashba system vanishes except at the origin in momentum space. Thus in the annular region, there is no contribution from the gauge curvature.

*Table 7.5* SHE in semiconductor spintronics and their respective universal conductivity

| | Semiconductor system | Spin Hall conductivity |
|---|---|---|
| 1. | Rashba 2DEG | $\sigma_{xy}^{z} = \frac{-e}{8\pi}$ |
| 2. | Luttinger (p-doped bulk) | $\sigma_{xy}^{z} = \frac{e\left(3k_F^H - k_F^L\right)}{12\pi^2}$ (Note: 'not universal' because of its $k_F$ dependence; H and L stand for heavy and light holes, respectively) |
| 3. | Cubic Dresselhaus (n-doped bulk) | $\sigma_{xy}^{z} = \frac{-ek_F}{12\pi^2}$ (Note: 'not universal' because of its $k_F$ dependence) |
| 4. | Rashba heavy holes | $\sigma_{xy}^{z} = \frac{9e}{8\pi}$ |

## 7.5.2 Topological anomalous Hall effects (local space)

We have provided a rather extensive discussion on gauge potential, curvature and Berry's phase in momentum space for numerous systems, e.g. spintronics, graphene, topological insulators. In the previous section, we showed the exact quantification of the conductivity which shows universal values under some conditions. In this section, we discuss a similar phenomenon but in real space.

One can reason that the Hall conductivity of the system is proportional to the average magnetic field felt by electron traveling in the local magnetic systems and thus:

$$\sigma_{xy} = \frac{1}{A}\iint \Omega_z^{rs}\, dx\, dy. \qquad [7.38]$$

One starts to imagine that, unlike the momentum space system where distribution of $\boldsymbol{b}(\boldsymbol{k})$ was predetermined by nature, in the local magnetic system it is possible to engineer distribution of $\boldsymbol{b}(\boldsymbol{r})$ or $(\boldsymbol{r})$. In full dimension, the curvature in a local magnetic system is $\Omega_z^{rs} = \frac{-\hbar}{2e}\boldsymbol{n}.(\partial_x \boldsymbol{n} \times \partial_y \boldsymbol{n})$. One can write the Hall conductivity as follows:

$$\sigma_{xy} = \frac{-1}{4\pi}\frac{\hbar}{eA}\iint \boldsymbol{n}.(\partial_x \boldsymbol{n} \times \partial_y \boldsymbol{n})\, dx \wedge dy \qquad [7.39a]$$

$$\sigma_{xy} = \frac{-1}{4\pi}\frac{\hbar}{eA}\iint \boldsymbol{n}.(\partial_r \boldsymbol{n} \times \partial_w \boldsymbol{n})\, dr \wedge dw \qquad [7.39b]$$

where $n = (\sin\theta\cos\phi, \sin\theta\sin\phi, \cos\theta)$. Instead of the $(x,y)$ coordinates, one uses the $(r,w)$ coordinates to track the magnetic moment where $x = r\sin w$ and $y = r\cos w$. On the other hand, $(\theta, \phi)$ are the spin coordinates of the local moment in spin or magnetic space.

**Exercise 7.6**

Show that:

(1) $\partial_r \boldsymbol{n} \times \partial_w \boldsymbol{n} = \boldsymbol{n} \sin\theta\,(\partial_r \theta\,\partial_w \phi - \partial_w \theta\,\partial_r \phi)$.

(2) $(\partial_r \theta\,\partial_w \phi - \partial_w \theta\,\partial_r \phi)\,\mathrm{d}r \wedge \mathrm{d}w = d\theta \wedge d\phi$.

Thus the Hall conductivity is:

$$\sigma_{xy} = \frac{-1}{4\pi}\left(\frac{h}{eA}\right)\iint \sin\theta\,(\partial_r \theta\,\partial_w \phi - \partial_w \theta\,\partial_r \phi)\,dr \wedge dw = \frac{-1}{4\pi}\left(\frac{h}{eA}\right)\iint \sin\theta\,d\theta \wedge d\phi. \quad [7.40]$$

The above shows that the Hall conductivity is topological in nature as it depends only on the solid angle covered by the range of mapping $n$: $R^2 \rightarrow S^2$. It measures the number of times the mapping wraps around the 2D unite sphere of $S^2$ as the electron traverses the 2D local magnetic system.

## 7.6 Spin torque induced by spin orbit coupling

Spin transport theories, including the phenomenological continuity of spin flux as well as microscopic approaches, have been developed to study CIMS and spin torque oscillations in a variety of structures[21–24] ranging from ferromagnetic spin valves and other nanostructures to Coulomb blockade transistors. This has led to renewed interest in the dynamics of local moment beyond the conventional Landau–Lifshitz–Gilbert (LLG) description. In this section, we describe theoretically an additional source of spin torque and local spin oscillations due to spin orbit coupling. Particular attention is given to Rashba spin orbit coupling (RSOC), which is gaining prominence in non-magnetic metals[25] and ferromagnetic metals,[26] as well as rare earth materials.[27] We show in this section that RSOC in ferromagnetic materials may lead to a modified LLG equation which physically leads to charge current induced spin transfer switching.[28–31]

In this section, we make use of the spin orbital gauge deployed in Chapter 5 to study SHE in Rashba 2DEG. In the presence of a local moment with smooth spatial variation, a separate local gauge transformation is required to describe the adiabatic alignment of electron spin along the moment texture. This transformation process aligns reference spin axis to the local moment, resulting in the SOC related gauge term of $\alpha\ U\ E_i \sigma_j \varepsilon_{ij\mu} U^\dagger$ and the chiral gauge of $\frac{-i\hbar}{e} U\partial_\mu U^\dagger$. The transformed Hamiltonian thus becomes:

$$H = \sum_{\mu}^{3} \frac{1}{2m}\left(p_\mu + e\left[\alpha\,U\,E_i \sigma_j \varepsilon_{ij\mu} U^\dagger - \frac{i\hbar}{e} U\partial_\mu U^\dagger\right]\right)^2 + \frac{eg\hbar}{4m}\sigma_z^r \left|M_z^r\right| \quad [7.41]$$

where in comparison to vacuum SOC, $\alpha$ simply takes on $\frac{\hbar}{4mc^2} = 3.2\times 10^{-22}\,s$. In the Rashba system, the material-dependent constant of $\langle \alpha E\rangle$ is determined from the relation of $e\langle \alpha E\rangle = m/\hbar\alpha_R$, where theoretical and experimental values of $\alpha_R$ for various material systems can be found in literatures. Note that superscript $r$ indicates the new reference frame. Physically, the $e\alpha\ UE_i\sigma_j\varepsilon_{ij\mu}U^\dagger$ term describes the momentum of an electron whose spin is aligned along the spin texture in the presence of the relativistic magnetic field due to SOC. The last term containing $\sigma_z^r|M_z^r|$ represents the Zeeman energy due to electron spin aligned along the local magnetization in the rotated frame where the local magnetization is the $z$ axis. Alignment of spin to the local moment in the present of SOC in ferromagnetic material with Rashba SOC is supported by experimental measurements.[26] Note that the line integral of $A_\mu^{Ch} = -\frac{i\hbar}{e}U\partial_\mu U^\dagger$ over the electron's trajectory is generally non-vanishing. It represents the phase acquired by the electron as it passes through the local magnetization and is related to the net chirality of the local magnetization under the Abelian (adiabatic) approximation, $A_\mu^{Ch} = \sigma_z a_\mu^{Ch}$, where $a_\mu^{Ch}$ is the top left diagonal element of the gauge field matrix $A_\mu^{Ch}$. Explicitly:

$$a_\mu^{Ch} = -\frac{i\hbar}{e}\left(\frac{1-\cos\theta}{2}\right)\frac{\partial\phi}{\partial \boldsymbol{n}}\cdot\frac{\partial \boldsymbol{n}}{\partial r_\mu} = a_{mon}^{v}(n)\partial_\mu n_v \qquad [7.42]$$

where it can be shown that $a_{mon}^{y}$ is a magnetic monopole in the space of local spin. The spin orbit potential is:

$$\boldsymbol{A}^{SO} = \alpha\sigma_r^z[(n_y E_z - n_z E_y)\,\boldsymbol{i} + (n_z E_x - n_x E_z)\boldsymbol{j} + (n_x E_y - n_y\ E_x)\,\boldsymbol{k}]. \quad [7.43]$$

The requisite conditions for the adiabatic approximation are:

1. An electron motion which is sufficiently slow to allow its spin to relax and align along the local spin texture. This condition can be met as the drift velocity in the magnetic system is fairly low.
2. The diagonal components of $A_\mu = A_\mu^{SO} + A_\mu^{Ch} = \alpha\ U\ E_i\sigma_j\varepsilon_{ij\mu}U^\dagger - \frac{i\hbar}{e}U\partial_\mu U^\dagger$ is sufficiently large in magnitude to ensure a large energy difference between the spin up and down eigenstates, and hence little mixing between the two eigenstates (i.e. relatively small non-diagonal components). This can be satisfied in systems with moderately strong magnetization.

In the adiabatic system where spin is constantly aligned to the local field, there is no probability of the spin assuming its other eigenstate. One can apply a continuous unitary transformation to the Hamiltonian such that the spin reference axis ($z$) in the rotated frame coincides with the local $b$ field direction. From the standpoint of electrodynamics, the gauge potentials $A_\mu^{SO}$ and $A_\mu^{Ch}$ due to SOC and chiral spin texture, respectively, give rise to an electromagnetic interaction between the current and the local spin, and results in an interaction energy density term:

$$\mathbb{E}_{int} = \langle\varphi_r|j_\mu A_\mu|\varphi_r\rangle$$
$$= \alpha\left[j_x\left(n_y E_z - n_z E_y + \frac{a_x^{Ch}}{\alpha}\right) + j_y\left(n_z E_x - n_x E_z + \frac{a_y^{Ch}}{\alpha}\right) + j_z\left(n_x E_y - n_y E_x + \frac{a_z^{Ch}}{\alpha}\right)\right] \quad [7.44]$$

where $a_\mu = a^v_{mon}\,\partial_\mu n_v, j_\mu$ is the charge current density and $A_\mu = A^S_{\mu O} + A^{Ch}_\mu$. It is worth noting that $\mathbb{E}_{int} = \langle \varphi_r | j_\mu A_\mu | \varphi_r \rangle$ implies carrying out $\mathbb{E}_{int} = \int \psi^\dagger \boldsymbol{A}.\nabla\, \psi - (\nabla\psi^\dagger).\, \boldsymbol{A}\, \psi\, d^3x$ in the adiabatic limit, i.e. considering spin aligned to moment only. In the presence of SOC, the gauge and hence the total anomalous velocity admits an additional term to become $A_\mu = \alpha\, U\, E_i \sigma_j \varepsilon_{ij\mu} U^\dagger - \frac{i\hbar}{e} U \partial_\mu U^\dagger$.

---

**Exercise 7.7**

Prove the following identities:

(1) $\varepsilon_{mkl}\,\varepsilon_{qnl} = \delta_{mq}\,\delta_{kn} - \delta_{mn}\,\delta_{kq}$

(2) $\delta(g(x)) = \sum_{a, g(a)=0,\, g'(a)\neq 0} \dfrac{\delta(x-a)}{|g'(a)|}$

The standard expression of the curvature of a field is:

$$\Omega_{\mu v} = \partial_\mu A_v - \partial_v A_\mu - \frac{ie}{\hbar}[A_\mu, A_v]$$

Show that the above is reduced to $-\frac{i\hbar}{e} U[\partial_\mu, \partial_v] U^\dagger$, and thus for any integrable function represented by $U$, this curvature term vanishes.

***Solution***

Make use of $\partial_\mu (U^\dagger U) = U^\dagger \partial_\mu U + (\partial_\mu U^\dagger) U$.

---

Since the gauge field derives its form from the electron spin relaxation to the local moment texture in the presence of SOC, the corresponding equation of motion (EOM) thus describes the above effect on the dynamics of the local magnetic moment. Generally, the local moment will adjust its orientation in order to achieve minimum energy. Thus, one would expect the local moment dynamics to be governed by the energy gradient with respect to a change in the local moment orientation. As in conventional LLG physics, we can regard the equilibrium state as one in which local moments are aligned along an effective magnetic field, which is given by the energy gradient with respect to the local moment, i.e.

$$\boldsymbol{H} = \frac{1}{\mu_0 M}\left[ j_x \left( \alpha \frac{\partial n_\mu}{\partial \boldsymbol{n}} E_v \varepsilon_{x\mu v} + \frac{\partial a^{Ch}_x}{\partial \boldsymbol{n}} \right) + j_y \left( \alpha \frac{\partial n_\mu}{\partial \boldsymbol{n}} E_v \varepsilon_{y\mu v} + \frac{\partial a^{Ch}_y}{\partial \boldsymbol{n}} \right) + j_z \left( \alpha \frac{\partial n_\mu}{\partial \boldsymbol{n}} E_v \varepsilon_{z\mu v} + \frac{\partial a^{Ch}_z}{\partial \boldsymbol{n}} \right) \right] \qquad [7.45]$$

where $\boldsymbol{n} = \boldsymbol{M}/M$ and $\mu_0 = 4\pi \times 10^{-7}$ T mA$^{-1}$. Note that the derivation applied to the gauge potential due to chiral spin texture is a functional derivative. By inspection the above shows cyclical nature and can be summarized as follows.

**Exercise 7.8**

Show by $\boldsymbol{H} = \frac{1}{\mu_0 M}\frac{\partial E_{\text{int}}}{\partial \boldsymbol{n}}$ the results of Eq. 7.45.

***Solution***

$$\boldsymbol{H} = \frac{1}{\mu_0 M}\frac{\partial \mathbb{E}int}{\partial \boldsymbol{n}}$$

$$= \frac{\alpha}{\mu_0 M}\left[j_x\left(\frac{\partial n_y}{\partial n_x}E_z - \frac{\partial n_z}{\partial n_x}E_y + \frac{1}{\alpha}\frac{\partial a_x^{Ch}}{\partial n_x}\right) + j_y\left(\frac{\partial n_z}{\partial n_x}E_x - \frac{\partial n_x}{\partial n_x}E_z + \frac{1}{\alpha}\frac{\partial a_y^{Ch}}{\partial n_x}\right)\right.$$

$$\left. + j_z\left(\frac{\partial n_x}{\partial n_x}E_y - \frac{\partial n_y}{\partial n_x}E_x + \frac{1}{\alpha}\frac{\partial a_z^{Ch}}{\partial n_x}\right)\right]\boldsymbol{i}$$

$$+ \frac{\alpha}{\mu_0 M}\left[j_x\left(\frac{\partial n_y}{\partial n_y}E_z - \frac{\partial n_z}{\partial n_y}E_y + \frac{1}{\alpha}\frac{\partial a_x^{Ch}}{\partial n_y}\right) + j_y\left(\frac{\partial n_z}{\partial n_y}E_x - \frac{\partial n_x}{\partial n_y}E_z + \frac{1}{\alpha}\frac{\partial a_y^{Ch}}{\partial n_y}\right)\right.$$

$$\left. + j_z\left(\frac{\partial n_x}{\partial n_y}E_y - \frac{\partial n_y}{\partial n_y}E_x + \frac{1}{\alpha}\frac{\partial a_z^{Ch}}{\partial n_y}\right)\right]\boldsymbol{j}$$

$$+ \frac{\alpha}{\mu_0 M}\left[j_x\left(\frac{\partial n_y}{\partial n_z}E_z - \frac{\partial n_z}{\partial n_z}E_y + \frac{1}{\alpha}\frac{\partial a_x^{Ch}}{\partial n_z}\right) + j_y\left(\frac{\partial n_z}{\partial n_z}E_x - \frac{\partial n_x}{\partial n_z}E_z + \frac{1}{\alpha}\frac{\partial a_y^{Ch}}{\partial n_z}\right)\right.$$

$$\left. + j_z\left(\frac{\partial n_x}{\partial n_z}E_y - \frac{\partial n_y}{\partial n_z}E_x + \frac{1}{\alpha}\frac{\partial a_z^{Ch}}{\partial n_z}\right)\right]\boldsymbol{k}.$$

If we consider the low-damping limit, the local moment will precess about the effective field, so that the general EOM can be written as $\frac{d\boldsymbol{M}}{dt} = \gamma \boldsymbol{M} \times \boldsymbol{H}$ where $\gamma$ is the gyromagnetic ratio (in units of $A^{-1}\ s^{-1}$). In fact, the precessional motion is analogous to the dynamical expression of $\frac{d\boldsymbol{S}}{dt} = \boldsymbol{S} \times \boldsymbol{H}$, for a spin $\boldsymbol{S}$ in the presence of an effective magnetic field $\boldsymbol{H}$, which arises due to the non-commutative spin algebra of $[\sigma_\mu, \sigma_\nu] = 2i\sigma_\kappa\, \varepsilon_{\mu\nu\kappa}$. In the classical description (of the micromagnetic method) where the spin operators $\sigma$ are replaced by magnetic moment vectors $\boldsymbol{M}$, such non-commutativity disappears. One can, however, use the well-based phenomenological approach to write:

$$\frac{d\boldsymbol{M}}{dt} = \frac{\gamma}{\mu_0}\boldsymbol{n} \times \boldsymbol{H} = \frac{\gamma}{\mu_0}\boldsymbol{n} \times \left(j_\lambda\left[\alpha\frac{\partial n_\mu}{\partial \boldsymbol{n}}E_\nu \varepsilon_{\lambda\mu\nu} + \frac{\partial A_\lambda}{\partial \boldsymbol{n}}\right]\right). \qquad [7.46]$$

To analyze the contribution from the chiral gauge component, one recalls $a_\mu^{Ch} = a_{mon}^\nu\, \partial_\mu\, n_\nu$ and notes the relations of $\varepsilon_{\alpha\beta\gamma}\frac{\partial a_{mon}^\beta}{\partial n_\gamma} = 2n_\alpha$, which merely reiterates that

$a_{mon}$ $(n)$ is a magnetic monopole in the space of the local magnetization as described earlier. Taking note of the fact that $\frac{\partial a_{\mu}^{Ch}}{\partial \boldsymbol{n}} = \frac{\partial a_{mon}^{v}}{\partial \boldsymbol{n}}(\partial_{\mu} n_{v})$, the EOM results in:

$$\frac{d\boldsymbol{M}}{dt} = \frac{\gamma}{\mu_0}\boldsymbol{n}\times j_{\lambda}\left(\alpha\frac{\partial n_{\mu}}{\partial \boldsymbol{n}}E_{v}\varepsilon_{\lambda\mu v}\right.$$

$$\left.-\frac{i\hbar}{e}\left[(n_y\partial_{\lambda}n_z - n_z\partial_{\lambda}n_y)\boldsymbol{i} + (n_z\partial_{\lambda}n_x - n_x\partial_{\lambda}n_z)\boldsymbol{j} + (n_x\partial_{\lambda}n_y - n_y\partial_{\lambda}n_x)\boldsymbol{k}\right]\right)$$

$$= \frac{\gamma}{\mu_0}\boldsymbol{n}\times j_{\lambda}\left(\alpha\frac{\partial n_{\mu}}{\partial \boldsymbol{n}}E_{v}\varepsilon_{\lambda\mu v} - \frac{i\hbar}{e}\boldsymbol{n}\times\partial_{\lambda}\boldsymbol{n}\right). \quad [7.47]$$

---

**Exercise 7.9**

Show that the EOM due to both SOC and local spin texture can be written in a more compact form of:

$$\frac{d\boldsymbol{M}}{dt} = \frac{\gamma}{\mu_0}j_{\lambda}n_A\left(\alpha E_{v}\varepsilon_{\lambda Bv} - \frac{i\hbar}{e}n_r\partial_{\lambda}n_s\varepsilon_{rsB}\right)e_C\,\varepsilon_{ABC}.$$

***Solution***

$$\frac{d\boldsymbol{M}}{dt} = \frac{\gamma}{\mu_0}j_{\lambda}n_A\left(\alpha\frac{\partial n_{\mu}}{\partial n_B}E_{v}\varepsilon_{\lambda\mu v} - \frac{i\hbar}{e}n_r\partial_{\lambda}n_s\varepsilon_{rsB}\right)e_C\,\varepsilon_{ABC}$$

where taking $\frac{\partial n_{\mu}}{\partial n_B} = \delta_{\mu B}$, yields the final form of the EOM.

---

The modified LLG equation in a continuously magnetic medium, taking into account the effect of spin-polarized current and SOC, is given by:

$$\frac{d\boldsymbol{M}}{dt} = \gamma(\boldsymbol{M}_f\times\boldsymbol{H}) + \alpha_d\boldsymbol{M}_f\times(\boldsymbol{M}_f\times\boldsymbol{H})$$

$$+ j_{\lambda}\left[a_C\boldsymbol{M}_f\times\left(\boldsymbol{M}_f\times\frac{\partial\boldsymbol{M}_f}{\partial\lambda}\right) + b_C\left(\boldsymbol{M}_f\times\frac{\partial\boldsymbol{M}_f}{\partial\lambda}\right) + d_C\boldsymbol{M}_f\times\frac{\partial}{\partial\boldsymbol{M}_f}M_{\mu}E_{v}\varepsilon_{\lambda\mu v}\right] \quad [7.48]$$

where $\alpha_d$ is the damping constant, $\alpha_C$, $b_C$ are the usual spin torque constants in a magnetic system (e.g. domain wall or magnetic medium with chiral local moment texture) and subscript '$C$' denotes a medium with continuous local spin variation. For comparison, the corresponding LLG equation for a discrete magnetic medium, in which current acquires the spin polarization of one medium (i.e. $\boldsymbol{M}_p$) and passes on to an adjacent medium (i.e. $\boldsymbol{M}_f$), is as follows:

$$\frac{d\boldsymbol{M}}{dt} = \gamma(\boldsymbol{M}_f \times \boldsymbol{H}) + \alpha_d \boldsymbol{M}_f \times (\boldsymbol{M}_f \times \boldsymbol{H}) + \left[ a_D \boldsymbol{M}_f \times (\boldsymbol{M}_f \times \boldsymbol{M}_p) + b_D(\boldsymbol{M}_f \times \boldsymbol{M}_p) + d_C \boldsymbol{M}_f \times \frac{\partial}{\partial \boldsymbol{M}_f} j_\lambda M_\mu E_\nu \varepsilon_{\lambda\mu\nu} \right] \quad [7.49]$$

where $a_D$ and $b_D$ are the corresponding spin torque constants and subscript 'D' denotes a medium with discrete local moment variation.

Physically, discrete magnetic systems include magnetic multilayers which are widely used in technologically important devices, e.g. spin valve and magnetic tunnel junction. In such multilayers, there are discrete changes in the magnetization direction across the interfaces. Typically, in such a system, the fixed magnetization layer $\boldsymbol{M}_p$ is used to generate spin current which exerts a spin torque on the free magnetization layer $\boldsymbol{M}_f$. It is important to note that, notwithstanding the different forms of the LLG equations in the continuous and discrete systems, the influence of spin orbit coupling on local spin dynamics occurs in both systems. It is worth noting in the discrete multilayer system that the spin orbit-induced local spin dynamics does not depend on the fixed magnetization of the adjacent layer. In short, spin injection is not needed to effect spin transfer in the free magnetization layer; thus, from the technological point of view, spin dynamics in magnetic system with SOC is suitable for single layer CIMS.

Figure 7.3a is a schematic diagram of a magnetic multilayer structure used conventionally in devices such as the spin valve recording head of a hard disk drive and the magnetic tunnel junction of magnetic random access memory (MRAM), while Fig. 7.3b illustrates a diluted magnetic semiconductor (DMS) in the form of a 2DEG structure of a III–V high-electron-mobility transistor (HEMT) device with large Rashba spin orbit coupling.

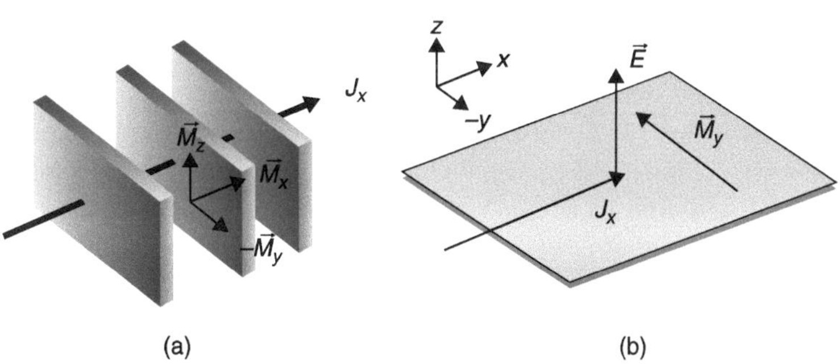


*7.3* (a) Magnetic multilayer with in-plane ($M_y$, $M_z$) or out-of-plane ($M_x$) local spin, and current passing in a perpendicular-to-plane direction through the multilayer. (b) 2DEG of a HEMT made of DMS material with in-plane local spin ($M_y$) and current passing in-plane along the *x* direction. The electric field due to structural inversion asymmetry aligns along the *z* direction, resulting in Rashba SOC.

In the above systems, the $\boldsymbol{H}$ field due to spin orbit coupling and $j_x$ is:

$$\boldsymbol{H}^{SO} = \frac{\alpha j_x}{\mu_0 M}\left[\frac{\partial}{\partial n_x}\boldsymbol{i} + \frac{\partial}{\partial n_y}\boldsymbol{j} + \frac{\partial}{\partial n_z}\boldsymbol{k}\right](n_y E_z - n_z E_y) \qquad [7.50]$$

which simplifies to $H_y^{SO} = \frac{j_x(\alpha E_z)}{\mu_0 M}$ and $H_z^{SO} = \frac{-j_x\langle \alpha E_y\rangle}{\mu_0 M}$. The spin orbit induced $\boldsymbol{H}^{so}$ can be treated like the externally applied magnetic field that might switch the local moment of a magnetic medium.

We will compare the strength of $\boldsymbol{H}^{so}$ for various SOC systems with fixed values of $j_x = 10^{11}$ A m$^{-2}$ and $M = 5 \times 10^{13}$ A m$^{-1}$. For a typical InAs/InGaAs Rashba 2DEG with $\alpha_R = 10^{-12}$ eVm, $k_F = 8 \times 10^6$ m$^{-1}$, the vacuum equivalent electric field can be worked out to be approximately $10^{13}$ V m$^{-1}$. The bare electron mass $m_0$ is considered in vacuum, but in 2DEG, the effective mass is $m_e = 0.05m_0$. The SOC energy of a specific material system is $\frac{\hbar e}{m}\langle \alpha E_z\rangle k_F = \alpha_R k_F$, and one works out $\langle \alpha E_z\rangle = 4.3 \times 10^{-10}$ T m, which, in turn gives $H_y^{SO} = \frac{j_x(\alpha E_z)}{\mu_0 M} = 6.5\times 10^3$ A m$^{-1}$. The switching fields for common ferromagnetic metals are given in Table 7.6.

Additionally, in some rare earth ferromagnetic metals (e.g. Gd), where the Rashba constant could be as high as $\alpha_R = 10^{-10}$ eVm, one finds that correspondingly $\langle \alpha E_z\rangle = 8.6 \times 10^{-7}$ T m. The corresponding $\boldsymbol{H}^{so} = 1.3 \times 10^7$ A m$^{-1}$ is large. Thus, for magnetic devices made from these rare earth elements, it may be possible to utilize the high $\boldsymbol{H}^{so}$ to achieve CIMS at a much reduced threshold current density. Table 7.7 gives a summary of the numerical estimates.

It is worth noting that the effect of spin transfer switching due to SOC elucidated here is in addition to the other previously described sources of spin transfer torque such as the $s$–$d$ coupling between local spin and the spin-polarized conduction electrons. A low magnetization for the local spin generates a higher $\boldsymbol{H}^{so}$. This is because the interaction energy is a function of the magnetization direction but not its magnitude. Thus, for a given current density, the resulting field $\boldsymbol{H}^{so}$ will be more efficient in switching a magnetic material with a lower magnetization.

We have modified the phenomenological LLG equation by incorporating the effect of spin orbit coupling. We obtained a general EOM for the local spin by formally deriving the SU(2) spin orbit gauge field arising due to spin orbit coupling, in addition to the previously derived topological $U(1) \otimes U(1)$ chiral gauge due to adiabatic relaxation of electron spin to local spin texture. The full

*Table 7.6* Switching fields for a few common ferromagnetic materials

| Fe | Ni | Co |
|---|---|---|
| 565 Oe<br>45×10$^3$ A m$^{-1}$ | 233 Oe<br>18.5×10$^3$ A m$^{-1}$ | 7429 Oe<br>591×10$^3$ A m$^{-1}$ |

*Table 7.7* Numerical estimates of the effective field ***H*** from the SOC effect and the spin precession frequency about the effective field for different SOC sources

| | External electric field (vacuum) | 2DEG with Rashba SOC $\alpha_R = 10^{-12}$ eVm | 2DEG with Rashba SOC $\alpha_R = 10^{-10}$ eVm |
|---|---|---|---|
| $\langle \alpha E \rangle = \frac{m}{\hbar e}\alpha_R (Tm)$ | $3.2 \times 10^{-15}$ $m = m_0$ | $4.3 \times 10^{-10}$ $m = 0.05m_0$ | $8.6 \times 10^{-7}$ $m = m_0$ |
| Vacuum equivalent electric field ($Vm^{-1}$) $m = m_0$ $\alpha = 3.2 \times 10^{-22}$ s | $10^7$ | $2.7 \times 10^{13}$ | $2.7 \times 10^{15}$ |
| Effective switching field $H^{SO}\left(\frac{A}{m}\right)$ | 0.051 | $6.5 \times 10^3$ | $1.3 \times 10^7$ |
| Simple relation between vacuum SOC and Rashba SOC | | $\frac{\hbar e}{4m^2c^2} pE = \alpha_R k \to E = \frac{m}{\alpha \hbar e}\alpha_R$ where $\alpha = \frac{\hbar}{4mc^2}$ | |

Note: Current density is $j_x = 10^{11}$ A $m^{-2}$ and the magnetization of local spin is $M = 5 \times 10^3$ A $m^{-1}$.

LLG equations were presented for both cases of continuous and discrete spatial variation of the local spin texture. The EOM in the LLG form will be applicable for micromagnetic or spin transport simulation across domain walls and in multilayer magnetic devices. We have also discussed the implications of our results in various spintronic applications; for example, SOC contribution to the spin torque will have direct relevance to CIMS operation in terms of the critical current density for switching or magnetization noise. In addition, the SOC has a substantial effect on the local spin trajectory and resonant frequency, which can be readily modified by voltage or current biasing. This tunability is particularly relevant to the generation of microwave oscillation in nanomagnetic devices.

## 7.7 Dirac string and monopole properties

To determine the vector potential for a regular monopole field is not a trivial task. In fact a single vector potential function which is regular everywhere on the $S_2$ manifold probably does not exist. We would like to remind readers that a $(\nabla_{\mathbf{B}} \times \boldsymbol{A}^{\boldsymbol{B}})$ based on one gauge expression (either north or south pole) will necessarily yield a vanishing surface integral over the entire $S^2$ manifold, due to the Dirac string. Whereas in modern understanding where $(\nabla_{\mathbf{B}} \times \boldsymbol{A}^{\boldsymbol{B}})$ is based on at least two gauge expressions, the curvature is regular everywhere even though the individual gauge expression is not, and the surface integral of such regular curvature will be a non-vanishing quantity as the string can be avoided. Here we will show using the south pole gauge that the Dirac magnetic field has a string. The vector form of the Dirac potential is:

$$\boldsymbol{A}^S = \frac{\hbar\,(\boldsymbol{b}\times\boldsymbol{k})}{2eb\,(b-\boldsymbol{b}.\,\boldsymbol{k})} = \frac{\hbar\, b_y\boldsymbol{i} - b_x\boldsymbol{j}}{2eb\,(b-b_z)}. \qquad [7.51]$$

At first sight, the curvature of the above potential may appear to be $\left(\nabla_{\mathbf{B}}\times\boldsymbol{A}^S\right) = \frac{\hbar b}{2eb^3}$. However, the expression is incomplete because it does not fully capture the fact that the Dirac potential is not a regular function; it is singular along $\theta = 0$ but regular along $\theta = \pi$. A proper approach is to first regularize the Dirac potential by inserting a $\epsilon^2$ to R so that $\theta = 0$ can be negotiated, and derive the regularized magnetic field. The regularized potential is:

$$\boldsymbol{A}^S = \frac{\hbar\,(\boldsymbol{b}\times\boldsymbol{k})}{2e\,R\,(R-\boldsymbol{b}.\,\boldsymbol{k})} \qquad [7.52]$$

where $R^2 = b^2_x + b^2_y + b^2_z + \epsilon^2$. The regularized magnetic field can be shown to be:

$$\boldsymbol{\Omega}(b,\in) = \frac{\hbar}{2e}\frac{\boldsymbol{b}}{R^3} - \frac{\hbar}{2e}\left(\frac{\in^2}{R^2(R-b_z)^2} + \frac{\in^2}{R^3(R-b_z)}\right)\boldsymbol{n}_k. \qquad [7.53]$$

Lifting regularization, which simply means presenting a modified expression that resides in both the regular and the singular regions, one obtains:

$$\lim_{\in\to 0}\boldsymbol{\Omega}\,(b,\in) = \frac{\hbar}{2e}\frac{\boldsymbol{b}}{b^3} - \frac{\hbar}{2e}4\pi\theta(b_z)\,\delta(b_x)\,\delta(b_y). \qquad [7.54]$$

Equation 7.54 consists of two parts, the regular magnetic monopole and the singular Dirac string. The curvature of Eq. 7.54 vanishes as $\oint\boldsymbol{\Omega}\cdot\boldsymbol{dS} = \oint\boldsymbol{\Omega}_r\,.\,\boldsymbol{dS} + \oint\boldsymbol{\Omega}_{string}\cdot\boldsymbol{dS} = 4\pi g - 4\pi g = 0$, noting here that the monopole has been separated into the regular and the string parts. This clearly shows that the south pole Dirac potential used above cannot be a correct representation of the vector potential for a magnetic monopole.

It is thus necessary to find a correct representation and this task was completed with the modern theory of fiber bundle. In this treatment, it is important to understand that there should exist at least two gauge expressions on the $S^2$ manifold in order to provide a non-vanishing $\oint(\nabla_{\mathbf{B}}\times\boldsymbol{A}^B).\boldsymbol{dS}$ over the manifold. In fact by Stokes' theorem, one sees that the surface integral of the monopole field is the line integral of the gauge field. To illustrate this more clearly, one resorts to the differential form which is related to the vector quantity as follows:

$$\tilde{A} = \frac{\hbar}{2e}(1-\cos\theta)\left(\frac{\partial\phi}{\partial\boldsymbol{b}}\right)\cdot\boldsymbol{db} = \frac{\hbar}{2e}(1-\cos\theta)d\phi \qquad [7.55]$$

$$d\tilde{A} = (\nabla\times\boldsymbol{A}).\boldsymbol{dS}. \qquad [7.56]$$

In Eq. 7.55, $\tilde{A}$ is expressed as a 1-form. Performing an exterior differentiation of $\tilde{A}$ following standard definitions in differential geometry will lead naturally to Eq. 7.56, which is the dot areal product of the curvature. $b$ is the parameter of an arbitrary space (e.g. the B space) and $d\boldsymbol{S}$ is in the same space too. In fact Eq. 7.56 is neither unique nor is it a complete form of the gauge potential. The gauge should be expressed as one of Eq 7.57, depending on the chart which covers the $S_b^2$ surface on which $\tilde{A}$ is defined, where $b$ is the radius of $S_b^2$. One example set of two charts that overlap but completely cover the $S_b^2$ manifold is:

$$\boldsymbol{A}^S = -\frac{\hbar}{2e}(1+\cos\theta)d\phi \quad \boldsymbol{A}^{\mathcal{N}} = \frac{\hbar}{2e}(1-\cos\theta)d\phi \qquad [7.57]$$

$dA^{\mathcal{N}}$ is defined everywhere on $S^2$ except the $-z$ axis, while $dA^S$ is defined everywhere except the $+z$ axis. Equation 7.57 is the differential form of the Dirac gauge potential in the spherical polar coordinates. For comparison in Cartesian coordinates, the differential form of the Dirac gauge is:

$$\boldsymbol{A}^S = \frac{\hbar}{2e}\left(\frac{-b_y}{b(b-b_z)}, \frac{b_x}{b(b-b_z)}, 0\right) \quad \boldsymbol{A}^{\mathcal{N}} = \frac{\hbar}{2e}\left(\frac{-b_y}{b(b+b_z)}, \frac{b_x}{b(b+b_z)}, 0\right) \qquad [7.58]$$

where $b$ is the required coordinates in Cartesian form. Since $\boldsymbol{A}^{\mathcal{N}}$ is related to $\boldsymbol{A}^S$ by $\boldsymbol{A}^{\mathcal{N}} = \boldsymbol{A}^S + d\theta$, it is obvious that $dA^S = dA^{\mathcal{N}}$ which shows $dA$ is unique. In summary, $A$ is not globally defined throughout $S_b^2$, otherwise $\int dA \neq 0$; in other words, if there is a vector $\tilde{A}$ such that its curvature has no singularity and $\tilde{A}$ is unique, then $\int dA \neq 0$. Now with the Dirac gauge potential defined on the north and south charts, one can then perform the surface integral of these gauge potentials as:

$$\oint \boldsymbol{\Omega}.d\boldsymbol{S} = \oint_N (\nabla \times \boldsymbol{A}^{\mathcal{N}}).d\boldsymbol{S} + \oint_S (\nabla \times \boldsymbol{A}^S).d\boldsymbol{S} = 2\pi g + 2\pi g = 4\pi g \qquad [7.59]$$

avoiding the string of each gauge potential.

## 7.8 Conclusion

We have provided a modern theoretical syllabus which covers recent developments in nanoelectronics, e.g. spintronics, graphene electronics, topological-based electronics, single-electronics, and others. Readers are constantly reminded that electrons in these systems are not only treated as charged particles but ones with internal degree of freedoms, e.g. spin, valley spin and pseudospin. In a nanoscale device, electrons propagate in an environment with periodic crystal, random impurity and device boundary potentials, subject to inter-particle interaction. Pure condensed matter physics focuses on the inter-particle interaction effect on phase transition. But nanoelectronics ought to focus on interaction effects on electron motion. Although inter-particle interaction has not been discussed in detail, the framework for future inclusion of these effects has been provided under the Kubo and the non-equilibrium Green's function (NEGF) formalisms in Chapters 3 and 4. The application of external fields is crucial for the operation of any nanodevice. It is thus important to consider the screening and the non-equilibrium physics. A large portion of Chapters 3 and 4 is devoted to discussing the non-equilibrium electron transport which takes into account its internal degree of freedom.

In Chapters 5 and 6, we introduce spintronics and graphene electronics since both areas are becoming technologically important. Chapter 7 discusses the new and important idea of topological and curvature effect. The current descriptions make no distinction between a bulk condensed matter and a nanoscale device channel, opening doors for future work. The recent discovery of a topological insulator is certain to further underpin the importance of topological physics.

Many physical phenomena discussed here are solely of nanodevice origin, e.g. single-electron transport, Rashba effects, NEGF transport, and monolayer and bilayer graphene physics. These physical effects cannot be found in natural materials, but exist only in nanoscale devices fabricated to nanometer scale and operated with external fields. This book should be expanded in the near future to include the topological insulators, plasmonics and metamaterials.

Lastly we hope this book will bring widespread awareness that there is now a need for theoretical physics to be defined and written for modern nanoelectronics. Nanoelectronics is a highly technologically relevant area. In fact, technological need is the only reason for its existence. The physics of nanoelectronics will be the prevailing physics of the twenty-first century. May this book prevail upon all young, creative individuals of today to participate in the future development of this beautiful physics.

## 7.9 References

[1] Bernevig B A and Zhang S C (2006), 'Quantum spin Hall effect', *Phys. Rev. Lett.* 96, 106802; Murakami S, Nagaosa N and Zhang S C (2003), 'Dissipationless quantum spin current at room temperature', *Science* 301, 1348; Berard A and Mohrbach H (2006), 'Spin Hall effect and Berry phase of spinning particles', *Phys. Lett. A* 352, 190; Murakami S (2007), 'Phase transition between the quantum spin Hall and insulator phases in 3D: Emergence of a topological gapless phase', *New J. Phys.* 9, 356; Sinova J, Culcer D, Niu Q, Sinitsyn N A, Jungwirth T *et al.* (2004), 'Universal intrinsic spin Hall effect' *Phys. Rev. Lett.* 92, 126603.

[2] Berger L (1996), 'Emission of spin waves by a magnetic multilayer traversed by a current', *Phys. Rev. B* 54, 9353; Slonczewski J C (1996), 'Current-driven excitation of magnetic multilayers', *J. Magn. Magn. Mater.* 159, L1.

[3] Waintal X, Myers E B, Brouwer P W and Ralph D C (2000), 'Role of spin-dependent interface scattering in generating current-induced torques in magnetic multilayers', *Phys. Rev. B* 62, 12317; Stiles M D and Zangwill A (2002), 'Noncollinear spin transfer in Co/Cu/Co multilayers', *J. Appl. Phys.* 91, 6812.

[4] Meier L, Salis G, Shorubalko I, Gini E, Schön S *et al.* (2007), 'Measurement of Rashba and Dresselhaus spin–orbit magnetic fields', *Nat. Phys.* 3, 650; Koga T, Nitta J, Akazaki T and Takayanagi H (2002), 'Rashba spin–orbit coupling probed by the weak antilocalization analysis in InAlAs/InGaAs/InAlAs quantum wells as a function of quantum well asymmetry', *Phys. Rev. Lett.* 89, 046801; Nitta J, Akazaki T, Takayanagi H and Enoki T (1997), 'Gate control of spin–orbit interaction in an inverted $In_{0.53}Ga_{0.47}As/In_{0.52}Al_{0.48}As$ heterostructure', *Phys. Rev. Lett.* 78, 1335.
[5] Yang C N and Mills R L (1954), 'Conservation of isotopic spin and isotopic gauge invariance', *Phys. Rev.* 96, 191; Chern S S (1992), 'Yang–Mills equations and Yang–Baxter equations', *Chin. J. Phys.* 30, 949.
[6] Shnir Y (2005), *Magnetic Monopoles*, Springer, Berlin.
[7] Karplus R and Luttinger J M (1954), 'Hall effect in ferromagnetics', *Phys. Rev.* 95, 1154; Sundaram G and Niu Q (1999), 'Wave-packet dynamics in slowly perturbed crystals: Gradient corrections and Berry-phase effects', *Phys. Rev. B* 59, 14915.
[8] Fujita T, Jalil M B A and Tan S G (2010), 'Unified description of intrinsic spin Hall effect mechanisms', *New J. Phys.* 12, 013016; Fujita T, Jalil M B A and Tan S G (2009), 'Topological spin Hall current in waveguided zinc-blende semiconductors with Dresselhaus spin-orbit coupling', *Ann. Phys.* 324, 2265.
[9] Bliokh K Y and Bliokh Y P (2005), 'Spin gauge fields: From Berry phase to topological spin transport and Hall effects', *Ann. Phys.* 319, 13; Tan S G, Jalil M B A, Fujita T (2010), 'Monopole and topological electron dynamics in adiabatic spintronic and graphene systems', *Ann. Phys*, 325, 1537.
[10] Bruno P, Dugaev V K and Taillefumier M (2004), 'Topological Hall effect and Berry phase in magnetic nanostructures', *Phys. Rev. Lett.* 93, 096806; Tatara G, Takayama T, Kohno H, Shibata J, Nakatani Y *et al.* (2006), 'Threshold current of domain wall motion under extrinsic pinning, β-term and non-adiabaticity', *J. Phys. Soc. Jpn.* 75, 064708.
[11] Tan S G, Jalil M B A, Liu X J and Fujita T (2008), 'Spin transverse separation in a two-dimensional electron-gas using an external magnetic field with a topological chirality', *Phys. Rev. B* 78, 245321; Bazaliy Y B, Jones B A and Zhang S C (1998), 'Modification of the Landau–Lifshitz equation in the presence of a spin-polarized current in colossal- and giant-magnetoresistive materials', *Phys. Rev. B* 57, R3213.
[12] Berry M V (1984), 'Quantal phase factors accompanying adiabatic changes', *Proc. R. Soc. London Ser. A* 392, 45.
[13] Simon B (1983), 'Holonomy, the quantum adiabatic theorem, and Berry's phase', *Phys. Rev. Lett.* 51, 2167.
[14] Wu T T and Yang C N (1975), 'Concept of nonintegrable phase factors and global formulation of gauge fields', *Phys. Rev. D* 12, 3845.
[15] Minami M (1979), 'Dirac's monopole and the Hopf map', *Prog. Theor. Phys.* 62, 1128.
[16] Pugh E M (1928), 'Hall effect and magnetic induction in a bar of electrolytic iron', *Phys. Rev.* 32, 824.
[17] Karplus R and Luttinger J M (1954), 'Hall effect in ferromagnetics', *Phys. Rev.* 95, 1154.
[18] Berger L (1970), 'Side jump mechanism for the Hall effect of ferromagnets', *Phys. Rev. B* 2, 4559.
[19] Moritz J, Rodmacq B, Auffret S and Dieny B (2008), 'Extraordinary Hall effect in thin magnetic films and its potential for sensors, memories and magnetic logic applications', *J. Phys. D: Appl. Phys.* 41, 135001.

[20] Onoda M and Nagaosa N (2002), 'Topological nature of anomalous Hall effect in ferromagnets', *J. Phys. Soc. Jpn.* 71, 19.
[21] Jalil M B A and Tan S G (2005), 'Spin transfer and current-induced switching in a ferromagnetic single-electron transistor', *Phys. Rev. B* 72, 214417;
[22] Myers E B, Ralph D C, Katine J A, Louie R N and Buhrman R A (1999), 'Current-induced switching of domains in magnetic multilayer devices', *Science* 285, 867.
[23] Jiang Y, Abe S, Ochiai T, Nozaki T, Hirohata A *et al.* (2004), 'Effective reduction of critical current for current-induced magnetization switching by a Ru layer insertion in an exchange-biased spin valve', *Phys. Rev. Lett.* 92, 167204.
[24] Huai Y, Albert F, Nguyen P, Pakala M and Valet T (2004), 'Observation of spin-transfer switching in deep submicron-sized and low-resistance magnetic tunnel junctions', *Appl. Phys. Lett.* 84, 3118; Meng H and Wang J P (2005), 'Spin transfer effect in magnetic tunnel junction with a nano-current-channel layer in free layer', *IEEE Trans. Magn.* 41, 2612.
[25] Lashell S, McDougall B A and Jensen E (1996), 'Spin splitting of an Au(111) surface state band observed with angle resolved photoelectron spectroscopy', *Phys. Rev. Lett.* 77, 3419; Henk J, Hoesch M, Osterwalder J, Ernst A and Bruno P (2004), 'Spin–orbit coupling in the L-gap surface states of Au(111): spin-resolved photoemission experiments and first-principles calculations', *J. Phys.: Condens. Matter* 16, 7581.
[26] Krupin O, Bihlmayer G, Starke K, Gorovikov S, Prieto J E *et al.* (2005), 'Rashba effect at magnetic metal surfaces', *Phys. Rev. B* 71, 201403 (R).
[27] Krupin O, Bihlmayer G, Döbrich K M, Prieto J E, Starke K *et al.* (2009), 'Rashba effect at the surfaces of rare-earth metals and their monoxides', *New J. Phys.* 11, 013035.
[28] Tan S G, Jalil M B A, Liu X J and Fujita T (2011), 'Spin dynamics under local gauge fields in chiral spin orbit coupling systems', *Ann. Phys.* 326, 207.
[29] Manchon A and Zhang S (2008), 'Theory of nonequilibrium intrinsic spin torque in a single nanomagnet', *Phys. Rev. B*, 78, 212405; Manchon A and Zhang S (2009), 'Theory of spin torque due to spin–orbit coupling', *Phys. Rev. B*, 79, 094422.
[30] Obata K and Tatara G (2008), 'Current-induced domain wall motion in Rashba spin–orbit system', *Phys. Rev. B* 77, 214429.
[31] Miron I M, Gaudin G, Auffret S, Rodmacq B, Schuhl A *et al.* (2010), 'Current-driven spin torque induced by the Rashba effect in a ferromagnetic metal layer', *Nat. Mater.* 9, 230; Miron I M, Zermatten P-J, Gaudin H, Auffret S, Rodmacq B *et al.* (2009), 'Domain wall spin torquemeter', *Phys. Rev. Lett.* 102, 137202.

## Appendix 7.A Mathematical properties of monopole fields

Although $A^{\mathcal{N}}$ and $A^S$ are by no means unique on their own, they uniquely determine a 1-form $\omega$ on the bundle space P which is a $S^3_R$, where $R$ is the radius of the 3-sphere. Here we provide the mathematical origin, the 1-form $\omega$ on $R_4$ which is given by:

$$\omega = i(-y_1 dx_1 + x_1 dy_1 - y_2 dx_2 + x_2 dy_2) \qquad [7.A.1]$$

which is well-defined on the $S^3_R . S^3_R$ and $S^2_R$ can be defined as:

$$S_R^3 = \{(x_1, y_1, x_2, y_2)\} \in \mathbb{R}_4 : x_1^2 + y_1^2 + x_2^2 + y_2^2 = 1\}$$
$$S_R^2 = \{(x, y, z)\} \in \mathbb{R}_3 : x^2 + y^2 + z^2 = 1\}. \quad [7.A.2]$$

One can re-parametrize such that $S_R^3 = \left\{(z_1, z_2) = \left(\cos\frac{\theta}{2}e^{i\varepsilon_1}, \sin\frac{\theta}{2}e^{i\varepsilon_2}\right); 0 \le \frac{\theta}{2} \le \frac{\pi}{2}, \varepsilon_1, \varepsilon_2 \in \mathbb{R}_4\right\}$, where $z_1 = x_1 + iy_1$, $z_2 = x_2 + iy_2$. It can then be seen that $(x_1, x_2, y_1, y_2) \equiv \left(\cos\frac{\theta}{2}\cos\varepsilon_1, \sin\frac{\theta}{2}\cos\varepsilon_2, \cos\frac{\theta}{2}\sin\varepsilon_1, \sin\frac{\theta}{2}\sin\varepsilon_2\right)$ and the constrain: $x_1^2 + y_1^2 + x_2^2 + y_2^2 = 1$ is obeyed.

Every point of $(z_1, z_2) = \left(\cos\frac{\theta}{2}e^{i\varepsilon_1}, \sin\frac{\theta}{2}e^{i\varepsilon_2}\right) \in S_R^3$ can then be mapped via:

$$\mathcal{P}(\theta, \varepsilon_1, \varepsilon_2) = (\sin\theta\cos(\varepsilon_1 - \varepsilon_2), \sin\theta\sin(\varepsilon_1 - \varepsilon_2), \cos\theta) \quad [7.A.3]$$

to a point on $S_b^2$ where $\mathcal{P}: S^3 \to S^2$, which is in fact a form of Hopf mapping to project the $\omega$ on $S_R^3$ to $S_b^2$. In other words, the Hopf mapping $\eta : S_R^3 \to S_b^2$ yields a 1-form gauge potential regular everywhere on $S_R^3$.

One then carries out the operation such that every point of $(z_1, z_2) \in S_R^3$ is mapped, i.e.

$$(z_1, z_2) \to (z_1, z_2) \times \frac{|z_1|}{z_1} = \left(|z_1|, \frac{z_2|z_1|}{z_1}\right) = \left(\cos\frac{\theta}{2}, \sin\frac{\theta}{2}e^{-i\phi}\right);$$
$$(z_1, z_2) \to (z_1, z_2) \times \frac{|z_2|}{z_2} = \left(\frac{z_1|z_2|}{z_2}, |z_2|\right) = \left(\cos\frac{0}{2}e^{i\phi}, \sin\frac{0}{2}\right). \quad [7.A.4]$$

The same point of $(z_1, z_2) \in S_R^3$ can be re-expressed in two different forms known as the sections, each corresponding to one part of the $S_b^2$. We note that:

$$\left(\cos\frac{\theta}{2}, \sin\frac{\theta}{2}e^{-i\phi}\right) \text{ corresponds to } U_N = S_b^2 - (0,0,-1)$$
$$\left(\cos\frac{\theta}{2}e^{i\phi}, \sin\frac{\theta}{2}\right) \text{ corresponds to } U_S = S_b^2 - (0,0,+1). \quad [7.A.5]$$

Equation 7.A.4 can also be written in formal mathematics as:

$$s_N(r \in U_N \text{ on } S^2) = \left(\cos\frac{\theta}{2}, \sin\frac{\theta}{2}e^{-i\phi}\right); s_S(r \in U_S \text{ on } S^2) = \left(\cos\frac{\theta}{2}e^{i\phi}, \sin\frac{\theta}{2}\right). \quad [7.A.6]$$

We have established the two sections of a particular point of $(z_1, z_2) \in S_R^3$, each corresponding to a chart on $S_b^2$. We will now work on the inclusion map ($\beth°$) of these points on $S_R^3$. One has $\beth°(z_1, z_2) = (x_1, x_2, y_1, y_2)$, which yields:

$$\beth°\left(\cos\frac{\theta}{2}e^{i\varepsilon_1}, \sin\frac{\theta}{2}e^{i\varepsilon_2}\right) = (x_1, x_2, y_1, y_2)$$
$$= \left(\cos\frac{\theta}{2}\cos\varepsilon_1, \sin\frac{\theta}{2}\cos\varepsilon_2, \cos\frac{\theta}{2}\sin\varepsilon_1, \sin\frac{\theta}{2}\sin\varepsilon_2\right). \quad [7.A.7]$$

The inclusion map performed on $\left(|z_1|, \frac{z_2|z_1|}{z_1}\right)$ and $\left(\frac{z_1|z_2|}{z_2}, |z_2|\right)$ yields:

$$\beth°S_N(r \in U_N \ on\ S^2) = \beth°\left(\cos\frac{\theta}{2}, \sin\frac{\theta}{2}e^{-i\phi}\right) = (x_1, y_1, x_2, y_2)$$
$$= \left(\cos\frac{\theta}{2}, 0, \sin\frac{\theta}{2}\cos\phi, -\sin\frac{\theta}{2}\sin\phi\right)$$
$$\beth°S_S(r \in U_S \ on\ S^2) = \beth°\left(\cos\frac{\theta}{2}e^{i\phi}\sin\frac{\theta}{2}\right) = (x_1, y_1, x_2, y_2)$$
$$= \left(\cos\frac{\theta}{2}\cos\phi, \cos\frac{\theta}{2}\sin\phi, \sin\frac{\theta}{2}, 0\right). \quad [7.A.8]$$

The above simply means that the $\omega$ defined in terms of $(x_1, y_1, x_2, y_2)$ can be re-expressed in terms of $(\theta, \phi)$ on a re-parametrized $S_R^3$. In summary, the gauge corresponding to the north and south charts can be derived as follows:

$$A^{\mathcal{N}} = \frac{\hbar}{2e}(\beth°S_N)^*\,\omega = \frac{\hbar}{2e}(-y_1dx_1 + x_1dy_1 - y_2dx_2 + x_2dy_2)$$
$$= \frac{\hbar}{2e}\left(-0.\,d\left(\cos\frac{\theta}{2}\right) + \cos\frac{\theta}{2}.d(0) + \sin\frac{\theta}{2}\sin\phi.\,d\left(\sin\frac{\theta}{2}\cos\phi\right)\right.$$
$$\left. + \sin\frac{\theta}{2}\cos\phi.\,d\left(-\sin\frac{\theta}{2}\sin\phi\right)\right) = \frac{\hbar}{2e}(1-\cos\theta)d\phi \quad [7.A.9]$$

$$A^S = \frac{\hbar}{2e}(\beth°S_S)^*\,\omega = \frac{\hbar}{2e}(-y_1dx_1 + x_1dy_1 - y_2dx_2 + x_2dy_2)$$
$$= \frac{\hbar}{2e}\left(-\cos\frac{\theta}{2}\sin\phi.\,d\left(\cos\frac{\theta}{2}\cos\phi\right)\right.$$
$$\left. + \cos\frac{\theta}{2}\cos\phi.\,d\left(\cos\frac{\theta}{2}\sin\phi\right) - 0.d\left(\sin\frac{\theta}{2}\right) + \sin\frac{\theta}{2}.\,d(0)\right)$$
$$= -\frac{\hbar}{2e}(1+\cos\theta)d\phi. \quad [7.A.10]$$

Noting the above, the gauge potential in the overlap regions are: $A^{\mathcal{N}} = A^S + d\left(2\frac{\hbar}{2e}\phi\right)$. The transition functions are:

$$g_{SN} = \frac{z_2/|z_2|}{z_1/|z_1|} = e^{-i\phi}; \qquad g_{NS} = \frac{z_1/|z_1|}{z_2/|z_2|} = e^{i\phi}. \qquad [7.A.11]$$

The topological magnetic charge is $\Omega = \int dA^{\mathcal{N}} = \int dA^S + dd\left(2\frac{\hbar}{2e}\phi\right) = \int dA^S$. It is thus apparent that one could avoid the singular part of $A^{\mathcal{N}}$ by switching to $A^S$ in the overlapping region, thus avoiding the conceptual difficulty of the Dirac string.

# Index

FSC
www.fsc.org
MIX
Papier | Fördert
gute Waldnutzung
FSC® C083411

Kolibri 360 GmbH
Ferdinand-Jühlke-Straße 7
99095 Erfurt, Deutschland
produktsicherheit@kolibri360.de